Nonrecursive Control Design for Nonlinear Systems

Based on the authors' recent advances, this book focuses on a class of nonlinear systems with mismatched uncertainties/disturbances and discusses their typical control problems. It aims to provide a comprehensive view of the nonrecursive control theory and application guidelines.

Nonrecursive synthesis of complex nonlinear systems not only greatly simplify the control design process, weaken the system assumptions, and reduce the conservatism of gain selection, but also realize the essential detachment of control law design and Lyapunov function-based stability analysis. Therefore, different from the classical recursive control design methods, it is of significance to study the synthesis of nonlinear systems from the perspective of a new nonrecursive control framework.

This book discusses the following typical control problems: theoretical background, homogeneous systems theory review, nonrecursive robust control design, nonrecursive adaptive control design, nonrecursive generalized dynamic predictive control, disturbance estimation and attenuation, nonrecursive stability analysis, implementation theory and real-life applications to series elastic actuators, DC microgrids, and permanent magnet synchronous motor (PMSM) systems under the proposed nonrecursive synthesis framework.

This book will be a great reference for scholars and students in the field of automation and control. It will also be a useful source for control engineers and those working on anti-disturbance control, nonlinear output regulation, nonsmooth control, and other related topics.

Chuanlin Zhang received the B.S. degree in mathematics and the Ph.D degree in control theory and control engineering from the School of Automation, Southeast University, Nanjing, China, in 2008 and 2014, respectively. He was a Visiting Ph.D. Student with the Department of Electrical and Computer Engineering, University of Texas at San Antonio, USA, from 2011 to 2012; a Visiting Scholar with the the Energy Research Institute, Nanyang Technological University, Singapore, from 2016 to 2017; and a Visiting Scholar with the Advanced Robotics Center, National University of Singapore, from 2017 to 2018. He has been with the College of Automation Engineering, Shanghai University of Electric Power, China since 2014, where he is currently a Professor and Director of Intelligent Autonomous Systems Lab. His research interests include nonlinear system control theory and applications for power systems.

Jun Yang received the B.Sc. degree in automation from the Department of Automatic Control, Northeastern University, Shenyang, China, in 2006, and the Ph.D. degree in control theory and control engineering from the School of Automation, Southeast University, Nanjing, China in 2011. He joined the Department of Aeronautical and Automotive Engineering at Loughborough University in 2020 as a Senior Lecturer. His research interests include disturbance observer, motion control, visual servoing, nonlinear control, and autonomous systems. He is a Fellow of IEEE and IET.

Nonrecursive Control Design for Nonlinear Systems

Theory and Applications

Chuanlin Zhang and Jun Yang

CRC Press is an imprint of the
Taylor & Francis Group, an **informa** business

Designed cover image: IR Stone

The book was supported in part by the National Natural Science Foundation of China under Grant 61503236, 61973080, 62173221; in part by the Program for Professor of Special Appointment (Eastern Scholar) at Shanghai Institutions of Higher Learning, in part by Shanghai Rising-Star program under Grant 20QA1404000, in part by the Natural Science Foundation of Shanghai under Grant 19ZR1420500.

First edition published 2023
by CRC Press
6000 Broken Sound Parkway NW, Suite 300, Boca Raton, FL 33487-2742

and by CRC Press
4 Park Square, Milton Park, Abingdon, Oxon, OX14 4RN

CRC Press is an imprint of Taylor & Francis Group, LLC

ISBN: 978-1-032-50599-2 (hbk)
ISBN: 978-1-032-50604-3 (pbk)
ISBN: 978-1-003-39923-0 (ebk)

DOI: 10.1201/9781003399230

Typeset in Nimbus Roman
by KnowledgeWorks Global Ltd.

Publisher's note: This book has been prepared from camera-ready copy provided by the authors.

To my parents, Wenhuan Zhang and Qingling Cheng, and my beloved wife, Esther Yanfei Jin, for their endless support and love.

Chuanlin Zhang

To my parents, Heyu Yang and Aiqing Li, my wife, Yanan Wei, and my son, Haoyu Yang, for their endless support and love.

Jun Yang

Contents

List of Figures

List of Tables

Preface

For nonlinear systems, finding an effective controller to keep the closed-loop control system stable at the working equilibrium is the primary goal of control. Moreover, it is also necessary to consider the performance indexes such as improving the robustness, convergence rate, regulation accuracy, and economy issue of the control system. For nearly half a century, there have been many systematic nonlinear controller design methods for different types of systems in the literature, which have attracted extensive attention from both the academic community and industrial practitioners, such as backstepping, sliding mode control, adding a power integrator, predictive control, disturbance/uncertainty estimation, and attenuation control. These nonlinear control technologies have already achieved rich research results in the theoretical field, and they have improved the performance of the control system from different points of view. Notably, for a wide class of nonlinear systems, control law derived from a recursive design procedure is one of the most common strategies, such as backstepping and adding a power integrator method. However, recursive design usually shows some disadvantages such as complex controller expression, large amount of calculation, and difficulty in selecting control parameters. Especially for high-order nonlinear systems, the phenomenon of "explosion of complexity" in control design is well acknowledged as a hurdle for practical implementation.

Considering the synthesis for linear systems in modern control theory, nonrecursive control design is a very common fashion and has gained successful applications. However, nonrecursive control design issue for nonlinear systems is indeed a non-trivial task. Therefore, a natural conjecture is:

For a class of n-order general nonlinear systems with the presence of uncertainties/disturbances, can a nonrecursive control design framework be proposed to greatly reduce the complexity of controller design and enhance the facility of engineering applications?

Focusing on the above problem, in recent years, based on several powerful tools such as homogeneous system theory, the authors have been continually working on how to discover a new nonrecursive synthesis framework for nonlinear systems to narrow the gap between advanced control theory and applications. Numerous research results on the nonrecursive control theory and application studies have been built. Various applications on the nonrecursive synthesis of complex nonlinear systems can not only greatly simplify the control design process, weaken the system assumptions and reduce the conservatism of gain selection, but also realize the essential detachment of control law design and Lyapunov function-based stability analysis. Therefore, different from the classical recursive control design method, it is of great theoretical and engineering significance to study the synthesis of nonlinear systems from the perspective of a new nonrecursive control framework.

Based on a series of recent advances achieved by the authors, this book will present the control theory and application communities with a comprehensive view of the nonrecursive control theory and application guidelines. It will not only enrich the existing nonlinear control theory but also depict an alternative practically oriented synthesis procedure with abundant practice experiences.

This book can be referred to as a reference for graduate students with a basic knowledge of classical control theory and some knowledge of the state-space control design methods and nonlinear systems. For control engineers, we have also provided a few practical case studies, which present the results of how we can implement the proposed methods into practices. Those examples are used to illustrate the successful practical application of the nonlinear control theory.

This book is structured as follows.

Chapter 1 reviews the recent advances in nonlinear control methods for nonlinear systems with the presence of uncertainties/disturbances. By presenting a few case studies, it is shown that even various existing methods have been investigated in the past few decades; however, those results may leave a big gap between theoretical justification and practical simplification requirements. Notably, with the famous backstepping control strategy it is easy to cause an "explosion of terms" phenomenon. Motivated by these facts, Chapter 1 will show a clear motivation of why the practitioners have a demand for simpler nonlinear controller construction procedures, especially for those systems requiring an accurate steady-state error and a strong disturbance/uncertainty attenuation ability.

Chapter 2 addresses the robust stabilization issue for a class of nonlinear systems with internal uncertainties. Different from existing recursive-based control strategies, this chapter presents a novel homogeneous controller construction framework. By opposing a series of nonlinearity constraints, one is now able to derive a nonrecursive stabilizing control law for a class of uncertain nonlinear systems without going through a series of virtual controller calculations. In addition, now the controller construction is essentially detached from the stability analysis. With an illustrative example, the efficacy is verified by numerical simulations.

Chapter 3 discusses the nonrecursive tracking control for nonlinear systems via both state-feedback and output feedback. In order to release the nonlinearity constraints generally opposed to the system in the existing nonsmooth control design framework, we develop a simple practically oriented finite-time control framework, to essentially remove any nonlinearity growth conditions under a less ambitious but very practical semi-global control objective. With abundant numerical simulations, the efficacy of the new nonrecursive robust tracking control design methods is illustrated.

Chapter 4 further addresses the adaptive control problem for a class of nonlinear systems with non-parameterized uncertainties. Noting that existing adaptive methods, such as adaptive backstepping, require very complex tuning functions to avoid the overparameterization problem, this chapter presents an alternative adaptive control design framework under a nonrecursive synthesis manner. By dividing the control objective into state feedback and output feedback cases, a unified smooth and nonsmooth control framework is presented. It is shown that under a semi-global control objective, one can unify the adaptive control within an identical design framework while the control performance is dependent on the selection of homogeneous degrees. At last, a few illustrative examples with numerical simulations demonstrate the effectiveness and theoretical justification.

Chapter 5 discusses the robust output regulation problem for a class of nonlinear systems perturbed by a series of mismatched disturbances. By integrating the output regulation theory and homogeneous domination method, now it is able to derive a very simple exact tracking controller without going through any complex calculations and virtual controller determination steps. The key factor in this algorithm is based on a delicate handling process of the nonlinearity and disturbance terms. For the mismatched disturbances, this book has also provided a few disturbance estimation tools, including higher-order sliding mode (HOSM) observer and nonsmooth extended state observer. Simulation comparison studies are conducted to show how the control parameters are configured and how can different homogeneous degree relates to the system control performance.

Chapter 6 further studies the nonrecursive adaptive output regulation problem for a class of nonlinear systems with the presence of mismatched disturbances. Different with the robust control method in Chapter 5, by adding a one-step dynamic gain tuning mechanism, the possible robustness redundancy issue caused by the gain selection guideline can be essentially avoided. Moreover, the adaptive output regulation strategy could also maintain an improved transient-time control performance for different variation levels of the disturbances. With rigorous stability proof, both semi-global attractivity and local asymptomatic/finite-time convergence rate can be ensured under the proposed controller. Illustrative examples and numerical simulation results have confirmed the performance improvement compared to the existing related robust control methods.

The above theoretical results have enriched the nonlinear control theory; however, the main motivation of this research is to facilitate practitioners and give them an alternative control design choice when they are dealing with some complicated real-life nonlinear systems. As such, the following chapters are dedicated to presenting several application examples to demonstrate the implementation simplicity and user-friendly features.

Chapter 7 presents the implementation of related nonrecursive smooth and nonsmooth controllers into the speed regulation or position control problems for permanent magnet synchronous motor (PMSM) systems. This type of system is generally employed to test the control performances of different control advances. By conducting the experimental tests on a laboratory platform, the control performances under different related controllers, the proposed nonsmooth controllers have shown their superiority in several control performance indexes. More specifically, we have also provided the parameter configuration details in order to give the engineers a typical experience if they are trying to apply the methods from this book to other real-life systems.

Chapter 8 shows the application of the nonrecursive controller design with a nonsmooth functional form to a series elastic actuator, which is widely applied in compliant robotic systems. First, the modeling procedure of the system is given to show the physical meaning of those control variables. Then by applying the robust control strategy into these practical systems, both simulation study and experimental verification are conducted. More specifically, what is interesting here is that a flexible homogeneous degree can make a serious impact on the control performance, hence providing the control engineering with a novel fractional power tuning procedure.

Chapter 9 gives a comprehensive application example of how we extend the theoretical method into the control practices of DC microgrid systems. These types of systems are strongly dependent on new nonlinear control methods owing to the fact that there are more and more constant power loads (CPLs) in the practical DC microgrids, and they normally have a negative impedance characteristic. By successfully applying our methods into the decentralized stabilization problem for DC microgrid systems, it is shown that a large-signal stability can be ensured by the proposed methods. Experimental results have validated our theoretical conjecture. Moreover, some extension results have also been presented to show the performance improvements and implementation facilities.

About the Authors

Chuanlin Zhang received the B.S. degree in mathematics and the Ph.D degree in control theory and control engineering from the School of Automation, Southeast University, Nanjing, China, in 2008 and 2014, respectively. He was a Visiting Ph.D. Student with the Department of Electrical and Computer Engineering, University of Texas at San Antonio, USA, from 2011 to 2012; a Visiting Scholar with the Energy Research Institute, Nanyang Technological University, Singapore, from 2016 to 2017; and a Visiting Scholar with the Advanced Robotics Center, National University of Singapore, from 2017 to 2018. He has been with the College of Automation Engineering, Shanghai University of Electric Power, China since 2014, where he is currently a Professor and Director of Intelligent Autonomous Systems Lab. His research interests include nonlinear system control theory and applications for power systems.

Jun Yang received the B.Sc. degree in automation from the Department of Automatic Control, Northeastern University, Shenyang, China, in 2006, and the Ph.D. degree in control theory and control engineering from the School of Automation, Southeast University, Nanjing, China in 2011. He joined the Department of Aeronautical and Automotive Engineering at Loughborough University in 2020 as a Senior Lecturer. His research interests include disturbance observer, motion control, visual servoing, nonlinear control, and autonomous systems. He is a Fellow of IEEE and IET.

Acknowledgments

The development of this book would not have been possible without the support and help from many collaborators. The authors are grateful to Prof. Shihua Li, Prof. Chunjiang Qian, Prof. Changyun Wen, Prof. Peng Wang, and Dr. Haoyong Yu for their long-lasting supervision and encouragement for the publication of the book, and are indebted to Dr. Yunda Yan, Dr. Jianliang Mao, Dr. Chenggang Cui, and Dr. Pengfeng Lin for their continuous support and credits to this book. The authors also sincerely thank Xin Dong, Xiaoyu Wang, Liwen Zhou, Zhongkun Cao, and Jing Ge for their efforts in the typesetting and proofreading of this book.

The book was supported in part by the National Natural Science Foundation of China under Grant 61503236, 61973080, 62173221; in part by the Program for Professor of Special Appointment (Eastern Scholar) at Shanghai Institutions of Higher Learning, in part by the Shanghai Rising-Star program under Grant 20QA1404000, and in part by the Natural Science Foundation of Shanghai under Grant 19ZR1420500.

Figures 4.5–4.7 © [2022] IEEE. Reprinted, with permission, from [C. Zhang and J. Yang, Nonsmooth adaptive control for uncertain nonlinear systems: A non-recursive design approach, IEEE Control Systems Letters, 2022, 6: 229–234.]

Figures 5.3–5.8 © [2020] IEEE. Reprinted, with permission, from [C. Zhang, J. Yang, Y. Yan, L. Fridman, and S. Li. Semi-global finite-time trajectory tracking realization for disturbed nonlinear systems via higher-order sliding modes. IEEE Transactions on Automatic Control, 2020, 60(5): 2185–2191.]

Figures 5.9 and 5.10, Figures 8.12–8.17 © [2018] IEEE. Reprinted, with permission, from [C. Zhang, Y. Yan, C. Wen, J. Yang, and H. Yu. A nonsmooth composite control design framework for nonlinear systems with mismatched disturbances: Algorithms and experimental tests. IEEE Transactions on Industrial Electronics, 2018, 65(11): 8828–8839.]

Figures 6.2–6.7, Figures 9.45 and 9.46, Tables 9.7 and 9.8 © [2022] IEEE. Reprinted, with permission, from [X. Dong, C. Zhang, T. Yang, and J. Yang. Nonsmooth dynamic tracking control for nonlinear systems with mismatched disturbances: Algorithm and practice, IEEE Transactions on Industrial Electronics, published online, DOI=10.1109/TIE.2022.3181367.]

Figures 8.1–8.11, Tables 8.1–8.3 © [2018] IEEE. Reprinted, with permission, from [C. Zhang, Y. Yan, A. Narayan, and H. Yu. Practically oriented finite-time control design and implementation: Application to series elastic actuator. IEEE Transactions on Industrial Electronics, 2018, 65(5): 4166–4176.]

Figures 9.1–9.13, Table 9.1 © [2019] IEEE. Reprinted, with permission, from [P. Lin, C. Zhang, P. Wang, and J. Xiao. A decentralized composite controller for unified voltage control with global system large-signal stability in DC microgrids. IEEE Transactions on Smart Grid. 2019, 10(5): 5075–5091.]

Figures 9.14–9.21, Table 9.2 © [2020] IEEE. Reprinted, with permission, from [C. Zhang, X. Wang, P. Lin, Peter X. Liu, Y. Yan, and J. Yang. Finite-time feedforward decoupling and precise decentralized control for DC microgrids towards large signal stability. IEEE Transactions on Smart Grid, 2020, 11(1): 391–402.]

Figure 9.3, Figures 9.22–9.34, Tables 9.3 and 9.4 © [2022] IEEE. Reprinted, with permission, from [X. Wang, X. Dong, X. Niu, C. Zhang, C. Cui, J. Huang, and P. Lin. Towards balancing dynamic performance and system stability for DC microgrids: A new decentralized adaptive control strategy. IEEE Transactions on Smart Grid, 2022, 13(5): 3439–3451.]

Figures 9.35–9.44, Tables 9.5 and 9.6 © [2020] IEEE. Reprinted, with permission, from [P. Lin, C. Zhang, J. Wang, C. Jin, and P. Wang. On autonomous large signal stabilization for islanded multi-bus DC microgrids: A uniform nonsmooth control scheme. IEEE Transactions on Industrial Electronics, 2020, 67(6): 4600–4612.]

Symbol Description

$\mathbb{R}^+$	a set of positive real numbers.
$\mathbb{R}^+_{\text{odd}}$	a set of ratios of two positive odd integers.
$\mathbb{N}_{i:j}$	defined mathematic symbol, which is $\mathbb{N}_{i:j} = \{i, i+1, \cdots, j\}$ for integers j and i satisfying $0 \leq i \leq j$.
$\mathbb{C}^i$	the set of all continuously differentiable functions whose ith time derivatives are continuous.
$\lfloor \cdot \rceil^a$	a $\mathbb{C}^0$ function defined by $\lfloor \cdot \rceil^a = \text{sign}(\cdot)\lvert \cdot \rvert^a$ with $a \in (0,1)$.
$\bar{x}_i$	defined mathematic symbol, which is $\bar{x}_i \triangleq (x_1, \cdots, x_i)^\top$.
$\bar{x}^{\tau}_{i\Delta^r}$	homogeneous vector defined by $\bar{x}^{\tau}_{i\Delta^r} \triangleq \left(x_1^{\frac{\tau}{r_1}}, x_2^{\frac{\tau}{r_2}}, \cdots, x_i^{\frac{\tau}{r_i}}\right)^\top$ for a constant τ, and the dilation weight $r = [r_1, r_2, \cdots, r_n]$.
$\lfloor x \rceil^{\tau}_{\Delta^r}$	homogeneous vector defined by $\lfloor x \rceil^{\tau}_{\Delta^r} = \left(\lfloor x_1 \rceil^{\frac{\tau}{r_1}}, \lfloor x_2 \rceil^{\frac{\tau}{r_2}}, \cdots, \lfloor x_n \rceil^{\frac{\tau}{r_n}}\right)^\top$
$\langle a \rangle_M$	a saturation function, for the positive constant threshold M, defined as $\langle a \rangle_M = \begin{cases} \text{sign}(a)M, \text{ if} \lvert a \rvert > M, \\ a, \text{ if} \lvert a \rvert \leq M. \end{cases}$

Chapter 1

Introduction

1.1 Overview of Nonlinear System Control

For different kinds of nonlinear systems with uncertainties and internal/external disturbances, the primary objective of system control design is to find an effective controller to keep the closed-loop control system stable at the operating point. Second, it is also necessary to consider improving the robustness, rapidity, accuracy, economy, and other performance indicators of the control system. More specifically, for a class of nonlinear systems with lower triangular structures, it is of special interest owing to its structural feature which supports a systematic controller design procedure. In this regard, many successful nonlinear controller design methods have been investigated in the literature. Back to the 1990s, *backstepping method*, which performs as a milestone tool bringing into the nonlinear control community, has stimulated a long-lasting bloom of research interests in nonlinear control systems [1, 32, 35, 97]. Therefore, different classes of nonlinear systems are extensively studied for different control issues. However, there are several clear drawbacks in most of the existing backstepping-based control approaches, which can be summarized in the following aspects.

- First, the nominal system considered should be restrained as a feedback linearizable structure, and hence system smoothness and triangular structure are always required.
- Second, the system internal nonlinearities are usually canceled recursively via feedback linearization; hence, the closed-loop system will behave only with linear or quasi-linear dynamics.

DOI: 10.1201/9781003399230-1

- Third, due to the recursive design, the virtual controllers usually contain abundant partial derivative terms that add much complexity and difficulty to be realized for practical implementations.

Therefore, backstepping-based control methods commonly have some drawbacks to hinder the industrial applications of the controller, especially, for high-order nonlinear systems, it is prone to the phenomenon of "explosion of complexity" in the control design.

On the one hand, aiming to release the "explosion of complexity" phenomenon of backstepping control methods, abundant extension results have been achieved in recent decades based on the core recursive synthesis idea of backstepping, such as Dynamic Surface Control (DSC) [74], Neural Network Control [50,52], other related intelligent control methods [25,98], etc.

On the other hand, for a class of inherent nonlinear systems, a common singularity problem of the backstepping design often leads to the failure of the recursive design method. Therefore, in order to improve the defects of the traditional backstepping method, by employing a "feedback domination" concept, rather than the "feedback cancellation" manner in backstepping design, a series of novel nonlinear design methods have been successively studied, aiming to solve the global asymptotic or finite-time control problems for a wider class of inherent nonlinear systems. As one representative method, adding a power integrator (API) method has successfully expanded the range of nonlinear systems that can be stabilized via recursive design by improving the singularity defect of adding a linear integrator in the backstepping method, see, for instances, [43,61] and references therein.

As mentioned above, there are numerous new control design methods have been reported in the literature, they are able to release the controller design complexity and extend the applicability of backstepping methods to some extent. However, it is worthy of pointing out here that these methods still inherit a recursive design procedure, leading to the fact that: 1) the virtual controllers are unavoidable and the controller design cannot be essentially detached from the stability analysis; 2) large computational burden of the allowable control gains or adaption functions; and 3) difficulty in selecting control parameters, especially for high-order systems.

Recall that the nonrecursive design manner is very common for the synthesis problems of linear systems and widely welcomed by control practitioners. However, most of the existing literature on advanced nonlinear control design may disobey the beauty of simplicity, and hence when control engineers try to implement those enormous nonlinear controllers, a series of costly calculations have to be tackled first in order to proceed. As a consequence, control engineers will finally return to simple PID controllers or state-feedback controllers via local linearization methods, etc. Therefore, for a class of n-order lower triangular nonlinear systems, designing a novel nonrecursive control design framework, which could significantly reduce the complexity of controller form, the difficulty of control gain selection, and thereby facilitate the implementation in industrial applications, will be of great significance in both control theory and industrial applications.

Since the 1980s, the homogeneous system theory has attracted extensive attention from the system and control community. Note that homogeneous dynamical systems, which take an intermediate between linear and nonlinear systems, have been extensively studied where many significant characteristics are revealed [21, 62]. These features have provided a promising conclusion that weighted homogeneity introduced into nonlinear system synthesis issues will significantly ease or facilitate the design procedure. Through the in-depth study of the homogeneous system theory, many new-fashioned nonlinear theoretical directions, such as adding a power integrator method [43], the homogeneous domination method [58], the higher-order sliding mode control algorithm [12] and others, have obtained fruitful research results.

A pioneer work [4] presents a novel simple finite-time controller and a non-recursive proof for the chain of integrators, and a later result [54] designs a full-order finite-time observer for the chain of integrators as well. A study [49] provides an explicit homogeneous feedback controller with the requirement that a control Lyapunov function exists for an affine control system and satisfies a homogeneous condition. Based on an adding a power integrator design, an interesting result in [55] reports a homogeneous domination technique that can handle a wide class of inherent nonlinear systems within one unified framework, while the controller can be designed in a nested form which will be much easier to be realized.

On the basis of the above-mentioned systematic recursive design methods and tools, for a class of lower-triangular nonlinear systems with matched/mismatched uncertainties/disturbances, numerous scholars have carried out related research from different perspectives of robust control and adaptive control. Some main research results are described as follows.

Robust control is a commonly widely applied strategy when the system is perturbed by mismatched disturbances. In [59], based on the homogeneous domination method, the authors propose a state feedback control law design method under an almost disturbance decoupling manner. Soon after, this robust control approach is generalized to the case of output feedback tracking in [20]. Regarding the disturbance processing, the above algorithms can usually be viewed as passive disturbance rejection techniques, and the regulation control objective of the system is achieved via a large control energy consumption. Another common design scheme is to combine the disturbance estimation and feedforward decoupling technology to handle the internal and external uncertainties/disturbances of the system, which is also called as "active disturbance attenuation control strategy". The advantage of this method lies in the real-time observation and compensation of disturbances. Therefore, the controller can relieve the problem of robustness redundancy while ensuring the control performance of the system. In the related literature, combining recursive control design techniques and active anti-disturbance methods, the robust control design problem for a class of mismatched disturbed systems has also been widely studied in recent years. Taking the finite-time control problem as an example, the works in [39, 86] realize the finite-time tracking control of a class of mismatched disturbed linear and nonlinear systems by combining the adding power integral method and the high-order sliding mode observation technique, respectively.

As is well acknowledged to all, in contrast to the robust control method, the adaptive control strategy can self-configure the control parameters under different working conditions, thereby improving the transient-time performance of the control system. In the existing literature, great progress has been made in the research on adaptive control of nonlinear systems with parametric uncertainties, such as the traditional adaptive backstepping method. However, it is noted that the non-parametric uncertainties (also called nonlinearly-parameterized uncertainties) exist more widely in practical nonlinear systems. In the past few years, for a class of nonlinear systems with non-parametric uncertainties, the adaptive control method has also received continuous attention from the control community. Notably, combined with the homogeneous system theory, the accompanying tuning functions in the adaptive backstepping method are extended to the nonsmooth case, and a category of nonsmooth adaptive control design methods can be obtained, such as [22, 44, 81]. However, such methods usually require a large number of mathematical scaling methods to estimate the nonlinear terms of the system, resulting in the self-tuning gain functions are possibly relatively conservative, and moreover, it is hard to construct these gain functions in practice.

In the literature, another indirect adaptive control design concept can also be found from the perspective of a self-tuned high gain design. For example, in [34], a class of feedback controllers with dynamic high gain has been designed to adaptively handle the nonparametric uncertainties. In [71] and [46], the recursive finite-time adaptive control law of this kind of system is developed by designing different dynamic gain self-renewal mechanisms and combining with the adding a power integrator method.

If we consider the application design and implementation issues, the advanced nonlinear control algorithms based on some of the methods mentioned above have also achieved some research results. From the perspective of theoretical research on sampling control implementation, it is of great significance for the realization of control algorithms to perform fixed-frequency or variable-frequency sampling analysis for nonlinear control systems. Taking the homogeneous control system as an example, the works [14, 68] analyze the quantitative relationship between the sampling period and the system stability for homogeneous control systems, and the realization mechanism of fixed frequency sampling is given. From the perspective of platform application, abundant application design schemes based on finite-time control algorithms for some specific practical engineering plants are reported, such as the spacecraft attitude control problem [95], the trajectory tracking control problem of surface vehicles [79], permanent magnet synchronous motor speed control problem [16], etc.

To sum up, although the research on nonlinear robust/adaptive control methodologies has achieved relatively fruitful results, it is worth noting that the current control design thoughts in the literature are mainly based on recursive design, hence it is similar to the backstepping method, the problems of complicated selection of control parameters, cumbersome recursive design process and a huge amount of calculation cannot be avoided in essence. Additionally, the existing nonsmooth control design methods usually require restrictive constraints on the nonlinear terms of the system, such as the commonly considered Hölder continuity requirements. As a consequence, practical implementations may still face a dilemma that the prior assumptions are very difficult to be verified, resulting in a questionable application potential.

The above problems also directly hinder the widespread application of these advanced nonlinear control algorithms in practical engineering. Oriented to practical engineering application, how to upgrade these deficiencies is also one of the topical issues in the field of nonlinear control theory research in recent years.

1.2 Overview of Nonrecursive Control Design for Nonlinear Systems

In order to simplify the nonlinear controller design procedure, reduce the difficulty of parameter selection, and relax the nonlinear constraints, some researchers have launched related research from the perspective of nonrecursive control design recently. The most immediate strength of this type of approach over the aforementioned recursive control methods is that the controller design can often be substantially separated from the Lyapunov function-based stability analysis.

For a chain of integrator systems, by introducing the homogeneous system theory, the work in [4] proposes a novel nonrecursive control design idea, meanwhile, provides a nonrecursive stability analysis approach. Later on, based on the result in [4], a class of homogeneous nonrecursive state observers are successively proposed in [54, 67], including the global/semi-global finite-time stability analysis. For a class of lower triangular systems with the presence of high-order nonlinearities, by introducing an enhanced homogeneous domination control strategy, a novel nonrecursive control method is proposed in [91], which realizes the design and integration of the controller, and is essentially detached from stability analysis based on the indirect Lyapunov method. This strategy can directly achieve the expression of the control law according to the structure of the control system, without designing a series of virtual controllers and recursive Lyapunov functions. The controller has the characteristics of simple controller form and easy selection of control gains. However, the global design still needs to oppose certain Hölder continuity constraints on the nonlinear terms of the system. Subsequently, based on the nonrecursive design framework of the homogeneous system approach, a class of output feedback controllers with a nonrecursive structure is achieved in [92] by setting a more practical control objective, namely semi-global stability. Referring to a recent result in [96], by utilizing homogeneous system theory, the authors present a nonrecursive method to design finite-time output feedback control laws for nonlinear systems, and apply it to the control issue for a class of single-degree-of-freedom manipulators. However, these methodologies usually have higher requirements for the accuracy of the control system model. When there exist internal parameter uncertainties (especially those non-parameterized uncertainties) and external disturbances in the system, a more conservative robust control strategy is usually adopted, that is, ensuring the steady-state error of the system to meet the design requirements via selecting a larger set of control gain parameters.

Furthermore, for a class of nonlinear systems with the presence of mismatched disturbances, rather than commonly seen matched disturbances, realizing precise decoupling of disturbances and developing higher performance output regulation schemes for such systems is never a trivial problem. In recent decades, there are also a number of results in the literature focusing on how to investigate the exact tracking control problem for such systems based on nonrecursive synthesis strategies. Aiming to solve the problem of nonsmooth output regulation for a class of nonlinear systems disturbed by time-varying mismatched disturbances, a finite-time output regulation control law for such systems is achieved [93]. Through employing a higher-order sliding mode observer for accurate disturbance estimation and subsequently decoupling the disturbances by the feedforward manner, a new framework of nonrecursive design is presented. However, in order to guarantee the absolute stability of the control system even in the sense of semi-global stability, the developed output regulation law has a certain degree of robustness redundancy, which may easily lead to problems such as undesirable transient-time performance of the control system under different initial conditions or different levels of disturbances.

For uncertain nonlinear systems, a typical handling strategy in the literature is to introduce a neural-network approximation or high-order sliding mode observation process to mitigate the complexity caused by recursive adaptive design [25,50,52,98]. In [98], by employing the neural network and dynamic surface control method, a novel adaptive nonrecursive design idea is proposed to avoid the common "explosion of complexity" problem in recursive approaches. In the case when the output of the system has certain constraints, integrating neural network and high-gain observation technology, a class of nonrecursive adaptive output feedback control law can be referred to [25]. Based on the homogeneous system theory, the nonrecursive robust control design method proposed in the work [91] is extended to the case of nonrecursive adaptive control design in [89], and a new type of continuously differentiable homogeneous adaptive control is proposed. Without recursively constructing a series of tuning functions and virtual controllers, the self-tuning gain mechanism is indirectly designed to obtain the updating configuration of the dynamic gain in the controller expression.

From the above literature review, it is clear to find out that: although nonrecursive control has been very common in the synthesis problems of linear systems, the research results of nonrecursive control problems for nonlinear systems in the literature are still very few, and the related design frameworks, analysis methods, and application design are not yet matured. In this regard, the initiative of this book is to summarize the existing related results from the authors to present a partial implementable nonrecursive control design framework.

1.3 Preliminaries

1.3.1 Introduction of nonsmooth control

For a general nonlinear dynamical system of the following form:

$$\dot{x} = f(t,x,u), \tag{1.1}$$

where $x \in \mathbb{R}^n$ is the system state, $u \in \mathbb{R}$ is the control input, and $f(\cdot) \in \mathbb{R}^n$ is a continuous nonlinear function.

If we consider the control design problem for system (1.1) under the framework of continuous-time design, it would be very straightforward to discuss about a significant control problem, i.e., the smoothness of the designed controller $u(t)$. From a mathematical point of view, the controller can be possibly designed as the following several manners:

$$(1)u(t) \in \mathbb{C}^0;\ (2)\,u(t) \in \mathbb{C}^1;\ (3)\,u(t) \in \mathbb{C}^l;\ l \geq 1 \in \mathbb{N};\ (4)\,u(t) \in \mathbb{C}^\infty.$$

Generally, one can define the controller design for nonlinear system (1.1) under the restriction that $u(t) \in \mathbb{C}^0$ as a typical nonsmooth control problem, which distinguishes itself essentially to continuously differentiable control ($u(t) \in \mathbb{C}^1, \mathbb{C}^l$), or smooth control ($\mathbb{C}^\infty$). Upon here, a naturally occurring problem is that:

For a given nonlinear system of the form (1.1)*, could the nonsmooth controller* ($u(t) \in \mathbb{C}^0$)*, continuously differentiable controller* ($u(t) \in \mathbb{C}^1, \mathbb{C}^l$) *or smooth controller* ($\mathbb{C}^\infty$) *be constructed in the meantime?*

From a theoretical point of view, indeed, what kind of influence will the controller with different smoothness have on the dynamic performance of the control system has aroused the attention from the system and control community for many decades.

Let's consider an illustrative simple 2-D nonlinear system of the form:

$$\begin{aligned}\dot{x}_1 &= x_2 + x_1^{2/3},\\ \dot{x}_2 &= u.\end{aligned} \tag{1.2}$$

For such system, a natural conjecture is to employ the famous backstepping controller to design a global nonlinear controller, however, owing to the fact that in the first-order subsystem, the existence of $x_1^{2/3}$ will directly lead to the insufficient smoothness of the virtual controller, therefore conventional backstepping method is not applicable for system (1.2). As a matter of fact, system (1.2) can not be stabilized by any continuously differentiable controller, and the system can only be stabilized by a continuously nondifferentiable controller, i.e., a $\mathbb{C}^0$ controller.

Further, we could also consider an illustrative real-life under-actuated mechanical system of the form [63]

$$\begin{aligned}\dot{x}_1 &= x_2,\\ \dot{x}_2 &= x_3^3 + \frac{g}{l}\sin x_1,\\ \dot{x}_3 &= x_4,\\ \dot{x}_4 &= u.\end{aligned} \tag{1.3}$$

According to the famous Brockett criterion [5], since the uncontrollable subsystem of the linearized system at the equilibrium point of the system is unstable, therefore, the system cannot be stabilized by any continuously differentiable feedback control law. Theoretically, systems that can be globally continuous and differentiable stabilized can also be globally nonsmooth stabilized, however in this case, the converse proposition is not true. Such underactuated mechanical systems cannot be stabilized by any continuously differentiable control law, therefore, nonsmooth control is the only choice for this kind of systems under the continuous-time control design requirements.

From the perspective of engineering application, the advanced control industry requires not only the stability of the system, but also the dynamic performance of the system to meet certain requirements. Among the dynamic performance indexes, the convergence rate and anti-disturbance ability of the closed-loop system are two key indexes. Compared with smooth control, the closed loop systems under nonsmooth control usually have faster convergence rate and better disturbance rejection performance, which has attracted extensive attentions on control application studies in recent years.

From the convergence point of view, the system states of nonsmooth control systems are able to converge to the equilibrium within a finite time, which is also called a finite-time control system [3].

For example, consider the following simple linear system:

$$\dot{x} = u. \tag{1.4}$$

It is very easy to design the following two controllers for system (1.4).

- $\mathbb{C}^0$ controller

$$u = -k\text{sign}(x)|x|^{\alpha},\ 0 < \alpha < 1,\ k > 0, \tag{1.5}$$

- $\mathbb{C}^\infty$ controller

$$u = -kx^{\beta},\ \beta \geq 1 \in \mathbb{N}_{\text{odd}},\ k > 0. \tag{1.6}$$

Under the smooth controller (1.6), without loss of generality, here we select $\beta = 1$, the closed-loop system

$$\dot{x} = -kx \tag{1.7}$$

has a solution of

$$x(t) = e^{-kt}x(0),$$

which implies that the closed-loop system (1.7) is globally asymptotically stable.

If we substitute the controller (1.5) into system (1.4), the closed-loop system gives:

$$\dot{x} = -k\text{sign}(x)|x|^{\alpha}. \tag{1.8}$$

Calculating the solution of (1.8) gives

$$x(t) = \begin{cases} \text{sign}(x(0))\left(|x(0)|^{1-\alpha} - k(1-\alpha)t\right)^{\frac{1}{1-\alpha}},\ 0 < t \leq \dfrac{|x(0)|^{1-\alpha}}{(1-\alpha)k}, \\ 0,\ t > \dfrac{|x(0)|^{1-\alpha}}{(1-\alpha)k}. \end{cases} \tag{1.9}$$

It can be concluded from (1.9) that, all the states of the nonsmooth control system (1.8) will converge to the origin within a finite time.

One can verify the fast convergence rate of a finite-time control system near the equilibrium point through a specific numerical simulation, depicted in Figures 1.1–1.2.

It is noted that under both the finite-time controller (1.5) and the smooth controller (1.6), the amplitude of the controller is almost at the same level for the sake of fair comparison. It can be clearly observed that, the nonsmooth control system (1.4)–(1.5) has shown a faster convergence rate than the smooth control system (1.4)–(1.6). The parameters are selected as $k = 1,\ \alpha = 0.5$ and $x(0) = 1$.

In practical systems, external disturbances are always inevitable for nonlinear systems. For a disturbed nonlinear system, the existence of nonvanishing disturbance will clearly affect the state convergence. Therefore, under a nonsmooth feedback control law, generally, the disturbances in the system cannot be completely attenuated. In other words, in this case, the finite-time convergence

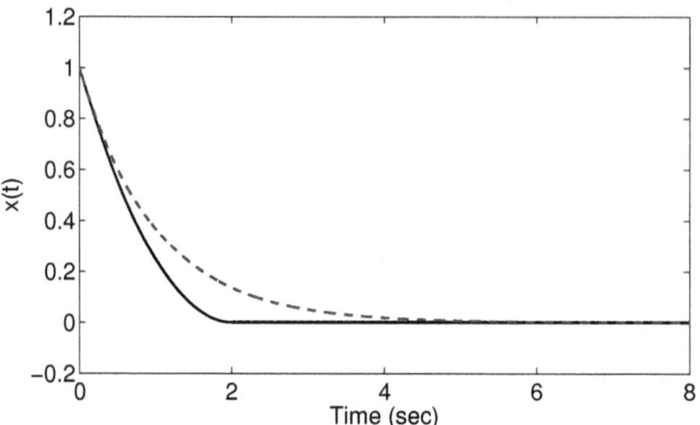

Figure 1.1 Response curves of system states.

characteristics of nonsmooth control systems will no longer be maintained. However, compared with smooth control system, nonsmooth control system normally performs better anti-disturbance performance. Consider system (1.4) again, and assume that the system has bounded external disturbance $d(t)$, that is:

$$\dot{x} = u + d(t), \tag{1.10}$$

where $d(t)$ satisfies $|d(t)| \leq d^*$. It can be simply concluded that under the nonsmooth controller (1.5), the steady state error of the closed-loop system (1.4)–(1.5) should be:

$$|x(\infty)| \propto \left(\frac{d^*}{k}\right)^{1/\alpha}. \tag{1.11}$$

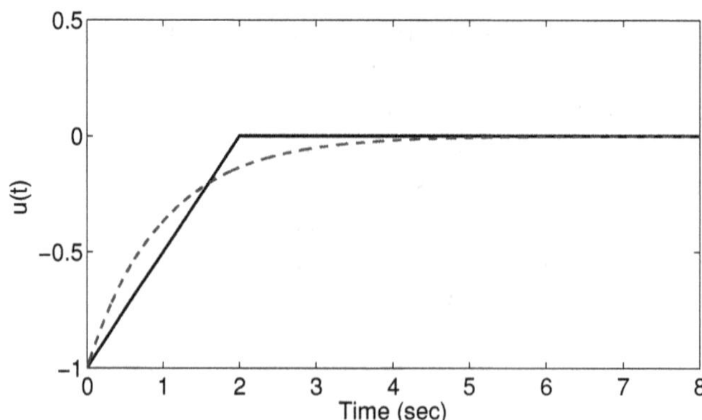

Figure 1.2 Response curves of the control inputs.

However, under the smooth controller (1.6), the steady state error of the closed-loop system (1.4)–(1.6) should be:

$$|x(\infty)| \propto \frac{d^*}{k}. \tag{1.12}$$

By comparing the formulas (1.11) and (1.12), one knows that under the smooth controller, in order to obtain a better anti-disturbance control performance, it would be necessary to continuously increase the controller gain k. One direct consequence is that the transient performance of the system might deteriorate. However, for the finite-time control system, in addition to adjusting the controller gain k, the engineers can also adjust an additional parameter, namely fractional power α.

Here, one can first adjust the gain k such that $k > d^*$, and then reduce the fractional power α, such that the boundary of the steady-state error $\left(\frac{d^*}{k}\right)^{1/\alpha}$ tends to zero, that is

$$\left(\frac{d^*}{k}\right)^{1/\alpha} << \frac{d^*}{k}.$$

The above analysis shows that the nonsmooth controller is able to significantly improve the anti-disturbance ability of the closed-loop system.

Next, we will verify through simulation that the nonsmooth control system(1.5)–(1.10) can show better disturbance attenuation ability than the smooth control system (1.6)–(1.10). Figures 1.3–1.4 show the state response and input response curves of the closed-loop system in the presence of external disturbances, respectively. Here, the parameters are selected as $k = 1,\ \alpha = 0.5,\ x(0) = 1$, and the external disturbance is set as $d(t) = 0.1\sin(t)$. It is obvious that the closed loop system under the finite-time controller has a better effect of restraining the external disturbance in the steady state.

1.3.2 Definitions and useful lemmas

Some definitions and useful lemmas for the sake of clear understanding of the deduction in the main theorems later on are presented below.

Definition 1.1 [29] For a fixed choice of coordinates $x = (x_1, \cdots, x_n) \in \mathbb{R}^n$, and positive real numbers $(r_1, r_2, \cdots, r_n) \triangleq r$, a one-parameter family of dilation is a map $\Delta^r : \mathbb{R}^+ \times \mathbb{R}^n \to \mathbb{R}^n$, defined by $\Delta_\varepsilon^r x = (\varepsilon^{r_1} x_1, \cdots, \varepsilon^{r_n} x_n)$.

- For a given dilation Δ^r and a real number k. A continuous function $V : \mathbb{R}^n \to \mathbb{R}$ is called Δ^r−homogeneous of degree k, denoted by $V \in \mathbb{H}_{\Delta^r}^k$ if $V \circ \Delta_\varepsilon^r = \varepsilon^k V$.
- A continuous vector field $f(x) = \sum f_j(x)(\frac{\partial}{\partial x_j})$ is Δ^r− homogeneous of degree τ, if $f_j \in \mathbb{H}_{\Delta^r}^{\tau + r_j},\ j \in \mathbb{N}_{1,n}$.

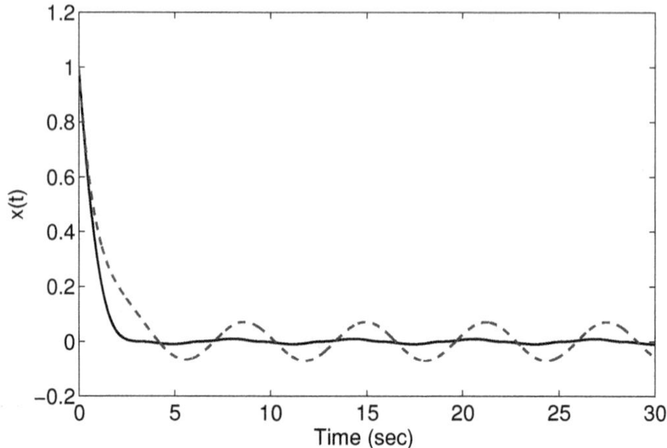

Figure 1.3 State response curves of the system (1.4) with the presence of external disturbance.

Definition 1.2 [29] For $p > 1, x \in \mathbb{R}^n$, the homogeneous p-norm of x regarding the dilation Δ^r is

$$\|x\|_{\Delta^r} \triangleq \left(\sum_{i=1}^{n} |x_i|^{\frac{p}{r_i}} \right)^{\frac{1}{p}}.$$

Without loss of generality, one could choose $p = 2$.

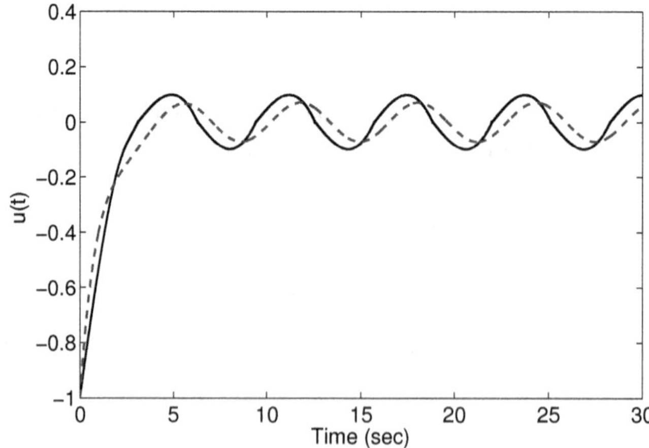

Figure 1.4 Input response curves with the presence of external disturbance.

Definition 1.3 A homogeneous vector $\bar{x}^{\tau}_{i\Delta^r}$ is denoted by $\bar{x}^{\tau}_{i\Delta^r} \triangleq \left(x_1^{\frac{\tau}{r_1}}, \cdots, x_i^{\frac{\tau}{r_i}}\right)^{\top}$, $x^{\tau}_{\Delta^r} = \bar{x}^{\tau}_{n\Delta^r}$ and $\lfloor x \rceil^{\tau}_{\Delta^r} = \left(\lfloor x_1 \rceil^{\frac{\tau}{r_1}}, \cdots, \lfloor x_n \rceil^{\frac{\tau}{r_n}}\right)^{\top}$.

Definition 1.4 [26] Consider a system

$$\dot{x} = f(x) + g(x)u, \; x \in \mathbb{R}^n, \tag{1.13}$$

with $x = 0$ as its equilibrium.

- System (1.13) is said to be *semi-globally stabilizable* if, for each (arbitrarily large) compact subset $\mathbb{P} \subset \mathbb{R}^n$, there exists a feedback law $u = u(x)$, which in general depends on $\mathbb{P}$, such that the equilibrium is locally asymptotically stable and $x(t_0) \in \mathbb{P} \Rightarrow \lim\limits_{t\to\infty} x(t) = 0$.
- A trajectory of system (1.13) is said to be *captured by the set* $\mathbb{Q}$ if it is defined for all $t \in [t_0, \infty)$, enters the set $\mathbb{Q}$ at some finite time T and remains in this set for all $t \geq T$.
- System (1.13) is said to be *semi-globally practically stabilizable* if, for any (arbitrarily large) set $\mathbb{P}$ and any arbitrarily small set $\mathbb{Q}$, there is a feedback law $u = u(x)$, which in general depends on both $\mathbb{P}$ and $\mathbb{Q}$, such that any trajectory with the initial condition in $\mathbb{P}$ is captured by the set $\mathbb{Q}$.

Lemma 1.1
[4] Let $V_1(x) \in \mathbb{H}^{\tau_1}_{\Delta^r}$ and $V_2(x) \in \mathbb{H}^{\tau_2}_{\Delta^r}$, respectively, then the following statements hold:

i) $V_1(x)V_2(x) \in \mathbb{H}^{\tau_1+\tau_2}_{\Delta^r}$.

ii) $\frac{\partial V_1(x)}{\partial x_i} \in \mathbb{H}^{\tau_1 - r_i}_{\Delta^r}$, $i \in \mathbb{N}_{1:n}$.

iii) If $V_1(x)$ is positive definite, then the following relation holds

$$\left(\min_{\{x:V_1(x)=1\}} V_2(x)\right) V_1^{\frac{\tau_2}{\tau_1}}(x) \leq V_2(x) \leq \left(\max_{\{x:V_1(x)=1\}} V_2(x)\right) V_1^{\frac{\tau_2}{\tau_1}}(x).$$

Lemma 1.2
[32] Consider the nonlinear system

$$\dot{x} = f(t, x, \eta). \tag{1.14}$$

Let $\mathcal{V} : [0, \infty) \times \mathbb{R}^n \to \mathbb{R}$ be a continuously differentiable function such that

$$\alpha_1(\|x\|) \leq \mathcal{V}(t, x) \leq \alpha_2(\|x\|),$$

$$\frac{\partial \mathcal{V}}{\partial t}+\frac{\partial \mathcal{V}}{\partial x}f(t,x,\eta)\leq -\Gamma(x),\ \forall \|x\|\geq \rho(\|\eta\|)>0,$$

for all $(t,x,\eta)\in[0,\infty)\times\mathbb{R}^n\times\mathbb{R}^m$, *where* α_1, α_2 *are class* $\mathcal{K}_\infty$ *functions,* ρ *is a class* $\mathcal{K}$ *function, and* $\Gamma(x)$ *is a continuous positive definite function on* $\mathbb{R}^n$. *Then system* (1.14) *is globally input-to-state stable (ISS).*

Lemma 1.3

[32] If the following conditions are satisfied:

- *system* $\dot{x}=f(t,x,\eta)$ *is globally ISS;*
- $\lim_{t\to\infty}\eta=0,$

then system (1.14) *is globally asymptotically stable.*

Lemma 1.4

[3] Consider a dynamical system $\dot{x}=f(x,t)$, $f(0,t)=0$. *Suppose there exist a* $\mathbb{C}^1$ *positive-definite and proper function* $V:\mathbb{R}^n\to\mathbb{R}^n$ *and real numbers* $\mu>0$ *and* $\iota\in(0,1)$, *such that* $\dot{V}+\mu V^{\iota}$ *is semi-negative definite, then the origin* $x=0$ *is a globally finite-time stable equilibrium with a settling time* $T\leq\frac{V^{1-\iota}(x_0,t_0)}{\mu(1-\iota)}$ *for any given initial condition* $x_0=x(t_0)$.

Lemma 1.5

[3] Let $\dot{x}=f(x)$, $x\in\mathbb{R}^n$ *where* f *is a continuous vector field homogeneous of degree* τ. *If the origin of* $\dot{x}=f(x)$ *is locally asymptotically stable, then it is also globally stable. If* τ *is negative definite, the origin of* $\dot{x}=f(x)$ *is globally finite-time stable.*

Lemma 1.6

[54] Consider the product space $\mathcal{X}\times\mathcal{Y}$ *where* $\mathcal{Y}$ *is a compact set. Let* $\mathbb{O}$ *be an open set of* $\mathcal{X}\times\mathcal{Y}$ *which contains a slice* $\{\mathbf{x}_0\}\times\mathcal{Y}$ *of* $\mathcal{X}\times\mathcal{Y}$, *then* $\mathbb{O}$ *will contain some tube* $\mathcal{W}\times\mathcal{Y}$ *about* $\{\mathbf{x}_0\}\times\mathcal{Y}$ *where* $\mathcal{W}$ *is a neighborhood of* $\mathbf{x}_0$ *in* $\mathcal{X}$.

Lemma 1.7

[83] Let $f:\ \mathbb{R}^n\to\mathbb{R}$ *be a* $\mathbb{C}^1$ *mapping and* $\Gamma_M=[-M,M]^n$ *is a compact set in* $\mathbb{R}^n$ *with* $M>0$ *being a real number. Then* $\forall\rho_i\in(0,1]$, $i\in\mathbb{N}_{1:n}$, *there exists a constant* $\gamma\geq1$ *depending on M, such that* $\forall x\in\Gamma_M$ *and* $\forall\hat{x}\in\Gamma_M$, *the following relation holds*

$$|f(x)-f(\hat{x})||_{\Gamma_M}\leq\gamma\sum_{i=1}^{n}|x_i-\hat{x}_i|^{\rho_i}.$$

Lemma 1.8

[34] Let $D=\mathrm{diag}\{d_1,d_2,\cdots,d_n\}$ *where* $d_i>0$, $i\in\mathbb{N}_{1:n}$ *is a constant and* $A\in\mathbb{R}^{n\times n}$, $B\in\mathbb{R}^n$ *be matrices in the controllable canonical form. Then there exist a*

positive definite, symmetrical matrix $P \in \mathbb{R}^{n\times n}$ *and a row vector* $K = [k_1, \cdots, k_n]$, *and two positive constants* ν_1 *and* ν_2, *such that*

$$(A-BK)^\top P + P(A-BK) \leq -I, \ \nu_1 I \leq DP + PD \leq \nu_2 I.$$

Lemma 1.9

(Barbalat's lemma) If $\lim_{t\to\infty} f(t) < \infty$ *exists and if* $\dot f$ *is uniformly continuous (or* $\ddot f$ *is bounded), then* $\lim_{t\to\infty} \dot f(t) = 0$.

Lemma 1.10

[44] For any real-valued continuous function $f(x,y)$ *where* $x \in \mathbb{R}^m$, $y \in \mathbb{R}^n$, *there exist two smooth scalar functions* $a(x) \geq 1$ *and* $b(y) \geq 1$ *such that*

$$|f(x,y)| \leq a(x)b(y).$$

Lemma 1.11

[83] Let c, d *be any real numbers, and* $p \in (0,1]$. *Then the following inequality holds:*

$$|c-d| \leq 2^{1-p}|\lfloor c\rceil^{1/p} - \lfloor d\rceil^{1/p}|^p.$$

Lemma 1.12

[60] Let c, d *be positive constants. Given any positive smooth function* $\pi(x,y)$, *the following inequality holds:*

$$|x|^c|y|^d \leq \frac{c}{c+d}\pi(x,y)|x|^{c+d} + \frac{d}{c+d}\pi^{-c/d}(x,y)|y|^{c+d}.$$

Lemma 1.13

[15] Assume $a \in [-M,M]$, $b \in \mathbb{R}$ *with the threshold* $M > 0$. *The following relation holds:*

$$|a - \langle b\rangle_M| \leq |a-b|.$$

Chapter 2

Nonrecursive Robust Stabilization for Uncertain Nonlinear Systems

By noting that classical nonlinear control design approaches usually follow a systematic recursive synthesis manner, such as popular backstepping and adding a power integrator methods, therefore an obvious long-term argument appears here naturally:

How can we ease the recursive design and propose simpler control laws with strong robustness while the implementation steps are user-friendly?

However, there are quite a few existing results on nonlinear control design may disobey the beauty of simplicity and hence when control engineers try to implement those enormous nonlinear controllers, a series of costly calculations (or, estimations) have been concurred first in order to proceed. As a consequence, control engineers will finally return to simple PID controllers or state-feedback controllers via local-linearization methods, etc.

In this chapter, we will investigate a simple one-step stabilization framework for a class of uncertain nonlinear systems. A novel global nonrecursive stabilization design framework is addressed for a class of inherent nonlinear systems with the presence of system uncertainties and external nonvanishing disturbances. By virtue of the facility that the weighted homogeneity brings into the system synthesis procedure, a nonrecursive design method is proposed to yield a globally

DOI: 10.1201/9781003399230-2

effective robust controller with its expression following a quasi-linear manner. By proceeding with a rigorous nonrecursive stability analysis framework, which covers both global asymptotical and finite-time convergence cases, the common recursively treated derivative items in backstepping based methods are totally avoided. Inspired by the homogeneous domination technique, a scaling gain performed as a bandwidth factor is introduced into the original system and hence the robustness of the controlled system can be adjusted to meet the practical performance requirements.

2.1 Problem Formulation

To begin with, we revisit the control problem of realizing the global stabilization for the following uncertain nonlinear system

$$\begin{cases} \dot{x}_i = x_{i+1} + f_i(t, \bar{x}_i, \theta, d), \ i \in \mathbb{N}_{1:n-1}, \\ \dot{x}_n = u + f_n(t, x, \theta, d), \end{cases} \tag{2.1}$$

where $\bar{x}_i = \mathrm{col}(x_1, x_2, \cdots, x_i)$, $x = \bar{x}_n$, and u are the system partial state vector, full state vector, and control input, respectively, $\theta \in \mathbb{R}^l$ is an unknown bounded parameter vector which could be either time-varying or constant, d represents the external mismatched disturbances which should be bounded, $f_i(\cdot)$, $i \in \mathbb{N}_{1:n}$ is a continuous nonlinear function satisfying a vanishing condition. The initial time is set as t_0 while the initial state vector is denoted by $x_0 \triangleq x(t_0)$.

The control objective of this chapter is to find a stabilizing controller of the following simple form

$$u(t) = u(x, \tau, K, L), \tag{2.2}$$

where τ is a homogeneous degree, K is the coefficient vector of a Hurwitz polynomial $s^n + k_n s^{n-1} + \cdots + k_2 s + k_1$ and $L \geq 1$ is a scaling gain such that:

- *Global Stability:* In the case when system (2.1) is not disturbed by external disturbances, the system states will converge to the origin asymptotically provided $\tau \geq 0$, or within a finite time provided $\tau < 0$.

- *Global Practical Stability:* In the case when system (2.1) suffers from nonvanishing bounded external disturbances, for any given tolerance $\varepsilon > 0$ and initial value $x(0)$, there exists a finite time $T^* > 0$, such that

$$\|x(t)\| \leq \varepsilon, \ \forall t > T^*.$$

2.2 Motivation—A Linear Case Study

In order to motivate the nonlinear controller design under a nonrecursive synthesis framework, we first present an illustrative linear case study to depict the main design idea of this chapter.

To proceed, a necessary assumption expressed as follows is required to be verified for system (2.1).

Assumption 2.2.1 *There exists a known constant* $\sigma > 0$, *such that*

$$|f_i(t,x,\theta,d)| \leq \sigma \sum_{j=1}^{i} |x_j|, \; i \in \mathbb{N}_{1:n}.$$

Subsequently, we will go through an explicit nonrecursive stability analysis procedure to obtain the global stabilizing controller and show that the control objective can be achieved with a unified framework of analysis.

First, we introduce the scaling gain L into system (2.1) by using a change of coordinates ζ defined by

$$\zeta = \mathbb{L}x, \; \mathbb{L} = \mathrm{diag}\{1/L^0, 1/L, \cdots, 1/L^{n-1}\}, \; u = L^n v. \tag{2.3}$$

Up to now, without any Lyapunov functions design and recursive deductions, a linear stabilizer is able to be created of the following form:

$$u = L^n v, \; v = -K\zeta. \tag{2.4}$$

Theorem 2.1

Consider the closed-loop system consisting of (2.1) *satisfying Assumption 2.2.1 and the linear controller* (2.4). *There exists a sufficiently large scaling gain L, such that the closed-loop system can be rendered globally asymptotically stable.*

Proof: With (2.3), system (2.1) is equivalent to the following one

$$\begin{cases} \dot{\zeta}_i = L\zeta_{i+1} + f_i(t,x,\theta,d)/L^{i-1}, \; i \in \mathbb{N}_{1:n-1}, \\ \dot{\zeta}_n = Lv + f_n(t,x,\theta,d)/L^{n-1}. \end{cases} \tag{2.5}$$

Build a Lyapunov function as $V(\zeta) = \zeta^\top P\zeta$, where P is a positive definite and symmetrical matrix that satisfies the following equation:

$$\mathcal{A}^\top P + P\mathcal{A} = -I, \tag{2.6}$$

with $\mathcal{A} = \begin{bmatrix} 0 & 1 & \cdots & 0 \\ \vdots & \vdots & \ddots & \vdots \\ 0 & 0 & \cdots & 1 \\ -k_1 & -k_2 & \cdots & -k_n \end{bmatrix}$.

Then, taking the time derivative of $V(\zeta)$ along (2.5) yields

$$\dot{V}(\zeta) = L\frac{\partial V(\zeta)}{\partial \zeta^{\top}}A\zeta + \frac{\partial V(\zeta)}{\partial \zeta^{\top}}F(\cdot), \tag{2.7}$$

where $F(\cdot) = \left(\frac{f_1(t,x,\theta,d)}{L^0}, \frac{f_2(t,x,\theta),d)}{L^1}, \cdots, \frac{f_n(t,x,\theta,d)}{L^{n-1}}\right)$.

By Assumption 2.2.1, the following inequalities can be obtained

$$\begin{aligned}\frac{f_i(\cdot)}{L^{i-1}} &\le \frac{\sigma}{L^{i-1}}\left(|x_1| + |x_2| + \cdots + |x_i|\right)\\ &= \frac{\sigma}{L^{i-1}}\left(|L^0\zeta_1| + |L^1\zeta_2| \cdots + |L^{i-1}\zeta_i|\right)\\ &\le \sigma(|\zeta_1| + |\zeta_2| + \cdots + |\zeta_i|)\\ &\le \sigma\sqrt{n}\|\zeta\|.\end{aligned} \tag{2.8}$$

Then the following relation holds:

$$\frac{\partial V(\zeta)}{\partial \zeta^{\top}}F(\cdot) \le c\|\zeta\|^2, \tag{2.9}$$

where $c \in \mathbb{R}^+$ is a constant.

Keeping (2.6) and (2.9) in mind, (2.7) can be rewritten as

$$\begin{aligned}\dot{V}(\zeta) &\le -L\|\zeta\|^2 + c\|\zeta\|^2\\ &\le -(L-c)\|\zeta\|^2.\end{aligned} \tag{2.10}$$

Now we are able to choose the scaling gain L under the following guideline:

$$L \ge c + 1, \tag{2.11}$$

such that (2.10) reduces to $\dot{V}(\zeta) \le -\|\zeta\|^2$. Then it is clear that the closed-loop system is globally asymptotically stable.

2.3 Nonrecursive Robust Stabilization—A Unified Framework

With the above illustration of a simple linear nonrecursive control design framework, it is shown that under a linear growth condition, we are able to derive a simple state feedback controller by utilizing a feedback domination design strategy. By further extending the main design conception, now it is made possible to derive a nonrecursive global stabilization design framework for nonlinear systems by taking advantage of the homogeneous system theory.

In this section, a novel nonrecursive control design framework is presented to stabilize the nonlinear system of the form (2.1). Firstly, by constructing a novel

Lyapunov function which adopts a quasi-linear system synthesis manner, the virtual controllers are not required to render recursive design steps in essence. Secondly, a simple control law, which consists of a scaling gain as a bandwidth factor and a quasi-linear state feedback law, can now be explicitly constructed without recursively determining procedures. Taking consideration of the case when the nonlinear system is subject to bounded mismatched disturbances, a sufficiently large value of the scaling gain can make the steady error as small as possible. Notably, without the presence of external disturbances, now both asymptotic and finite-time stabilization results can be established under a unified framework whereas the sign of the homogeneous degree performs as a flexibly adjustable factor.

2.3.1 Global unified nonrecursive control design framework

In order to realize a global control objective, Assumption 2.2.1 on the system (2.1) is modified into the following form:

Assumption 2.3.1 *There exist a constant* $\tau > -\frac{1}{n}$ *and two known constants* $\sigma > 0,\ \gamma \geq 0$ *such that*

$$|f_i(t,x,\theta,d| \leq \sigma \sum_{j=1}^{i} |x_j|^{\frac{r_i+\tau}{r_j}} + \gamma,\ i \in \mathbb{N}_{1:n},$$

where $r \triangleq (r_1, r_2, \cdots, r_n)$ *with* $r_1 = 1,\ r_i = 1 + (i-1)\tau,\ i \in \mathbb{N}_{1:n}$ *are positive real numbers.*

Remark 2.1 Assumption 2.3.1 can be regarded as a generalized version of the assumption proposed in [55] where the external disturbance items are taken into consideration. In the case when $\gamma = 0$, the system nonlinearities can be seen as a special vanishing case, i.e., $f_i(0) = 0$. More specifically, in the case when $\tau = 0$, Assumption 2.3.1 covers a wide class of linear growth nonlinearity assumptions.

The following lemma is necessary for subsequent homogeneous controller design.

Lemma 2.1
Consider the following chain of integrators

$$\dot{x}_i = x_{i+1},\ i \in \mathbb{N}_{1:n-1},\ \dot{x}_n = v, \tag{2.12}$$

under a homogeneous control law of the following form

$$v = -K\lfloor x \rceil_{\Delta^r}^{r_n+\tau}. \tag{2.13}$$

There exist two constants $\varsigma_1 \in (0, \frac{1}{n})$ *and* $\varsigma_2 \in (0, +\infty)$, *such that the closed-loop system (2.12)–(2.13) is globally finite-time stable for* $\tau \in (-\varsigma_1, 0)$ *or globally asymptotically stable for* $\tau \in [0, \varsigma_2)$.

Proof: With the dilation mapping $\Delta_\varepsilon^r x$, one can easily obtain the following relations

$$\begin{cases} x_{i+1} \circ \Delta_\varepsilon^r = \varepsilon^{r_{i+1}} x_{i+1} = \varepsilon^{r_i+\tau} x_{i+1},\ i \in \mathbb{N}_{1:n-1}, \\ v \circ \Delta_\varepsilon^r = \varepsilon^{r_n+\tau} v, \end{cases} \tag{2.14}$$

which conclude that the closed-loop system (2.12)–(2.13) is Δ^r−homogeneous of degree τ.

Inspired by the finite-time homogeneous analysis method in [4] and [54], construct a homogeneous Lyapunov function candidate in the following form

$$V(\tau,x) = \left(\lfloor x \rceil_{\Delta^r}^{\kappa-\frac{\tau}{2}}\right)^\top P \lfloor x \rceil_{\Delta^r}^{\kappa-\frac{\tau}{2}}, \tag{2.15}$$

with P satisfying (2.6).

With the denotation $\kappa \geq \max\{r_1, r_i+\tau\}_{i\in\mathbb{N}_{1:n}}$ in mind, we can conclude that

$$V(\tau,x) \in \mathbb{C}^1 \cap \mathbb{H}_{\Delta^r}^{2\kappa-\tau}.$$

Define a set $\mathbb{S} = \{x \in \mathbb{R}^n : V(0,x) = 1\}$ which is a compact set of $\mathbb{R}^n$, and a continuous function φ as

$$\begin{cases} \varphi : \left(-\dfrac{1}{n}, +\infty\right) \times \mathbb{S} \to \mathbb{R}, \\ \varphi(\tau,x) = \dfrac{\partial V(\tau,x)}{\partial x}\dot{x}\Big|_{(2.12)-(2.13)}. \end{cases} \tag{2.16}$$

Considering the case when $\tau = 0$, it is clear that the closed-loop system (2.12)–(2.13) reduces to $\dot{x} = \mathcal{A}x$. Note that $\mathcal{A}$ is Hurwitz. Hence, we can conclude that $\varphi(0,x) = -\|x\|^2$. This relation concludes $\forall x \in \mathbb{S},\ \varphi(0,x) < 0$, that is, $\varphi(\{0\} \times \mathbb{S}) \subset (-\infty, 0)$.

By utilizing the Tube Lemma (Lemma 1.6), there exist two constants $\varsigma_1 > 0$ and $\varsigma_2 > 0$, such that

$$\varphi((-\varsigma_1, \varsigma_2) \times \mathbb{S}) \subset (-\infty, 0).$$

This concludes that for $\tau \in (-\varsigma_1, \varsigma_2)$, the closed-loop system is locally stable. It then follows directly from Lemma 1.5 that the closed-loop system is also globally stable.

In addition, note that $\frac{\partial V(\tau,x)}{\partial x}\dot{x} \in \mathbb{H}_{\Delta^r}^{2\kappa}$. Using Lemma 1.1, the following relation can be obtained:

$$\frac{\partial V(\tau,x)}{\partial x}\dot{x} \leq -cV^{\frac{2\kappa}{2\kappa-\tau}}(\tau,x),\ \tau \in (-\varsigma_1, \varsigma_2), \tag{2.17}$$

where $c > 0$ is a constant dependent of τ. Moreover, the closed-loop system is globally asymptotically stable in the case when $\tau \in [0, \varsigma_2)$, or globally finite-time stable in the case when $\tau \in (-\varsigma_1, 0)$. ■

Different from the smooth controller expression (2.4), a unified nonlinear stabilizing controller can be developed in the following form:

$$u = -L^n K \lfloor \zeta \rceil_{\Delta^r}^{r_n+\tau}. \tag{2.18}$$

At this stage, we are able to straightforwardly present the following theorem.

Theorem 2.2

Consider the closed-loop system consisting of system (2.1) satisfying Assumption 2.3.1 and the controller (2.18) with a sufficient large scaling gain L determined by

$$Lc - \tilde{c}\left(L^{1-\frac{1}{\max\limits_{i\in\mathbb{N}_{1:n}}\{r_i\}}} + 1\right) \geq 1, \tag{2.19}$$

where c, $\tilde{c}$ are positive constants. Then the following statements hold:

- *In the case when $\gamma = 0$, the closed-loop system is globally asymptotically stable provided $\tau \in [0, \varsigma_2)$ or globally finite-time stable provided $\tau \in (-\varsigma_1, 0)$.*
- *In the case when $\gamma > 0$, for any given tolerance $\varepsilon > 0$ and initial value $x(0)$, there exists a finite time $T^* > 0$, such that $\|x\| \leq \varepsilon$, $\forall t > T^*$.*

Proof: Construct a Lyapunov function

$$V(\tau,\zeta) = \left(\lfloor \zeta \rceil_{\Delta^r}^{\kappa-\frac{\tau}{2}}\right)^{\top} P \lfloor \zeta \rceil_{\Delta^r}^{\kappa-\frac{\tau}{2}},$$

where $\kappa \geq \max\{r_1, r_i+\tau\}_{i\in\mathbb{N}_{1:n}}$ and P satisfies the relation (2.6), it is clear that $V \in \mathbb{H}_{\Delta^r}^{2\kappa-\tau}$.

Different from the linear controller design analysis, by Assumption 2.3.1, the following inequalities can be obtained:

$$\begin{aligned} \frac{f_i(\cdot)}{L^{i-1}} &\leq \frac{\sigma}{L^{i-1}}\left(|x_1|^{\frac{r_i+\tau}{r_1}} + \cdots + |x_i|^{\frac{r_i+\tau}{r_i}}\right) + \frac{\gamma}{L^{i-1}} \\ &= \frac{\sigma}{L^{i-1}}\left(|L^0\zeta_1|^{\frac{r_i+\tau}{r_1}} + \cdots + |L^{i-1}\zeta_i|^{\frac{r_i+\tau}{r_i}}\right) + \frac{\gamma}{L^{i-1}}. \end{aligned} \tag{2.20}$$

Note that for $j \in \mathbb{N}_{1:i}$, the following relation can be derived:

$$\begin{aligned} (j-1)\frac{r_i+\tau}{r_j} - (i-1) &= \frac{(j-1)r_i + (j-1)\tau - (i-1)r_j}{r_j} \\ &= \frac{(j-i) + (j-1)(i-1)\tau + (j-1)\tau - (i-1)(j-1)\tau}{r_j} \\ &\leq \frac{(j-1)\tau}{r_j} = \frac{r_j - 1}{r_j} \\ &\leq 1 - \frac{1}{\max\limits_{j\in\mathbb{N}_{1:i}}\{r_j\}} < 1. \end{aligned} \tag{2.21}$$

Denote $\bar{f}_i(\bar{\zeta}_i) = |\zeta_1|^{\frac{r_i+\tau}{r_1}} + \cdots + |\zeta_i|^{\frac{r_i+\tau}{r_i}}$. With (2.21) in mind, now we have

$$\frac{f_i(\cdot)}{L^{i-1}} \leq \sigma L^{1-\frac{1}{\max_{i\in\mathbb{N}_{1:n}}\{r_i\}}} \bar{f}_i(\bar{\zeta}_i) + \frac{\gamma}{L^{i-1}}. \tag{2.22}$$

Keeping Lemmas 1.1–1.12 in mind and from (2.22), there exist constants $\bar{c} > 0$ and $\tilde{c} > 0$, such that the following relations hold:

$$\begin{aligned}
\sum_{i=1}^{n} \frac{\partial V(\tau,\zeta)}{\partial \zeta_i}\frac{\tilde{f}_i}{L^{i-1}} &\leq \sum_{i=1}^{n} \sigma L^{1-\frac{1}{\max_{i\in\mathbb{N}_{1:n}}\{r_i\}}} \left|\frac{\partial V(\tau,\zeta)}{\partial \zeta_i}\right| \bar{f}_i(\bar{\zeta}_i) + \sum_{i=1}^{n}\left|\frac{\partial V(\tau,\zeta)}{\partial \zeta_i}\right| \frac{\gamma}{L^{i-1}} \\
&\leq \bar{c} L^{1-\frac{1}{\max_{i\in\mathbb{N}_{1:n}}\{r_i\}}} V^{\frac{2\kappa}{2\kappa-\tau}}(\tau,\zeta) + \sum_{i=1}^{n}\left|\frac{\partial V(\tau,\zeta)}{\partial \zeta_i}\right| \left(\frac{\gamma^{\frac{1}{r_i+\tau}}}{L^{\frac{i-1}{r_i+\tau}}}\right)^{r_i+\tau} \\
&\leq \tilde{c}\left(L^{1-\frac{1}{\max_{i\in\mathbb{N}_{1:n}}\{r_i\}}} + 1\right) V^{\frac{2\kappa}{2\kappa-\tau}}(\tau,\zeta) + \sum_{i=1}^{n} \frac{\tilde{c}\gamma^{\frac{2\kappa}{r_i+\tau}}}{L^{\frac{2\kappa(i-1)}{r_i+\tau}}} \\
&:= \tilde{c}\left(L^{1-\frac{1}{\max_{i\in\mathbb{N}_{1:n}}\{r_i\}}} + 1\right) V^{\frac{2\kappa}{2\kappa-\tau}}(\tau,\zeta) + \Gamma.
\end{aligned} \tag{2.23}$$

From Lemma 2.1, calculating the derivative of $V(\tau,\zeta)$ along the closed-loop system (2.5)–(2.18) gives

$$\begin{aligned}
\dot{V}(\tau,\zeta) &= \frac{\partial V(\tau,\zeta)}{\partial \zeta} L(\zeta_2,\cdots,\zeta_n,v)^{\top} + \sum_{i=1}^{n}\frac{\partial V(\tau,\zeta)}{\partial \zeta_i}\frac{f_i(\cdot)}{L^{i-1}} \\
&\leq -\left(Lc - \tilde{c}\left(L^{1-\frac{1}{\max_{i\in\mathbb{N}_{1:n}}\{r_i\}}} + 1\right)\right) V^{\frac{2\kappa}{2\kappa-\tau}}(\tau,\zeta) + \Gamma.
\end{aligned} \tag{2.24}$$

By noting that $L \geq 1$, following the guideline (2.19), (2.24) will be yield to

$$\dot{V}(\tau,\zeta) \leq \begin{cases} -V^{\frac{2\kappa}{2\kappa-\tau}}(\tau,\zeta),\ \gamma = 0; \\ -V^{\frac{2\kappa}{2\kappa-\tau}}(\tau,\zeta) + \Gamma,\ \gamma > 0. \end{cases} \tag{2.25}$$

- In the case when $\gamma = 0$ and $\tau \in [0, \varsigma_2)$, we know that $\frac{2\kappa}{2\kappa-\tau} \geq 1$, (2.25) will lead to a conclusion that the closed-loop system is globally asymptotically stable. If $\tau \in (-\varsigma_1, 0)$, then $\frac{2\kappa}{2\kappa-\tau} \in (0,1)$. Hence the finite-time stabilization result can be concluded straightforwardly from Lemma 1.4.
- In the case when $\gamma > 0$, one can conclude from (2.25) and Lemma 1.1 that there exists a constant $\hat{c} > 0$ and a finite-time T^*, such that

$$\|\zeta\|_{\Delta^r} \leq \hat{c} V^{\frac{1}{2\kappa-\tau}} \leq \hat{c}(2\Gamma)^{\frac{1}{2\kappa}},\ \forall t \geq T^* > 0. \tag{2.26}$$

It is clear that Γ is a bounded constant and can be made arbitrarily small by tuning the scaling gain L to be sufficiently large. Hence the practical stability of the closed-loop system can be achieved.

This completes the proof of Theorem 2.2. ■

Remark 2.2 Without external disturbances, the proposed nonrecursive design framework shows us a novel handling strategy to achieve both global asymptotical and finite-time stabilization results. By taking advantage of the weighted homogeneity for the considered inherent nonlinear system, now the controller has several new features. First, the control law is a result by totally avoiding the costly calculations of a series of partial derivative terms and hence the expression is very simple, as depicted by (2.18). Second, without the virtual controller design in recursive approaches, the stability analysis can be done within only one step and hence is easier to be realized. Third, the control gains now can be simply selected as the coefficients of a Hurwitz polynomial which will significantly ease the practical implementations.

Remark 2.3 As is well acknowledged in practical plants that external disturbances are always inevitable, the proposed method presents us with an alternative robust control law design procedure. By introducing into the controller a constant bandwidth factor, the robustness margin of the controlled system can be artificially tuned and the steady error can be practically adjusted to meet certain control requirements.

2.3.2 Extension to non-triangular systems

Note that Assumption 2.3.1 can be classified as a nonlinear lower-triangular structural condition. However, the proposed generalized control scheme can also be applicable to solve a class of non-triangular nonlinear systems which can be treated as a direct application of Theorem 2.2.

Assumption 2.3.2 *Assume there exist constants $\nu > 0$, $\sigma > 0$, and $\gamma \geq 0$ such that the following relation holds for system* (2.1)

$$|f_i(t,\zeta(t),L,\theta(t),d(t))/L^{i-1}| \leq \sigma L^{1-\nu}\|\zeta\|_{\Delta^r}^{r_i+\tau} + \gamma/L^{i-1},\ i \in \mathbb{N}_{1:n}.$$

Based on the above assumption, one can obtain the following theorem.

Theorem 2.3

Under Assumption 2.3.2, there exists a sufficient large scaling gain $L \geq 1$, such that the closed-loop system (2.1)–(2.18) *can be rendered globally practically stable provided $\gamma > 0$ and globally stable provided $\gamma = 0$.*

Proof: The proof of Theorem 2.3 can be straightforwardly derived following a similar analysis with the proof of Theorem 2.2 and hence is omitted here. ■

2.3.3 Numerical simulation

In this subsection, a numerical example and its control performance simulation is provided in order to illustrate the effectiveness of the proposed nonrecursive control law design.

Example 2.3.1 *Consider the following nonlinear system:*

$$\begin{aligned}\dot{x}_1 &= x_2 + d_1(t),\\ \dot{x}_2 &= x_3 + \theta(t)\ln(1+x_2^2),\\ \dot{x}_3 &= u + d_2(t),\end{aligned} \tag{2.27}$$

where $|\theta(t)| \le 2$ *is an uncertain time-varying parameter and* $d_1(t)$ *and* $d_2(t)$ *are two external disturbances satisfying* $|d_1| \le 0.5$ *and* $|d_2| \le 3$*, respectively.*

Set $r_1 = 1$, $r_2 = 1+\tau$, $r_3 = 1+2\tau$, and $\kappa = 1$. With the relation $\ln(1+x_2^2) \le 2|x_2|^{\frac{1+2\tau}{1+\tau}}$ with $\tau \le 0$, it is concluded that Assumption 2.3.1 is satisfied. By using the proposed method, we can design the following stabilizing control law:

$$u(t) = -L^3 K\left(\lfloor x_1\rceil^{\frac{r_3+\tau}{r_1}}, \lfloor x_2/L\rceil^{\frac{r_3+\tau}{r_2}}, \lfloor x_3/L^2\rceil^{\frac{r_3+\tau}{r_3}}\right)^{\top}. \tag{2.28}$$

In the following simulations, we choose $\tau = -2/15$. k_1, k_2, k_3 can be chosen as the coefficients of a Hurwitz polynomial $s^3 + k_3 s^2 + k_2 s + k_1$. As a consequence, the parameters are selected as $K = [5, 25, 30]$.

In the case when $d_1(t) = d_2(t) = 0$, we choose the scaling gain as $L = 3$ according to the guideline (2.19). As shown in Figure 2.1(a), the states response curves of the system (2.27) under the stabilizing control (2.28) can approach the origin in satisfactory control performance. The time history of the control signal is shown in Figure 2.1(b).

In the case when $d_1(t) = 0.5$ and $d_2(t) = 3\cos(10t)$, it is clear that only practical control result can be achieved by the proposed method. In the simulation, we show that by choosing different values of the scaling gain L, different steady error of the system output (i.e., $y = x_1$) is performed. As one can observe from Figure 2.2(a), by increasing the scaling gain while the control gain K is set as the same value, a smaller steady-error is obtained. In Figure 2.2(b), the control signals are given which concludes that a larger control energy is needed to maintain the controlled system with better robustness.

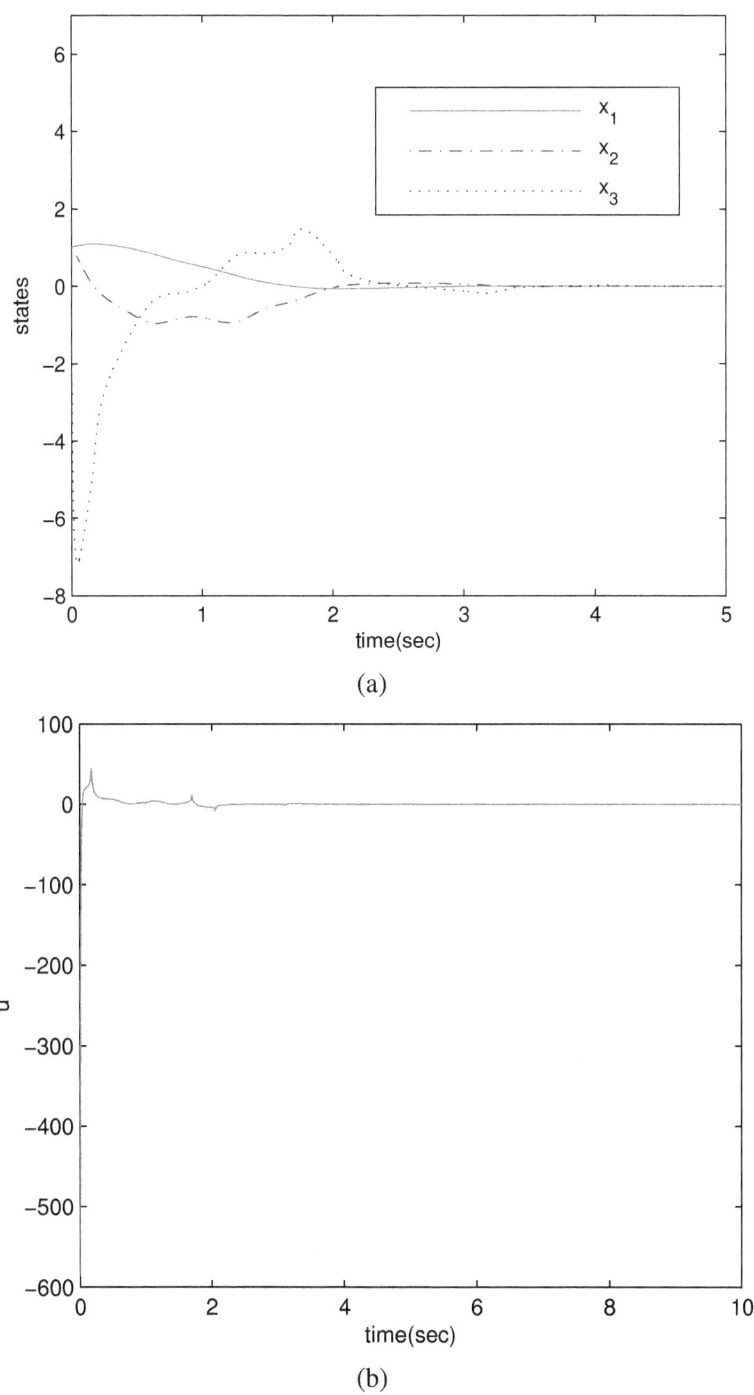

Figure 2.1 Control performance. (a) State response curves; (b) Time history of the control input.

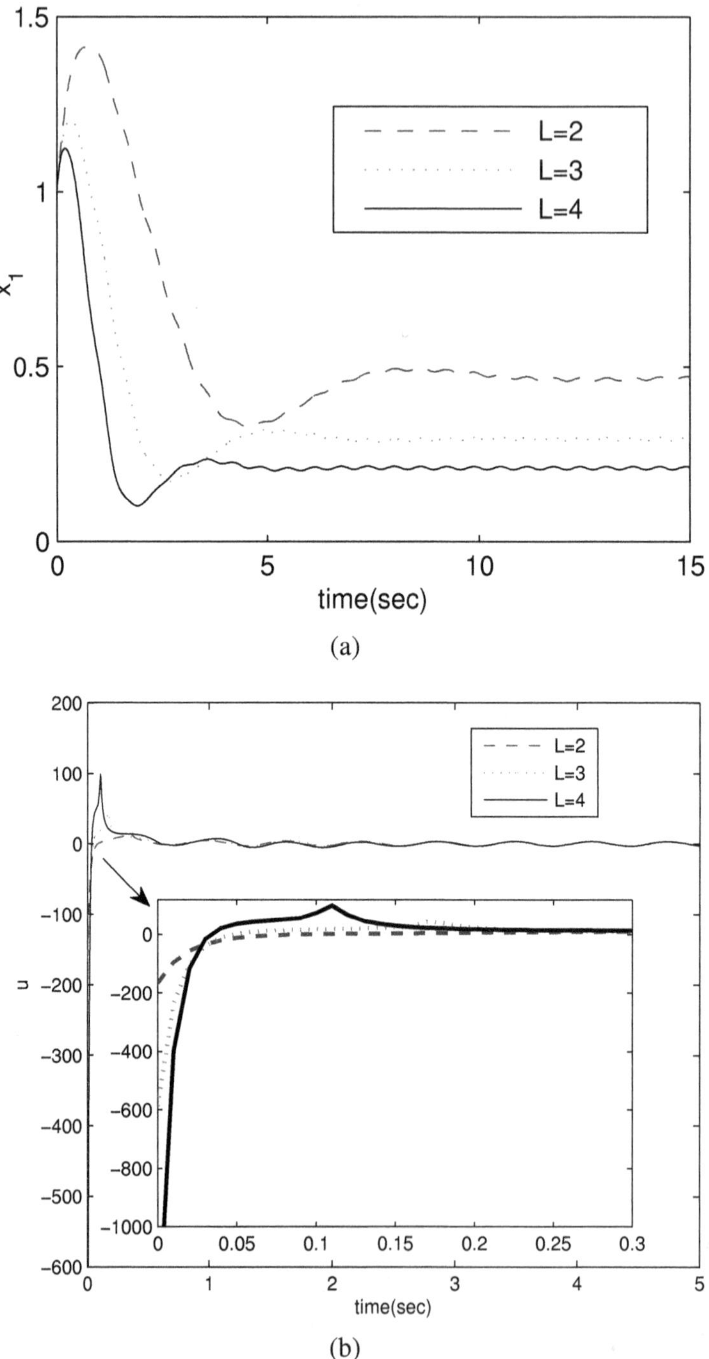

(a)

(b)

Figure 2.2 (a) Output response curves under different scaling gains; (b) Time histories of the control input under different scaling gains.

Chapter 3

Practically Oriented Finite-Time Tracking Control for Nonlinear Systems

Chapter 2 has presented a nonrecursive stabilization framework for a class of nonlinear systems, however, aiming for designing a globally convergent controller, certain nonlinearity constraints are essential and should be verified in priority. As a matter of fact, it is well known that a common nonlinear growth constraint of the considered system is essentially required for global continuous finite-time stabilization design. For general real-life plants, it is clearly of suspicion that most of the nonlinear hypothesis could be widely satisfied. Hence, a key question arises that for a general class of nonlinear systems, how to essentially relax the pre-requirement of the system nonlinearity constraints, such that a practically oriented finite-time controller can be widely applied.

In this chapter, aiming to facilitate the implementation of practical systems, we propose a practically oriented finite-time control design strategy for a class of lower-triangular nonlinear systems under a less ambitious but more practical control objective, namely, semi-global rather than restrictive global control. Firstly, a delicate coordinates transform is presented by exactly calculating the steady-state generators. Secondly, we show that by going through a nonrecursive design that the controller scheme design can be essentially detached from its stability analysis, which is inevitable in the existing recursive design approaches. Thirdly,

DOI: 10.1201/9781003399230-3

a rigorous semi-global attractivity and local finite-time convergence provide the theoretical justifications of the proposed method.

Further, consider that controller design via state feedback design relies on a restrictive condition that all the states are assumed to be measurable. Due to the fact that a dynamic tracking algorithm is imperatively required for physical realization when part of the system states are unmeasurable, in the latter part of this chapter, a generalized nonrecursive output feedback tracking control design framework is also presented, which will not only generate a concise design procedure but also simple dynamic control algorithms. As an additional compensator to the existing state feedback law, a unified high-gain observer is exploited which can reduce to a conventional high-gain observer as one special case and also performs a finite-time state observation under a negative homogeneous degree. In order to obtain a generalized tracking control result by covering a wider class of nonlinear systems and relaxing the implementable requirements in essence, we further propose a novel tracking control design framework with bounded trajectories without the prior nonlinearity constraints by trading off the control objective from global stability to a semi-global one. By modifying the dynamic compensators proposed in the global control design section, a delicate semi-global stability analysis is provided to ensure the semi-global attractivity and exact tracking realization, respectively.

Thereafter, as an illustration of the practical nature of this proposed method, we show that by utilizing an almost identical nonrecursive design procedure, a practical tracking control law can also be obtained for a class of more general nonlinear systems with mismatched disturbances. What is worthy of pointing out here that by distributing each state a tracking task of their reference nominal trajectory respectively, rather than only regulating the single output to its reference signal by neglecting the other states, a relatively more exquisite practical tracking control law can be obtained.

Compared with the existing related finite-time control design results for lower-triangular nonlinear systems, the main distinguishable improvements of this chapter can be stated as follows.

- Firstly, under a semi-global control infrastructure, we show that the existing essentially required nonlinearity growth constraints, which are generally restrictive to be verified, can now be fully removed.

- Secondly, the involved adjustable homogeneous degree, which could significantly affect the control performance, has been endowed with much flexibility within a tunable region.

- Thirdly, the proposed novel nonrecursive synthesis approach could render a finite-time control scheme to be designed in a very simple expression, whereas its control gain selection follows a conventional pole placement manner. More specifically, the obtained controller can reduce to its linear feedback controller counterpart simply by assigning the homogeneous degree to zero.

3.1 Practically Oriented Finite-Time Control via State Feedback

3.1.1 Problem formulation

In what follows, a practically oriented finite-time controller is designed in the nonrecursive manner first. To begin with, the finite-time control problem for a class of lower-triangular nonlinear systems is revisited:

$$\begin{cases} \dot{x}_i = x_{i+1} + f_i(\bar{x}_i),\ i \in \mathbb{N}_{1:n-1}, \\ \dot{x}_n = u + f_n(x), \\ y = x_1, \end{cases} \tag{3.1}$$

where $\bar{x}_i = \text{col}(x_1, x_2, \cdots, x_i)$, $i \in \mathbb{N}_{1:n-1}$, and $x = \bar{x}_n$ are the system partial and full-state vectors, respectively; y is the system output; u is the system input and $f_i(\cdot)$, $i \in \mathbb{N}_{1:n}$ is a known smooth nonlinear function.

Assumption 3.1.1 *The output reference signal, denoted by y_r and its n-th order derivative are assumed to be piecewise continuous, known and bounded.*

In what follows, for $x(0) \in \mho_x \triangleq [-\rho, \rho]^n$ where $\rho \in \mathbb{R}_+$ is any given constant which could be arbitrarily large, the control objective is to find, if possible, a $\mathbb{C}^0$ controller of the following state feedback form

$$u = u(\bar{y}_r, L, K, r, x) \in \mathbb{C}^0, \tag{3.2}$$

where $\bar{y}_r = (y_r, y_r^{(1)}, \cdots, y_r^{(n)})^\top$, $L \geq 1$ is a sufficiently large scaling gain, $K = [k_1, k_2, \cdots, k_n]$ is the coefficient vector of a Hurwitz polynomial $p(s) = s^n + k_n s^{n-1} + \cdots + k_2 s + k_1$ and the dilation weight is defined by $r = (1, 1+\tau, \cdots, 1+(n-1)\tau)$ with a homogeneous degree $\tau \in (-\frac{1}{n}, 0)$, such that the following statements hold:

- All the states in the closed-loop system (3.1)–(3.2) are semi-globally uniformly bounded;
- There exists a finite-time instant $T > 0$, such that $y - y_r = 0$, $\forall t \geq T$.

3.1.2 One-step controller construction

To achieve a realizable control scheme, firstly, we define $\bar{x}_i^* = (x_1^*, x_2^*, \cdots, x_i^*)^\top$, $i \in \mathbb{N}_{1:n+1}$ as an auxiliary variable vector where x_i^* is determined by the following steady-state generators:

$$\begin{cases} x_1^* = y_r, \\ x_i^* = \dfrac{dx_{i-1}^*}{dt} - f_{i-1}(\bar{x}_{i-1}^*),\ i \in \mathbb{N}_{2:n+1}. \end{cases} \tag{3.3}$$

Second, denote $z = (z_1, z_2, \cdots, z_n)^\top$ where

$$z_i = (x_i - x_i^*)/L^{i-1},\ i \in \mathbb{N}_{1:n},\ v = (u - x_{n+1}^*)/L^n. \tag{3.4}$$

In what follows, we show that, without going through the procedure of recursive stability analysis, a simple practically oriented finite-time controller can be explicitly pre-built of the following form:

$$\begin{aligned} v &= -k_1 \lfloor z_1 \rceil^{\frac{1+n\tau}{1}} - k_2 \lfloor z_2 \rceil^{\frac{1+n\tau}{1+\tau}} - \cdots - k_n \lfloor z_n \rceil^{\frac{1+n\tau}{1+(n-1)\tau}} \\ &\triangleq -K \lfloor z \rceil_{\Delta^r}^{1+n\tau}, \\ u &= L^n v + x_{n+1}^*. \end{aligned} \tag{3.5}$$

Remark 3.1 The proposed regulation methodology presents us with an alternative, but more practical control design procedure owing to the semi-global control infrastructure. More distinguishably, the common requirements of various nonlinearity growth conditions on the system nonlinearities, which are always employed in the existing finite-time control related literatures, are essentially relaxed. Hence any nonlinear systems of the form (3.1) can be finite-timely controlled by the proposed controller (3.5) where the main difference relies on the determination guideline of the bandwidth factor L.

3.1.3 Semi-global stability analysis

Theorem 3.1

Consider the closed-loop system consisting of (3.1) satisfying $x(0) \in \mho_x$ and the finite-time control law (3.5) with a sufficiently large bandwidth factor L which is dependent on ρ. Then the following statements hold.

- *Any trajectory starting from the compact set $\mho_x$ will converge to its equilibrium $x = x^*$.*
- *There exists a finite time instant $T > 0$ such that $\forall t \geq T$, $y(t) = y_r$.*

Proof: Based on the relations (3.3) and (3.4), the z-dynamical system can be directly expressed as:

$$\begin{cases} \dot{z}_i = L z_{i+1} + (f_i(\bar{x}_i) - f_i(\bar{x}_i^*))/L^{i-1},\ i \in \mathbb{N}_{1:n-1}, \\ \dot{z}_n = L v + (f_n(x) - f_n(\bar{x}_n^*))/L^{n-1}. \end{cases} \tag{3.6}$$

Construct a simplified Lyapunov function $V(z)$ of the following form:

$$V(z) = \left(\lfloor z \rceil_{\Delta^r}^{1-\frac{\tau}{2}}\right)^\top P \lfloor z \rceil_{\Delta^r}^{1-\frac{\tau}{2}}, \tag{3.7}$$

where P is a positive definite and symmetrical matrix satisfying $\Lambda^\top P + P\Lambda = -I$ with Λ being a companion matrix of K. From the definition of K, it is obvious that Λ is Hurwitz.

Firstly, the following statement holds:

$$V(z) \in \mathbb{C}^1 \cap \mathbb{H}_{\Delta^r}^{2-\tau}. \tag{3.8}$$

By Lemma 2.1, one can obtain that $\frac{\partial V(z)}{\partial z^T}\left(z_2, \cdots, z_n, -K\lfloor z \rceil_{\Delta^r}^{r_n+\tau}\right)^\top$ is negative definite for a homogeneous degree $\tau \in (-\varepsilon, 0)$. With $V(z) \in \mathbb{H}_{\Delta^r}^{2-\tau}$ and $\left(z_2, \cdots, z_n, -K\lfloor z \rceil_{\Delta^r}^{r_n+\tau}\right)^\top \in \mathbb{H}_{\Delta^r}^{\tau}$ in mind, using Lemma 1.1, the following relations can be achieved for a constant $\alpha \in \mathbb{R}_+$

$$\begin{aligned}\frac{\partial V(z)}{\partial z^\top} L\left(z_2, \cdots, z_n, -K\lfloor z \rceil_{\Delta^r}^{r_n+\tau}\right)^\top &\le -\alpha L \|z\|_{\Delta^r}^2 \\ &\le -\alpha L V^{\frac{2}{2-\tau}}(z),\ \tau \in (-\varepsilon, 0).\end{aligned}$$

From the definitions of x_i^* and y_r, there exists a constant $\bar{\rho} > 0$ such that

$$\max_{i \in \mathbb{N}_{1:n}} \{\sup_{t \ge 0}\{x_i^*(t)\}\} \le \bar{\rho}.$$

Then for a given compact set $\mho_x$, define a level set

$$\Omega = \left\{ z \in \mathbb{R}^n | V(z) \le \sup_{z \in \mho_z \triangleq [-(\rho+\bar{\rho}), (\rho+\bar{\rho})]^n} V(z) \right\}.$$

Let $N = \sup_{z \in \Omega} \|z\|_\infty$ where $\|\cdot\|_\infty$ stands for L_∞ norm of vectors and $\mho_N = [-N, N]^n$. It is not difficult to conclude that

$$\mho_x \subset \mho_z \subset \Omega \subset \mho_N.$$

The time derivative of $V(z)$ along the closed-loop system (3.5)–(3.6) gives

$$\begin{aligned}\dot{V}(z) &= \frac{\partial V(z)}{\partial z^\top} L\left(z_2, \cdots, z_n, -K\lfloor z \rceil_{\Delta^r}^{1+n\tau}\right)^\top + \sum_{i=1}^n \frac{\partial V(z)}{\partial z_i} \frac{f_i(\bar{x}_i) - f_i(\bar{x}_i^*)}{L^{i-1}} \\ &\le -\alpha L V^{\frac{2}{2-\tau}}(z) + \sum_{i=1}^n \frac{\partial V(z)}{\partial z_i} \frac{f_i(\bar{x}_i) - f_i(\bar{x}_i^*)}{L^{i-1}}.\end{aligned} \tag{3.9}$$

Recalling f_i, $i \in \mathbb{N}_{1:n}$ is a smooth function, by utilizing Mean-Value Theorem, we have

$$f_i(\bar{x}_i) - f_i(\bar{x}_i^*) \leq \bar{\gamma}_i(\bar{x}_i, \bar{x}_i^*)\Big(|x_1 - x_1^*| + |x_2 - x_2^*| + \cdots + |x_i - x_i^*|\Big),$$

where $\bar{\gamma}_i(\bar{x}_i, \bar{x}_i^*)$ is a $\mathbb{C}^0$ nonnegative function. $\forall x \in \mho_N$, it is clear that there exists a constant $\tilde{\gamma}_i$ dependent on N, such that $\bar{\gamma}_i(\bar{x}_i, \bar{x}_i^*) \leq \tilde{\gamma}_i$.

In what follows, two cases are discussed below.

1) In the case when $|x_j - x_j^*| \geq 1$, $\forall j \in \mathbb{N}_{1:i}$, by noting that $|x_j| \leq N$ and $|x_j^*| \leq \bar{\rho}$, we know that

$$|x_j - x_j^*| \leq N + \bar{\rho} \leq (N + \bar{\rho})|x_j - x_j^*|^{\frac{1+i\tau}{1+(j-1)\tau}}.$$

2) In the case when $|x_j - x_j^*| < 1$, $\forall j \in \mathbb{N}_{1:i}$, by noting $\frac{1+i\tau}{1+(j-1)\tau} \leq 1$, we know that

$$|x_j - x_j^*| \leq |x_j - x_j^*|^{\frac{1+i\tau}{1+(j-1)\tau}}.$$

By summarizing the above two cases, the following relation holds with a constant $\check{\gamma}_i = \tilde{\gamma}_i \max\{N + \bar{\rho}, 1\}$:

$$(f_i(\bar{x}_i) - f_i(\bar{x}_i^*))/L^{i-1} \leq \check{\gamma}_i \frac{\left(|L^0 z_1|^{\frac{1+i\tau}{1}} + |L^1 z_2|^{\frac{1+i\tau}{1+\tau}} + \cdots + |L^{i-1} z_i|^{\frac{1+i\tau}{1+(i-1)\tau}}\right)}{L^{i-1}}, \; x \in \mho_N.$$

Noting that $L \geq 1$, the following relation:

$$(j-1)\frac{1+i\tau}{1+(j-1)\tau} - (i-1) \in (-\infty, 0), \; \forall j \in \mathbb{N}_{1:i},$$

implies the fact that $L^{(j-1)\frac{1+i\tau}{1+(j-1)\tau} - (i-1)} < 1$. Hence one can obtain

$$(f_i(\bar{x}_i) - \phi_i(\bar{x}_i^*))/L^{i-1} \leq \check{\gamma}_i \|z\|_{\Delta^r}^{1+i\tau}. \tag{3.10}$$

It is clear that $V(z) \in \mathbb{H}_{\Delta^r}^{2-\tau}$ and $\|z\|_{\Delta^r}^{1+i\tau} \in \mathbb{H}_{\Delta^r}^{1+i\tau}$. With Lemma 1.1 in mind, there exists a constant $\tilde{\alpha} \in \mathbb{R}_+$ which is dependent on N but independent of L, such that the following relations hold:

$$\begin{aligned}\sum_{i=1}^{n} \frac{\partial V(z)}{\partial z_i}(f_i(\bar{x}_i) - f_i(\bar{x}_i^*))/L^{i-1}\Big|_{\mho_N} &\leq \sum_{i=1}^{n} \check{\gamma}_i \left|\frac{\partial V(z)}{\partial z_i}\right| \|z\|_{\Delta^r}^{1+i\tau} \\ &\leq \tilde{\alpha} V^{\frac{2}{2-\tau}}(z).\end{aligned} \tag{3.11}$$

Substituting the relation (3.11) into (3.9) yields

$$\dot{V}(z)\big|_{\Omega} \leq -(\alpha L - \tilde{\alpha}) V^{\frac{2}{2-\tau}}(z). \tag{3.12}$$

Now one can select a sufficiently large scaling gain $L \geq 1$ to satisfy the following guideline:

$$\alpha L - \tilde{\alpha} \geq 1, \tag{3.13}$$

which leads to

$$\dot{V}(z)\big|_{\Omega} \leq -V^{\frac{2}{2-\tau}}(z). \tag{3.14}$$

In what follows, we will use a contradiction argument to prove that under the guideline (3.13), for any non-zero initial states satisfying $x(0) \in \mho_x$, all the trajectories of $z(t)$ will stay in Ω forever.

If the above statement is not true, the trajectory of $z(t)$ will escape the set Ω within a finite-time. Due to the fact that $z(0) \in \mho_N$, it yields

$$\begin{aligned}\dot{V}(z(0)) &\leq -(\alpha L - \tilde{\alpha})V^{\frac{2}{2-\tau}}(z(0))\\ &\leq -V^{\frac{2}{2-\tau}}(z(0)) < 0.\end{aligned} \tag{3.15}$$

Hence there must exist two time instants $t_2 > t_1 > 0$, such that

$$\begin{aligned}&i)\ \dot{V}(z(t_1)) < 0,\\ &ii)\ V(z(t_2)) = V(z(t_1)),\\ &iii)\ \dot{V}(z(t_2)) > 0.\end{aligned} \tag{3.16}$$

It is clear that (3.14) still holds for $t \in [t_1, t_2]$. Then the following relations hold:

$$V(z(t_2)) - V(z(t_1)) = \int_{t_1}^{t_2} \dot{V}(z(s))ds \leq -\int_{t_1}^{t_2} V^{\frac{2}{2-\tau}}(z(s))ds. \tag{3.17}$$

By the relation ii) of (3.16) and the fact that $V(z(s)) > 0,\ s \in [t_1, t_2]$, (3.17) leads to an obvious contradiction, expressed as

$$0 \geq \int_{t_1}^{t_2} V^{\frac{2}{2-\tau}}(z(s))ds > 0.$$

Hence, we can arrive at the following conclusion:

$$\forall x(0) \in \mho_x \Rightarrow z(0) \in \mho_z \Rightarrow z(t) \in \Omega,\ \forall t \geq 0,$$

which implies that the set Ω is an invariance set.

Further, with Lemma 1.4 in mind and owing to the fact that $\tau \in (-\varepsilon, 0)$, hence $0 < \frac{2}{2-\tau} < 1$. Then the following relation

$$\dot{V}(z)\big|_{\Omega} \leq -V^{\frac{2}{2-\tau}}(z)$$

leads to a straightforward conclusion that there exists a finite time instant $T > 0$, such that $y(t) - y_r = 0$, $t \in [T, \infty)$.

This completes the proof of Theorem 3.1. ■

Remark 3.2 In practice, different from local control strategies, the compact set $\mho_x$ of the initial states can be defined under a worst-case study. One can determine a proper bandwidth factor L via a practical "trial and error" way, which can be stated as follows. The control gain K can be simply selected following the pole placement manner first. By setting a large L to guarantee the stability, then one can tune L to be smaller and smaller while testing the gap between current control performance and pre-given performance indexes until satisfactory response curves appear.

3.2 Practically Oriented Finite-Time Control via Output Feedback

Considering the case when parts of the states in system (3.1) are very difficult to be measured, output feedback strategy could show its great potential for practical implementations. In this section, we first present a global output feedback control design procedure based on a homogeneous growth constraint for the system nonlinearities. Then, by proposing a novel saturated observer and controller, explicit semi-global stability analysis is presented, which is able to extend the nonrecursive output feedback design to a wide range of nonlinear systems.

3.2.1 Global finite-time output feedback control

In what follows, an explicit step-by-step controller construction procedure to derive a simple globally stable dynamic tracking law for system (3.1) will be presented.

3.2.1.1 Change of coordinates

Firstly, define the following change of coordinates with the denotation of $\bar{z}_i = (z_1, z_2, \cdots, z_i)^\top$ and $z = \bar{z}_n$:

$$\begin{cases} z_i = \dfrac{x_i - x_i^*}{L^{i-1}}, \ i \in \mathbb{N}_{1:n-1}, \\ v = \dfrac{u - u^*}{L^n}, \end{cases} \tag{3.18}$$

where x_i^* is defined in (3.3), $L \geq 1$ is a scaling gain and $\tilde{f}_i$ is determined by

$$\tilde{f}_0 = 0, \ \tilde{f}_i = f_i(x_1^*, x_2^*, \cdots, x_i^*), \ i \in \mathbb{N}_{1:n}. \tag{3.19}$$

Then system (3.18) is equivalent to the following z-dynamical system

$$\begin{cases} \dot{z}_i = Lz_{i+1} + (f_i - \tilde{f}_i)/L^{i-1}, \ i \in \mathbb{N}_{1:n-1}, \\ \dot{z}_n = Lv + (f_n - \tilde{f}_n)/L^{n-1}, \end{cases} \tag{3.20}$$

which can be rewritten as the following compact form:

$$\begin{aligned} \dot{z} &= L(z_2, \cdots, z_n, v)^\top + \left((f_1 - \tilde{f}_1)/L^0, \cdots, (f_n - \tilde{f}_n)/L^{n-1}\right)^\top \\ &\triangleq L\psi(z, v) + \left[\{(f_i - \tilde{f}_i)/L^{i-1}\}_{i\in\mathbb{N}_{1:n}}\right]_{n\times 1}. \end{aligned} \tag{3.21}$$

3.2.1.2 *Nonsmooth observer design*

In what follows, for the dynamic system (3.20), a full-order state observer can be constructed in the following form:

$$\begin{cases} \dot{\hat{z}}_i = L\hat{z}_{i+1} + (\hat{f}_i - \tilde{f}_i)/L^{i-1} + L\ell_i\lfloor z_1 - \hat{z}_1\rceil^{r_{i+1}}, i \in \mathbb{N}_{1:n-1}, \\ \dot{\hat{z}}_n = Lv + (\hat{f}_n - \tilde{f}_n)/L^{n-1} + L\ell_n\lfloor z_1 - \hat{z}_1\rceil^{r_{n+1}}, \end{cases} \tag{3.22}$$

where $H \triangleq (\ell_1, \ell_2, \cdots, \ell_n)^\top$ is the observer gain vector and ℓ_i, $i \in \mathbb{N}_{1:n}$ is the corresponding coefficient of a Hurwitz polynomial $p(s) = s^n + \ell_n s^{n-1} + \cdots + \ell_1$, $\hat{x}_i = L^{i-1}\hat{z}_i + x_i^*$, and $\hat{f}_i$ is defined by $\hat{f}_i = f_i(\hat{\bar{x}}_i)$, $i \in \mathbb{N}_{1:n}$.

Remark 3.3 The observer (3.22) can also be regarded as a nonsmooth extension of the well-known high-gain observers reported in [31] whereas the difference relies on the integration of homogeneous dilation weights $r_1, r_2, \cdots, r_{n+1}$ with $\tau < 0$. As a matter of fact, (3.22) reduces to a conventional high-gain observer in the case when $\tau = 0$.

With (3.21) and (3.22), the error dynamics can be derived as

$$\begin{cases} \dot{e}_i = Le_{i+1} + (f_i - \hat{f}_i)/L^{i-1} - L\ell_i\lfloor e_1\rceil^{r_{i+1}}, i \in \mathbb{N}_{1:n-1}, \\ \dot{e}_n = (f_n - \hat{f}_n)/L^{n-1} - L\ell_n\lfloor e_1\rceil^{r_{n+1}}, \end{cases} \tag{3.23}$$

with the denotation of $e_i = z_i - \hat{z}_i$, $i \in \mathbb{N}_{1:n}$ and $e = (e_1, e_2 \cdots, e_n)^\top$.

(3.23) can also be rewritten as the following compact form:

$$\begin{aligned} \dot{e} &= L\begin{bmatrix} e_2 - \ell_1\lfloor e_1\rceil^{r_2} \\ \vdots \\ e_n - \ell_{n-1}\lfloor e_1\rceil^{r_n} \\ -\ell_n\lfloor e_1\rceil^{r_{n+1}} \end{bmatrix} + \begin{bmatrix} (f_1 - \hat{f}_1)/L^0 \\ \vdots \\ (f_{n-1} - \hat{f}_{n-1})/L^{n-2} \\ (f_n - \hat{f}_n)/L^{n-1} \end{bmatrix} \\ &\triangleq Lf(\tau, H, e) + \left[\{(f_i - \hat{f}_i)/L^{i-1}\}_{i\in\mathbb{N}_{1:n}}\right]_{n\times 1}. \end{aligned} \tag{3.24}$$

3.2.1.3 Output feedback controller construction

At this stage, one can design a state feedback controller for system (3.21) of the following form provided all the states are measurable:

$$v_c = -K\lfloor z \rceil_{\Delta^r}^{r_{n+1}}, \tag{3.25}$$

where $K \triangleq (k_1, \cdots, k_n)$ is the coefficient vector of a Hurwitz polynomial $s^n + k_n s^{n-1} + \cdots + k_2 s + k_1$.

In order to present a physically realizable tracking control law, by using the estimated full states information from the observer (3.22) to replace the states used in (3.25), now we can construct a globally stabilizing output feedback control law for system (3.21) of the following form:

$$v(t) = -K\lfloor \hat{z} \rceil_{\Delta^r}^{r_{n+1}}, \tag{3.26}$$

which implies that u should be expressed as

$$u(t) = -L^n K\lfloor \hat{z} \rceil_{\Delta^r}^{r_{n+1}} + u^*. \tag{3.27}$$

3.2.1.4 Explicit global stability analysis

In order to proceed with the global stability analysis, a prior constrain condition of the nonlinearity growth rates, well known as a generalized Hölder continuous condition, is required.

Assumption 3.2.1 *Given any $\bar{x}_i$, $\hat{\bar{x}}_i \in \mathbb{R}^i$, and $\tilde{\bar{x}}_i = \bar{x}_i - \hat{\bar{x}}_i$, there exists a known constant vector $\bar{\gamma}_i = (\gamma_{i,1}, \cdots, \gamma_{i,i})$ such that*

$$|f_i(\bar{x}_i) - f_i(\hat{\bar{x}}_i)| \leq |\bar{\gamma}_i \lfloor \tilde{\bar{x}}_i \rceil_{\Delta^r}^{r_i+\tau}|, \ i \in \mathbb{N}_{1:n}.$$

Remark 3.4 Assumption 3.2.1 can be seen as a homogeneous growth condition which covers both vanishing and nonvanishing cases. Assumption 3.2.1 reduces to the widely known global Lipschitz continuous condition in the case when $\tau = 0$, and becomes a Hölder continuous condition in the case when $\tau < 0$.

By combining (3.21) and (3.24) together, denoting $\mathbb{Z} \triangleq (z^\top, e^\top)^\top$, we can obtain the following closed-loop system:

$$\begin{aligned}
\dot{\mathbb{Z}} &= L\left(\begin{bmatrix} A & 0 \\ 0 & A \end{bmatrix}\mathbb{Z} + \begin{bmatrix} B \\ 0 \end{bmatrix} v - \begin{bmatrix} 0_{n\times 1} \\ \{\ell_i \lfloor e_1 \rceil^{r_{i+1}}\}_{i\in\mathbb{N}_{1:n},\ n\times 1} \end{bmatrix}\right) \\
&\quad + \begin{bmatrix} [\{(f_i - \tilde{f}_i)/L^{i-1}\}_{i\in\mathbb{N}_{1:n}}]_{n\times 1} \\ [\{(f_i - \hat{f}_i)/L^{i-1}\}_{i\in\mathbb{N}_{1:n}}]_{n\times 1} \end{bmatrix} \\
y &= Cz + y_r,
\end{aligned} \tag{3.28}$$

with A, B, C representing the corresponding matrices of a controllable canonical system.

The nominal system of (3.28) can be rewritten in the following form:

$$\begin{aligned}\dot{\mathbb{Z}} &= L\left(\begin{bmatrix} A & 0 \\ 0 & A \end{bmatrix}\mathbb{Z} + \begin{bmatrix} B \\ 0 \end{bmatrix} v - \begin{bmatrix} 0_{n\times 1} \\ \{\ell_i \lfloor e_1 \rceil^{r_{i+1}}\}_{i\in\mathbb{N}_{1:n},\ n\times 1} \end{bmatrix}\right) \\ &\triangleq L\Psi(\tau,\mathbb{Z},v). \end{aligned} \tag{3.29}$$

Denote an extended homogeneous dilation in the following form

$$\Delta_{\bar{r}} = \Big(\underbrace{r_1,\cdots,r_n}_{\text{for } z},\underbrace{r_1,\cdots,r_n}_{\text{for } e}\Big)$$

for the vector field $\Psi(\tau,\mathbb{Z},v)$. By a simple process of verification, we know that the closed-loop system (3.26)–(3.29) is $\Delta_{\bar{r}}$-homogeneous of degree τ.

Now we are ready to present the following theorem.

Theorem 3.2

Consider system (3.1) satisfying Assumptions 3.1.1–3.2.1. There exists a sufficiently large scaling gain $L \geq 1$ such that the dynamic output feedback tracking controller of the form (3.22)–(3.27) will render the output y of system (3.1) globally exactly track its desired value y_r asymptotically provided $\tau = 0$ or within a finite time provided $\tau < 0$.

Proof: Construct a homogenous, positive definite and proper Lyapunov function of the following form:

$$V(\tau,\mathbb{Z}) = \left(\lfloor \mathbb{Z} \rceil_{\Delta^{\bar{r}}}^{1-\frac{\tau}{2}}\right)^{\top} P \lfloor \mathbb{Z} \rceil_{\Delta^{\bar{r}}}^{1-\frac{\tau}{2}}, \tag{3.30}$$

where P is a positive definite matrix satisfying $\mathcal{A}^{\top}P + P\mathcal{A} = -I$, with $\mathcal{A} = \begin{bmatrix} A-BK & BK \\ 0 & A-HC \end{bmatrix}$. From its construction, it is easy to verify that $V(\tau,\mathbb{Z}) \in \mathbb{C}^1 \cap \mathbb{H}_{\Delta^{\bar{r}}}^{2-\tau}$.

Calculating the time derivative of the Lyapunov function $V(\tau,\mathbb{Z})$ along the closed-loop system (3.26)–(3.28) gives

$$\dot{V}(\tau,\mathbb{Z}) = \frac{\partial V(\tau,\mathbb{Z})}{\partial \mathbb{Z}^{\top}} L\Psi(\tau,\mathbb{Z},v) + \frac{\partial V(\tau,\mathbb{Z})}{\partial \mathbb{Z}^{\top}} \begin{bmatrix} \left[\{(f_i - \tilde{f}_i)/L^{i-1}\}_{i\in\mathbb{N}_{1:n}}\right]_{n\times 1} \\ \left[\{(f_i - \hat{f}_i)/L^{i-1}\}_{i\in\mathbb{N}_{1:n}}\right]_{n\times 1} \end{bmatrix}. \tag{3.31}$$

Then, according to Lemma 1.1, we arrive at

$$\frac{\partial V(\tau,\mathbb{Z})}{\partial \mathbb{Z}^{\top}} L\Psi(\tau,\mathbb{Z},v) \leq -cLV^{\frac{2}{2-\tau}}(\tau,\mathbb{Z}),\ \tau \in (-\varsigma, 0], \tag{3.32}$$

where $\varsigma \in [0, \frac{1}{n})$ and $c > 0$ are all constants.

With Assumption 3.2.1 and $L \geq 1$ in mind, the following relations hold naturally:

$$\begin{aligned}\frac{f_i-\tilde{f}_i}{L^{i-1}} &\leq \frac{|f_i-\tilde{f}_i|}{L^{i-1}}\\ &\leq \frac{\max_{j\in\mathbb{N}_{1:i}}\{\gamma_{ij}\}}{L^{i-1}}\left(|x_1-x_1^*|^{\frac{r_i+\tau}{r_1}}+|x_2-x_2^*|^{\frac{r_i+\tau}{r_2}}+\cdots+|x_i-x_i^*|^{\frac{r_i+\tau}{r_i}}\right)\\ &= \frac{\max_{j\in\mathbb{N}_{1:i}}\{\gamma_{ij}\}}{L^{i-1}}\left(|L^0z_1|^{\frac{r_i+\tau}{r_1}}+|Lz_2|^{\frac{r_i+\tau}{r_2}}+\cdots+|L^{i-1}z_i|^{\frac{r_i+\tau}{r_i}}\right).\end{aligned} \tag{3.33}$$

Similarly, the following relations can also be deduced:

$$\begin{aligned}\frac{f_i-\hat{f}_i}{L^{i-1}} &\leq \frac{\max_{j\in\mathbb{N}_{1:i}}\{\gamma_{ij}\}}{L^{i-1}}\left(|x_1-\hat{x}_1|^{\frac{r_i+\tau}{r_1}}+|x_2-\hat{x}_2|^{\frac{r_i+\tau}{r_2}}+\cdots+|x_i-\hat{x}_i|^{\frac{r_i+\tau}{r_i}}\right)\\ &= \frac{\max_{j\in\mathbb{N}_{1:i}}\{\gamma_{ij}\}}{L^{i-1}}\left(|L^0e_1|^{\frac{r_i+\tau}{r_1}}+|L^1e_2|^{\frac{r_i+\tau}{r_2}}+\cdots+|L^{i-1}e_i|^{\frac{r_i+\tau}{r_i}}\right).\end{aligned} \tag{3.34}$$

Note that for $j \in \mathbb{N}_{1:i}$, we have

$$\begin{aligned}(j-1)\frac{r_i+\tau}{r_j}-(i-1) &= \frac{(j-i)+(j-1)(i-1)\tau+(j-1)\tau-(i-1)(j-1)\tau}{r_j}\\ &\leq \frac{(j-1)\tau}{r_j}\\ &= \frac{r_j-1}{r_j} \leq 0.\end{aligned} \tag{3.35}$$

With (3.35) and the fact that $L \geq 1$ in mind, now it comes to a conclusion that the following relations can be achieved:

$$\begin{cases}\left|\dfrac{f_i-\tilde{f}_i}{L^{i-1}}\right| \leq \gamma\left(|z_1|^{\frac{r_i+\tau}{r_1}}+\cdots+|z_i|^{\frac{r_i+\tau}{r_i}}\right),\\ \left|\dfrac{f_i-\hat{f}_i}{L^{i-1}}\right| \leq \gamma\left(|e_1|^{\frac{r_i+\tau}{r_1}}+\cdots+|e_i|^{\frac{r_i+\tau}{r_i}}\right),\end{cases} \tag{3.36}$$

where $\gamma = \max_{i\in\mathbb{N}_{1:n},j\in\mathbb{N}_{1:i}}\{\gamma_{ij}\}$.

Recall that $V(\tau,\mathbb{Z}) \in \mathbb{H}_{\Delta^{\bar{r}}}^{2-\tau}$ and $\bar{f}_i(\cdot) \in \mathbb{H}_{\Delta^{\bar{r}}}^{r_i+\tau}$. With Lemma 1.1 in mind, we know that $\frac{\partial V(\tau,\mathbb{Z})}{\partial z_i} \in \mathbb{H}_{\Delta^{\bar{r}}}^{2-\tau-r_i}$ and $\frac{\partial V(\tau,\mathbb{Z})}{\partial e_i} \in \mathbb{H}_{\Delta^{\bar{r}}}^{2-\tau-r_i}$, hence there exist two constants $\breve{c} > 0$ and $\bar{c} > 0$, such that the following relations hold:

$$\begin{aligned}
&\frac{\partial V(\tau,\mathbb{Z})}{\partial \mathbb{Z}^{\top}}\begin{bmatrix}\left[\{(f_i-\tilde{f}_i)/L^{i-1}\}_{i\in\mathbb{N}_{1:n}}\right]_{n\times 1}\\ \left[\{(f_i-\hat{f}_i)/L^{i-1}\}_{i\in\mathbb{N}_{1:n}}\right]_{n\times 1}\end{bmatrix}\\
&=\sum_{i=1}^{n}\frac{\partial V(\tau,\mathbb{Z})}{\partial z_i}\frac{f_i-\tilde{f}_i}{L^{i-1}}+\sum_{i=1}^{n}\frac{\partial V(\tau,\mathbb{Z})}{\partial e_i}\frac{f_i-\hat{f}_i}{L^{i-1}}\\
&\leq \gamma\left(\sum_{i=1}^{n}\left|\frac{\partial V(\tau,\mathbb{Z})}{\partial z_i}\right|\bar{f}_i(\bar{z}_i)+\sum_{i=1}^{n}\left|\frac{\partial V(\tau,\mathbb{Z})}{\partial e_i}\right|\bar{f}_i(\bar{e}_i)\right)\\
&\leq \breve{c}\|\mathbb{Z}\|_{\Delta^{\bar{r}}}^{2}\\
&\leq \bar{c}V^{\frac{2}{2-\tau}}(\tau,\mathbb{Z}).
\end{aligned} \tag{3.37}$$

At this stage, after presenting the selection guideline

$$cL-\bar{c}\geq 1, \tag{3.38}$$

(3.31) reduces to

$$\dot{V}(\tau,\mathbb{Z})\leq -V^{\frac{2}{2-\tau}}(\tau,\mathbb{Z}). \tag{3.39}$$

Now we will deduce the exact tracking realization results with the following analysis of two different cases.

Case 1: If $\tau=0$, it is clear that the controller (3.27) reduces to a linear controller and the conclusion of asymptotical stability of the closed-loop system can be directly obtained from (3.39), then we have

$$\lim_{t\to\infty}(y(t)-y_r)=0.$$

Case 2: If $\tau\in(-\varsigma,0)$, we know that $0<\frac{2}{2-\tau}<1$. By Lemma 1.4, it is straightforward from (3.39) that

$$y(t)-y_r=0,\ \forall t\geq T\triangleq \frac{V^{1-\frac{2}{2-\tau}}(\tau,\mathbb{Z}(t_0))}{(1-\frac{2}{2-\tau})}.$$

This completes the proof of Theorem 3.2. ■

3.2.2 Semi-global finite-time output feedback control

In order to obtain a global control result via output feedback, Assumption 3.2.1, which to some extent is restrictive to be verified, is strongly depended. However, by a trade-off of the control target, a less ambitious semi-global stabilization result is considered, which could facilitate practical implementations. Now the exact tracking objective of system (3.1) can be realized under only a smooth assumption of the system nonlinearities, which will cover much more general nonlinear systems.

3.2.2.1 Saturated observer design

Noting that the peaking phenomenon, which is revealed in high-gain observers, will possibly destabilize the control system, hence under a semi-global control objective, the observer (3.22) is modified as the following saturated form:

$$\begin{cases} \dot{\hat{x}}_i = \hat{x}_{i+1} + \langle \hat{f}_i \rangle_M + L^i \ell_i \lfloor x_1 - \hat{x}_1 \rceil^{r_{i+1}}, \ i \in \mathbb{N}_{1:n-1}, \\ \dot{\hat{x}}_n = u + \langle \hat{f}_n \rangle_M + L^n \ell_n \lfloor x_1 - \hat{x}_1 \rceil^{r_{n+1}}, \end{cases} \tag{3.40}$$

where $\langle \hat{f}_i \rangle_M = f_i\big(\langle \hat{x}_1 \rangle_M, \langle \hat{x}_2 \rangle_M, \cdots, \langle \hat{x}_i \rangle_M\big)$, $i \in \mathbb{N}_{1:n}$, and the saturation threshold M will be assigned later.

Remark 3.5 Inspired by [30, 76, 83], a saturation function is employed in the high-gain observer (3.40) to bound the state and estimation trajectories. The main difference is the introduced flexibility of a homogeneous degree τ, which could significantly affect the convergence speed. In addition, the gain tuning mechanism can follow a simple pole placement manner while the bandwidth factor will be explicitly determined in the stability analysis later.

Correspondingly with the same definition of e, the error dynamics should be modified from (3.24) as

$$\dot{e} = Lf(\tau, H, e) + \left[\{(f_i - \langle \hat{f}_i \rangle_M)/L^{i-1}\}_{i \in \mathbb{N}_{1:n}}\right]_{n \times 1}. \tag{3.41}$$

3.2.2.2 Saturated dynamic control law design

A saturated output feedback controller is then constructed as

$$\begin{aligned} v_s(t) &= -\sum_{i=1}^{n} k_i \lfloor (\langle \hat{x}_i \rangle_M - x_i^*)/L^{i-1} \rceil^{\frac{r_{n+1}}{r_i}}, \\ u_s(t) &= -L^n v_s(t) + u^*. \end{aligned} \tag{3.42}$$

3.2.2.3 Explicit semi-global stability analysis

The semi-global output feedback control problem for system (3.1) can now be depicted by the following theorem.

Theorem 3.3
Consider the closed-loop system consisting of system (3.1) satisfying Assumption 3.1.1 and a dynamic output feedback tracking controller of the form (3.40)–(3.42). There exists a sufficiently large scaling gain L, such that the following relations hold:

- *All the states of the closed-loop system starting from the compact set $\Gamma_x \times \Gamma_{\hat{x}} \triangleq [-r, r]^n \times [-r, r]^n \subset \mathbb{R}^n \times \mathbb{R}^n$ are uniformly bounded.*

- *The output $y(t)$ of system (3.1) can be rendered exactly track its reference value y_r.*

Proof: Construct two positive definite and proper Lyapunov functions of the following forms:

$$\begin{cases} W(\tau,z) = \left(\lfloor z\rceil_{\Delta^r}^{1-\frac{\tau}{2}}\right)^\top Q_1 \lfloor z\rceil_{\Delta^r}^{1-\frac{\tau}{2}}, \\ U(\tau,e) = \left(\lfloor e\rceil_{\Delta^r}^{1-\frac{\tau}{2}}\right)^\top Q_2 \lfloor e\rceil_{\Delta^r}^{1-\frac{\tau}{2}}, \end{cases} \tag{3.43}$$

where $\Delta_r \triangleq (r_1, r_2, \cdots, r_n)$, Q_1, Q_2 are two positive definite and symmetrical matrices satisfying

$$(A-BK)^\top Q_1 + Q_1(A-BK) = -I,$$
$$(A-HC)^\top Q_2 + Q_2(A-HC) = -I. \tag{3.44}$$

From the construction of W and U, it is easy to verify that $W(\tau,z) \in \mathbb{C}^1 \cap \mathbb{H}_{\Delta^r}^{2-\tau}$ and $U(\tau,e) \in \mathbb{C}^1 \cap \mathbb{H}_{\Delta^r}^{2-\tau}$. It is clear that by using an almost identical analysis with (3.32), the following relations can be achieved for $\tau \in (-\varsigma, 0]$ and a constant $\alpha > 0$:

$$\dot{W}(\tau,z) \leq -\alpha L W^{\frac{2}{2-\tau}} + \frac{\partial W}{\partial z_n} L(v_s - v_c) + \frac{\partial W}{\partial z^\top}\left[\{(f_i - \tilde{f}_i)/L^{i-1}\}_{i\in\mathbb{N}_{1:n}}\right]_{n\times 1},$$
$$\dot{U}(\tau,e) \leq -\alpha L U^{\frac{2}{2-\tau}} + \frac{\partial U}{\partial e^\top}\left[\{(f_i - \langle\hat{f}_i\rangle_M)/L^{i-1}\}_{i\in\mathbb{N}_{1:n}}\right]_{n\times 1}.$$

With Assumption 3.1.1 in mind, it is clear that $\tilde{f}_i$, $i \in \mathbb{N}_{1:n}$ are all bounded, hence there must exist a constant $\bar{r} > 0$ such that the following relation holds:

$$\max_{i\in\mathbb{N}_{1:n}}\{\sup_{t\geq t_0} x_i^*\} \leq \bar{r}. \tag{3.45}$$

For a given compact set $\Gamma_x = [-r, r]^n$, define another compact set $\Gamma_z \triangleq [-(r + \bar{r}), r + \bar{r}]^n$ and the following level set:

$$\Omega_z = \left\{ z \in \mathbb{R}^n | W(\tau,z) < r_0 + 1,\ r_0 \geq \sup_{z\in\Gamma_z} W(\tau,z) \right\}.$$

Let $M \triangleq \sup_{z\in\Omega_z} \|z\|_\infty$ being the saturation threshold and $\Gamma_M = [-M, M]^n$. Note that $f_i \in \mathbb{C}^1$, $x \in \Gamma_M$ and also the relation (3.45). Hence by using the Mean-Value Theorem, we are able to obtain

$$\begin{aligned} f_i - \tilde{f}_i =& f_i(x_1, \cdots, x_i) - f_i(x_1^*, \cdots, x_i^*) \\ \leq& \hat{\gamma}_1\left(|x_1 - x_1^*| + |x_2 - x_2^*| + \cdots + |x_i - x_i^*|\right), \end{aligned}$$

where $\hat{\gamma}_1$ is a bounded constant dependent on Γ_M.

1) In the case when $|x_j - x_j^*| \geq 1$, $\forall j \in \mathbb{N}_{1:i}$, by noting that $|x_j| \leq M$ and $|x_j^*| \leq \bar{r}$, the following relations hold:

$$|x_j - x_j^*| \leq M + \bar{r} \leq (M+\bar{r})|x_j - x_j^*|^{\frac{r_i+\tau}{r_j}}.$$

2) In the case when $|x_j - x_j^*| < 1$, $\forall j \in \mathbb{N}_{1:i}$, by noting $\frac{r_i+\tau}{r_j} \leq 1$, we have $|x_j - x_j^*| \leq |x_j - x_j^*|^{\frac{r_i+\tau}{r_j}}$. By summarizing the above two cases, the following inequalities can be obtained with a constant $\gamma_1 = \hat{\gamma}_1 \max\{M+\bar{r}, 1\}$:

$$\begin{aligned}\frac{f_i - \tilde{f}_i}{L^{i-1}}\Big|_{\Gamma_M} &\leq \frac{\gamma_1}{L^{i-1}}\left(|x_1 - x_1^*|^{\frac{r_i+\tau}{r_1}} + |x_2 - x_2^*|^{\frac{r_i+\tau}{r_2}} + \cdots + |x_i - x_i^*|^{\frac{r_i+\tau}{r_i}}\right) \\ &= \frac{\gamma_1}{L^{i-1}}\left(|L^0 z_1|^{\frac{r_i+\tau}{r_1}} + |L^1 z_2|^{\frac{r_i+\tau}{r_2}} + \cdots + |L^{i-1} z_i|^{\frac{r_i+\tau}{r_i}}\right).\end{aligned}$$

Similarly with the above analysis and by using Lemma 1.13, there must exist a constant $\gamma_2 > 0$, such that

$$\begin{aligned}\frac{f_i - \hat{f}_i}{L^{i-1}}\Big|_{\Gamma_M \times \mathbb{R}^n} &\leq \frac{\gamma_2}{L^{i-1}}\left(|x_1 - \langle\hat{x}_1\rangle_M|^{\frac{r_i+\tau}{r_1}} + |x_2 - \langle\hat{x}_2\rangle_M|^{\frac{r_i+\tau}{r_2}} + \cdots + |x_i - \langle\hat{x}_i\rangle_M|^{\frac{r_i+\tau}{r_i}}\right) \\ &\leq \frac{\gamma_2}{L^{i-1}}\left(|x_1 - \hat{x}_1|^{\frac{r_i+\tau}{r_1}} + |x_2 - \hat{x}_2|^{\frac{r_i+\tau}{r_2}} + \cdots + |x_i - \hat{x}_i|^{\frac{r_i+\tau}{r_i}}\right) \\ &= \frac{\gamma_2}{L^{i-1}}\left(|e_1|^{\frac{r_i+\tau}{r_1}} + |L^1 e_2|^{\frac{r_i+\tau}{r_2}} + \cdots + |L^{i-1} e_i|^{\frac{r_i+\tau}{r_i}}\right).\end{aligned}$$

With (3.35) in mind, now it comes to a conclusion that the following relations can be achieved:

$$\begin{cases}\dfrac{f_i - \tilde{f}_i}{L^{i-1}}\Big|_{\Gamma_M} \leq \gamma_1\left(|z_1|^{\frac{r_i+\tau}{r_1}} + \cdots + |z_i|^{\frac{r_i+\tau}{r_i}}\right); \\ \dfrac{f_i - \hat{f}_i}{L^{i-1}}\Big|_{\Gamma_M \times \mathbb{R}^n} \leq \gamma_2\left(|e_1|^{\frac{r_i+\tau}{r_1}} + \cdots + |e_i|^{\frac{r_i+\tau}{r_i}}\right).\end{cases} \tag{3.46}$$

Then there exist constants $\bar{\alpha}_1 > 0$, $\bar{\alpha}_2 > 0$, such that the following relations hold:

$$\begin{aligned}&\frac{\partial W(\tau,z)}{\partial z^T}\left[\{(f_i - \tilde{f}_i)/L^{i-1}\}_{i\in\mathbb{N}_{1:n}}\right]_{n\times 1}\Big|_{\Gamma_M} \leq \bar{\alpha}_1 W^{\frac{2}{2-\tau}}; \\ &\frac{\partial U(\tau,e)}{\partial e^T}\left[\{(f_i - \hat{f}_i)/L^{i-1}\}_{i\in\mathbb{N}_{1:n}}\right]_{n\times 1}\Big|_{\Gamma_M \times \mathbb{R}^n} \leq \bar{\alpha}_2 U^{\frac{2}{2-\tau}},\end{aligned} \tag{3.47}$$

and denote $\bar{\alpha} = \max\{\bar{\alpha}_1, \bar{\alpha}_2\}$.

By comparing (3.25) and (3.42), using Lemmas 1.11 and 1.13, the following relations can be obtained:

$$\begin{aligned}v_s - v_c\Big|_{\Gamma_M \times \mathbb{R}^n} &\leq \sum_{i=1}^{n} k_i \left|\left\lceil \frac{x_i - x_i^*}{L^{i-1}} \right\rfloor^{\frac{r_{n+1}}{r_i}} - \left\lceil \frac{\langle\hat{x}_i\rangle_M - x_i^*}{L^{i-1}} \right\rfloor^{\frac{r_{n+1}}{r_i}}\right| \\ &\leq \sum_{i=1}^{n} k_i 2^{1-\frac{r_{n+1}}{r_i}} |e_i|^{\frac{r_{n+1}}{r_i}}.\end{aligned} \tag{3.48}$$

Then by applying Lemmas 1.1 and 1.12, we have

$$\frac{\partial W(\tau,z)}{\partial z_n}L(v_s-v_c)\Big|_{\Gamma_M\times\mathbb{R}^n}\leq\left|\frac{\partial W(\tau,z)}{\partial z_n}\right|L\sum_{i=1}^{n}k_i2^{1-\frac{r_{n+1}}{r_i}}|e_i|^{\frac{r_{n+1}}{r_i}}$$
$$\leq L\hat{\alpha}\|z\|_{\Delta^r}^{2-\tau-r_n}\sum_{i=1}^{n}|e_i|^{\frac{r_{n+1}}{r_i}}\leq\frac{\alpha}{2}LW^{\frac{2}{2-\tau}}+\tilde{\alpha}LU^{\frac{2}{2-\tau}}, \tag{3.49}$$

where $\hat{\alpha}>0$, $\tilde{\alpha}>0$ are constants.

Inspired by the semi-global design tool proposed in [77], choose a Lyapunov candidate function of the following form:

$$V(\tau,\mathbb{Z})=\frac{r_0W(\tau,z)}{r_0+1-W(\tau,z)}+\frac{\beta\mu U(\tau,e)}{\mu+1-U(\tau,e)}, \tag{3.50}$$

where two design constants β and μ are defined as:

$$\beta\geq\frac{\tilde{\alpha}(r_0^2+\mu^2+r_0+1)^2(\mu+1)}{\mu\alpha r_0(r_0+1)}+\frac{1}{2},$$

$$\mu\geq\sup_{e\in\Gamma_e\triangleq[-2(r+\bar{r}),2(r+\bar{r})]^n}U(\tau,e).$$

Correspondingly, we can define a level set depicted below:

$$\Omega=\{(z,e)\in\mathbb{R}^{2n}|V(\tau,z,e)\leq r_0^2+\beta\mu^2+1\}.$$

From its construction, the defined sets satisfy the following relations:

$$\begin{cases}\Gamma_x\subset\Gamma_z\subset\Omega_z\subset\Gamma_M,\\ \Gamma_x\times\Gamma_{\hat{x}}\subset\Omega\subset\Gamma_M\times\mathbb{R}^n.\end{cases} \tag{3.51}$$

Moreover, for each scaling gain $L\geq1$, the Lyapunov function

$$V(\tau,\mathbb{Z}):\Omega_z\times\{e\in\mathbb{R}^n|U(\tau,e)<\mu+1\}\to\mathbb{R}_{\geq0}$$

is positive define on $\Omega\setminus\{\mathbf{0}\}$ and is proper on Ω.

From the relation (3.51), with the help of (3.47) and (3.49), calculating the time derivative of the Lyapunov function $V(\tau,\mathbb{Z})$ along the closed-loop system (3.21)–(3.41)–(3.42) gives

$$\begin{aligned}\dot{V}(\tau,\mathbb{Z})\Big|_{\Omega} &= \frac{r_0(r_0+1)}{(r_0+1-W)^2}\dot{W}(\tau,z)+\frac{\mu(\mu+1)\beta}{(\mu+1-U)^2}\dot{U}(\tau,e)\\ &\le \frac{r_0(r_0+1)}{(r_0+1-W)^2}\left(-\left(\frac{1}{2}\alpha L-\bar{\alpha}\right)W^{\frac{2}{2-\tau}}+\tilde{\alpha}LU^{\frac{2}{2-\tau}}\right)\\ &\quad -\frac{\mu(\mu+1)}{(\mu+1-U)^2}(\alpha\beta L-\bar{\alpha}\beta)U^{\frac{2}{2-\tau}}. \end{aligned} \tag{3.52}$$

Note that for $\mathbb{Z}\in\Omega$, the following inequalities can be obtained.

$$\begin{aligned}\frac{r_0}{r_0+1}&\le\frac{r_0(r_0+1)}{(r_0+1-W)^2}\le\frac{(r_0^2+\mu^2+r_0+1)^2}{r_0(r_0+1)},\\ \frac{\mu}{\mu+1}&\le\frac{\mu(\mu+1)}{(\mu+1-U)^2}\le\frac{(r_0^2+\mu^2+\mu+1)^2}{\mu(\mu+1)}.\end{aligned} \tag{3.53}$$

Hence (3.52) reduces to

$$\begin{aligned}\dot{V}(\tau,\mathbb{Z})\Big|_{\Omega} &\le -\left(\frac{1}{2}\frac{\alpha r_0}{r_0+1}L-\frac{\bar{\alpha}(r_0^2+\mu^2+r_0+1)^2}{r_0(r_0+1)}\right)W^{\frac{2}{2-\tau}}\\ &\quad -\left(\frac{\mu\alpha\beta}{\mu+1}-\frac{\tilde{\alpha}(r_0^2+\mu^2+r_0+1)^2}{r_0(r_0+1)}\right)LU^{\frac{2}{2-\tau}}\\ &\quad +\frac{\bar{\alpha}\beta(r_0^2+\mu^2+\mu+1)^2}{\mu(\mu+1)}U^{\frac{2}{2-\tau}}.\end{aligned} \tag{3.54}$$

Given the following guideline:

$$\begin{aligned}L\ge\max&\left\{\frac{2\bar{\alpha}(r_0^2+\mu^2+r_0+1)^2+2(r_0+1)r_0}{\alpha r_0^2},\right.\\ &\left.\frac{2\bar{\alpha}\beta(r_0^2+\mu^2+\mu+1)^2+2(\mu+1)\mu}{\alpha\mu^2},1\right\},\end{aligned} \tag{3.55}$$

and with the definition of β in mind, the relation (3.54) reduces to

$$\dot{V}(\tau,\mathbb{Z})\big|_{\Omega}\le-\left(W^{\frac{2}{2-\tau}}+U^{\frac{2}{2-\tau}}\right). \tag{3.56}$$

In what follows, provided the scaling gain L is set according to (3.55), we will prove that for any non-zero initial states satisfying $(x(t_0),\hat{x}(t_0))\in\Gamma_x\times\Gamma_{\hat{x}}$, the state trajectories will be guaranteed to stay in the compact set Ω forever.

If the above statement is not true, due to the fact that the initial states $(z(t_0),e(t_0))\in\Omega\subset\Gamma_M\times\mathbb{R}^n$ which yields $\dot{V}(\tau,\mathbb{Z}(t_0))<0$, there must exist two

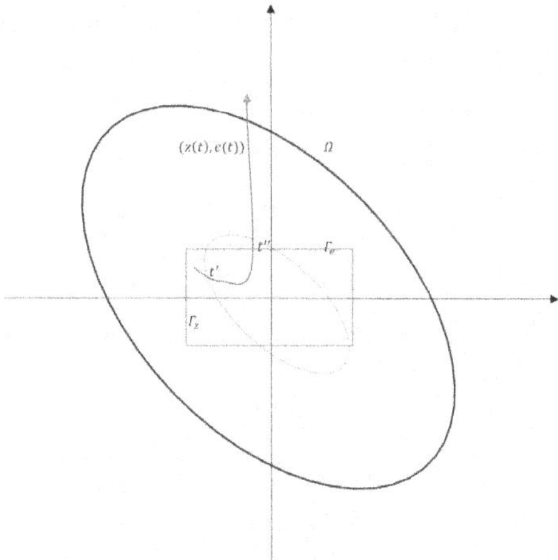

Figure 3.1 Sketch depiction of the state trajectories and the relation (3.57).

time instants $t^{'} > t_0$, $t^{''} > t^{'}$, as depicted by Figure 3.1, such that the following relations hold:

$$\begin{cases} i)\ \dot{V}(\tau,\mathbb{Z}(t^{'})) < 0; \\ ii)\ V(\tau,\mathbb{Z}(t^{''})) = V(\tau,\mathbb{Z}(t^{'})); \\ iii)\ \dot{V}(\tau,\mathbb{Z}(t^{''})) > 0. \end{cases} \tag{3.57}$$

It is clear that (3.47) and (3.49) still hold for $t \in [t^{'},t^{''}]$, hence with L satisfying (3.55), one can obtain the following relations from (3.56):

$$\begin{aligned} V(\tau,\mathbb{Z}(t^{''})) - V(\tau,\mathbb{Z}(t^{'})) &= \int_{t^{'}}^{t^{''}} \dot{V}(\tau,\mathbb{Z}(s))ds \\ &\leq -\int_{t^{'}}^{t^{''}} (W^{\frac{2}{2-\tau}} + U^{\frac{2}{2-\tau}})ds. \end{aligned} \tag{3.58}$$

With the relation ii) of (3.57), (3.58) will clearly lead to a contradiction that

$$0 \geq \int_{t^{'}}^{t^{''}} (W^{\frac{2}{2-\tau}} + U^{\frac{2}{2-\tau}})ds > 0.$$

Hence, a conclusion arrives at

$$(x(t_0),\hat{x}(t_0)) \in \Gamma_x \times \Gamma_{\hat{x}} \Rightarrow \mathbb{Z}(t) \in \Omega \Rightarrow \lim_{t\to\infty} \mathbb{Z}(t) = 0,\ \forall t \geq t_0, \tag{3.59}$$

which concludes that the exact tracking realization can be achieved under the semi-global control objective. Moreover, it is clear from the relation (3.56) that the output y will exactly track its desired value y_r asymptotically provided $\tau = 0$ or within a finite time provided $\tau < 0$.

This completes the proof of Theorem 3.3. ■

Remark 3.6 Due to the careful selection of the scaling gain L to dominate the nonlinear perturbation terms, the known boundary of the initial states (i.e., located in a compact set) and the fact that a series of saturation functions are employed to avoid the destabilization of peaking phenomenon, now we can obtain the leverage to avoid the finite time escape issues.

Remark 3.7 Note that by trading off the control target from global control to a more practical semi-global one, now the tracking control law can be designed as a finite-time dynamic compensator without the pre-verification of Assumption 3.2.1 or other restrictive nonlinear growth constraints utilized in many existing literatures, such as [20, 61], etc. Moreover, by recalling that the homogeneous degree τ in existing global design methods is always restricted to certain values owing to the exact verification of the nonlinearity growth assumptions, now the homogeneous degree could be flexibly tuned to meet certain pre-required convergence rate.

3.2.2.4 Extension to practical tracking control

Aiming for illustrating the nonrecursive output feedback design to achieve exact tracking control, system (3.1) is assumed to satisfy a certainty hypothesis. However, for a wider class of general nonlinear systems with the presence of mismatched external disturbances, the proposed method will also lead to an easily implementable practical tracking controller and the robustness of closed-loop system can be enhanced by increasing the scaling gain L. More specifically, consider the following nonlinear system:

$$\begin{cases} \dot{x}_i(t) = x_{i+1}(t) + f_i(\bar{x}_i(t)) + d_i(t), \ i \in \mathbb{N}_{1:n-1}, \\ \dot{x}_n(t) = u(t) + f_n(x(t)) + d_n(t), \\ y(t) = x_1(t), \end{cases} \tag{3.60}$$

where $d_i(t),\ i \in \mathbb{N}_{1:n}$ represents a mismatched disturbance term satisfying the following assumption.

Assumption 3.2.2 *There exists a constant $D > 0$, such that $|d_i(t)| \leq D,\ i \in \mathbb{N}_{1:n}$.*

In what follows, we will show that by using the same design architecture, a general practical tracking control result for system (3.60) can be achieved to enrich the practical nature of the proposed nonrecursive method, which can be summarized by the following theorem.

Theorem 3.4

Consider the closed-loop system consisting of (3.60) satisfying Assumptions 3.1.1–3.2.2 and the output feedback tracking control law (3.40)–(3.42). For any given tolerance $\sigma > 0$, there exists a proper scaling gain $L \geq 1$ and a finite-time T^*, such that the closed-loop system is semi-globally practically stable and $|y(t) - y_r| \leq \sigma, \forall t \geq T^*$.

Proof: Using the same coordinates transformation as (3.18), the z-dynamical system transformed from (3.60) should be modified to the following form:

$$\dot{z} = L\psi(z,v) + \left[\{(f_i - \hat{f}_i + d_i(t))/L^{i-1}\}_{i\in\mathbb{N}_{1:n}}\right]_{n\times 1}. \tag{3.61}$$

Henceforth, with the same observer as (3.40), the error dynamics now should be presented to the following form:

$$\dot{e} = Lf(\tau,H,e) + \left[\{(f_i - \langle\hat{f}_i\rangle_M + d_i(t))/L^{i-1}\}_{i\in\mathbb{N}_{1:n}}\right]_{n\times 1}. \tag{3.62}$$

With Assumption 3.2.2 in mind and using the same Lyapunov function as (3.43), the following relations can be derived directly by applying Lemmas 1.1 and 1.12:

$$\begin{aligned}\frac{\partial W(\tau,z)}{\partial z^\top}\left[\frac{d_1}{L^0},\cdots,\frac{d_n}{L^{n-1}}\right]^\top &\leq \sum_{i=1}^{n}\breve{\alpha}\|z\|_{\Delta^r}^{2-\tau-r_i}\left(\frac{D^{\frac{1}{\tau+r_i}}}{L^{\frac{i-1}{\tau+r_i}}}\right)^{\tau+r_i}\\ &\leq W^{\frac{2}{2-\tau}}(\tau,z) + \hat{\alpha}\sum_{i=1}^{n}\frac{D^{\frac{2}{\tau+r_i}}}{L^{2(i-1)/(\tau+r_i)}}\\ &\triangleq W^{\frac{2}{2-\tau}}(\tau,z) + D^*,\end{aligned} \tag{3.63}$$

where $\breve{\alpha} > 0$ and $\hat{\alpha} > 0$ are constants. Similarly, we can also have

$$\frac{\partial U(\tau,e)}{\partial e}\left[\frac{d_1}{L^0},\cdots,\frac{d_n}{L^{n-1}}\right]^\top \leq U^{\frac{2}{2-\tau}}(\tau,e) + D^*. \tag{3.64}$$

Thereafter, the following semi-global practical stability analysis is almost identical to the proof of Theorem 3.3, whereas the difference mainly relies on the additional terms in (3.63)–(3.64). Consequently, with the relation (3.54) in mind, the derivative of $V(\tau,\mathbb{Z})$ along the closed-loop system (3.26)–(3.61) gives

$$\begin{aligned}\dot{V}(\tau,\mathbb{Z})\big|_\Omega \leq &-\left(\frac{1}{2}\frac{\alpha r_0}{r_0+1}L - \frac{(\bar{\alpha}+1)(r_0^2+\mu^2+r_0+1)^2}{r_0(r_0+1)}\right)W^{\frac{2}{2-\tau}}\\ &-\left(\frac{\mu\alpha\beta}{\mu+1} - \frac{\tilde{\alpha}(r_0^2+\mu^2+r_0+1)^2}{r_0(r_0+1)}\right)LU^{\frac{2}{2-\tau}}\\ &+\frac{(\bar{\alpha}+1)\beta(r_0^2+\mu^2+\mu+1)^2}{\mu(\mu+1)}U^{\frac{2}{2-\tau}}\\ &+\left(\frac{(r_0^2+\mu^2+r_0+1)^2}{r_0(r_0+1)} + \frac{\beta(r_0^2+\mu^2+\mu+1)^2}{\mu(\mu+1)}\right)D^*.\end{aligned} \tag{3.65}$$

Following the guideline (3.55) which yields a proper choice of L, (3.65) will reduce to

$$\dot{V}(\tau,\mathbb{Z})\big|_{\Omega} \leq -(W^{\frac{2}{2-\tau}}+U^{\frac{2}{2-\tau}}) + \left(\frac{(r_0^2+\mu^2+r_0+1)^2}{r_0(r_0+1)}+\frac{\beta(r_0^2+\mu^2+\mu+1)^2}{\mu(\mu+1)}\right)D^*. \tag{3.66}$$

For a given arbitrary small constant $\nu > 0$, define a set

$$\Omega_1 = \{\mathbb{Z}\in\mathbb{R}^{2n}|V(\tau,\mathbb{Z})\leq \nu\}.$$

From the derivation of D^* in (3.63) which implies the fact that D^* can be made arbitrary small by tuning L to be sufficiently large, then there must exist a sufficiently large positive constant $L^*(\nu)\geq 1$, such that if $L\geq L^*(\nu)$, for any $\mathbb{Z}\in\Omega\setminus\Omega_1$, the following inequality holds:

$$\dot{V}(\tau,\mathbb{Z})\big|_{\Omega\setminus\Omega_1} \leq -\frac{1}{2}\left(W^{\frac{2}{2-\tau}}+U^{\frac{2}{2-\tau}}\right). \tag{3.67}$$

Hence we know that Ω is an invariance set and any trajectories starting from Ω will be captured by the set Ω_1. Therefore, for any given tolerance constant $\sigma > 0$, by using (3.67), there must exist a time instant T^* such that the following relation can be achieved:

$$|y(t)-y_r| \leq \|z\|_{\Delta^r} \leq \sigma,\ \forall t\geq T^* > t_0, \tag{3.68}$$

which concludes that the practical tracking for system (3.60) in the presence of bounded mismatched disturbances is realized. This completes the proof of Theorem 3.4. ■

Remark 3.8 Compared to many related existing practical tracking control methods, such as [20, 61, 80], etc., the proposed practical tracking control law can now be realized in a one-step manner where the expression of the form (3.27) will be much more concise. More significantly, in the presence of mismatched disturbances and nonvanishing nonlinearities, the proposed practical tracking law can render every states practically track their nominal reference values respectively, rather than practically regulate the states $(x_1-y_r, x_2, \cdots, x_n)$ to the origin via a relatively larger control energy, which is a common approach proposed in [20, 61].

Remark 3.9 In practice, a set of performance specifications is usually given before the controller design. It is worthy of pointing out that by following a similar technique utilized in [24], the selection of the control parameters can also be guided without much effort to meet the requirements of pre-specified performance indexes.

3.2.2.5 Numerical simulations

In what follows, a numerical example is provided to illustrate the design procedure and verify the effectiveness of the proposed exact tracking and practical tracking controllers.

Example 3.2.1 *Consider the following 3-D nonlinear system*

$$\begin{cases} \dot{x}_1 = x_2 + x_1^{5/3} + d_1 \\ \dot{x}_2 = x_3 + 2\cos(x_2) + d_2 \\ \dot{x}_3 = u + x_1 \ln(1 + x_2^2) + d_3 \\ y = x_1, \end{cases} \tag{3.69}$$

with its control objective being to semi-globally exactly (practically) track a reference time-varying signal $y_r = 3 + 2\sin(2t) + \cos(t)$.

Utilizing the proposed semi-global control method with a series of non-recursive calculations of the auxiliary variables as

$$\begin{aligned} & y_r = 3 + 2\sin(2t) + \cos(t),\ y_r^{(1)} = 4\cos(2t) - \sin(t), \\ & y_r^{(2)} = -8\sin(2t) - \cos(t),\ y_r^{(3)} = -16\cos(2t) + \sin(t); \\ & x_1^* = y_r,\ x_2^* = y_r^{(1)} - y_r^{5/3},\ x_3^* = x_2^{*(1)} - 2\cos(x_2^*); \\ & u^* = x_3^{*(1)} - x_1^* \ln(1 + x_2^{*2}); \\ & \tilde{f}_1 = y_r^{5/3},\ \tilde{f}_2 = 2\cos(y_r^{(1)} - y_r^{5/3}),\ \tilde{f}_3 = y_r \ln(1 + (y_r^{(1)} - y_r^{5/3})^2), \end{aligned} \tag{3.70}$$

we can achieve the tracking output feedback control law, which is explicitly shown in the following forms:

$$\begin{cases} \dot{\hat{z}}_1 = L\hat{z}_2 + \langle \hat{x}_1 \rangle_M^{5/3} - \tilde{f}_1 + L\ell_1 \lfloor z_1 - \hat{z}_1 \rceil^{1+\tau}, \\ \dot{\hat{z}}_2 = L\hat{z}_3 + \dfrac{2\cos(\langle \hat{x}_2 \rangle_M) - \tilde{f}_2}{L} + L\ell_2 \lfloor z_1 - \hat{z}_1 \rceil^{1+2\tau}, \\ \dot{\hat{z}}_3 = Lv + \dfrac{\langle \hat{x}_1 \rangle_M \ln(1 + \langle \hat{x}_2 \rangle_M^2) - \tilde{f}_3}{L^2} + L\ell_3 \lfloor z_1 - \hat{z}_1 \rceil^{1+3\tau}; \end{cases} \tag{3.71}$$

$$\begin{cases} v = -k_1 \lfloor \langle \hat{x}_1 \rangle_M - x_1^* \rceil^{1+3\tau} - k_2 \left\lfloor \dfrac{\langle \hat{x}_2 \rangle_M - x_2^*}{L} \right\rceil^{\frac{1+3\tau}{1+\tau}} - k_3 \left\lfloor \dfrac{\langle \hat{x}_3 \rangle_M - x_3^*}{L^2} \right\rceil^{\frac{1+3\tau}{1+2\tau}}, \\ u = L^3 v + u^*. \end{cases} \tag{3.72}$$

In what follows, two cases are studied to show that a unified control framework can achieve both exact tracking and practical tracking results.

Case 1: Exact Tracking Provided $d_i = 0,\ i = 1, 2, 3.$

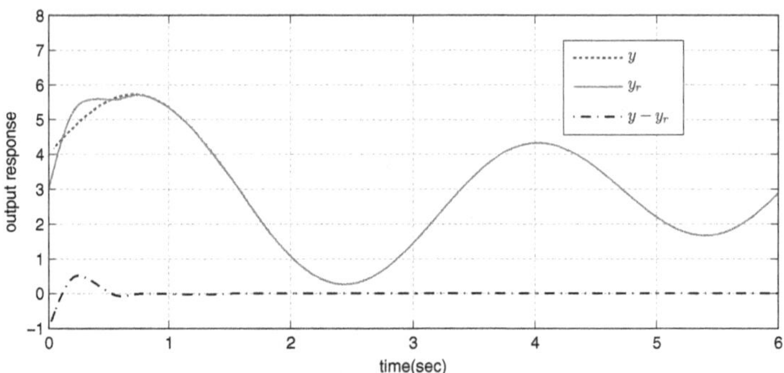

Figure 3.2 Output tracking performance (Case 1).

In the following simulation, we choose the homogeneous degree as $\tau = -0.21$, the control and observer gain vectors are set as $K = [27,27,9]$, $H = [15,75,125]$ which render $A-BK$ and $A-HC$ both Hurwitz, the saturation threshold is set as $M = 100$, the scaling gain is chosen as $L = 3.2$. To conduct the simulation, the initial value vector is set as $(x_1(0),x_2(0),x_3(0),\hat{z}_1(0),\hat{z}_2(0),\hat{z}_3(0)) = (3,6,17,-1.8,2.5,2)$. From Figure 3.2, the exact tracking control objective is realized by the proposed semi-global output feedback control law. By selecting a negative homogeneous degree, the tracking task can be achieved within a finite time. The states x_1, x_2, and x_3 can be precisely estimated by the finite-time observer whereas the observation curves are shown in Figure 3.3. The time history of the control input is given in Figure 3.4.

Case 2: Practical Tracking Provided $d_i \neq 0,\ i = 1,2,3.$

In the following simulations, assume $d_1(t) = d_3(t) = 5\sin(10t) + 4$ and $d_2(t) = 2\sin(10t) - 4$. The control parameters are chosen the same as Case 1 except that the scaling gain L is tuned to have different values in order to show different practical tracking performances. Figure 3.5 shows that under the same tracking control law, a practical tracking control performance can be obtained and moreover, the steady-state error can be adjusted to be small enough by tuning the scaling gain to be a larger value. It can be clearly observed from Figure 3.5 that with an increasing scaling gain (i.e., L is tuned from 4 to 5), the steady-state error can be reduced and the convergence speed can be made faster as well. However on the other side, a trade-off is the consumption of a larger control energy as depicted in Figure 3.7. Figure 3.6 shows that practical observation results can be achieved for system (3.69) in the presence of nonvanishing mismatched disturbances.

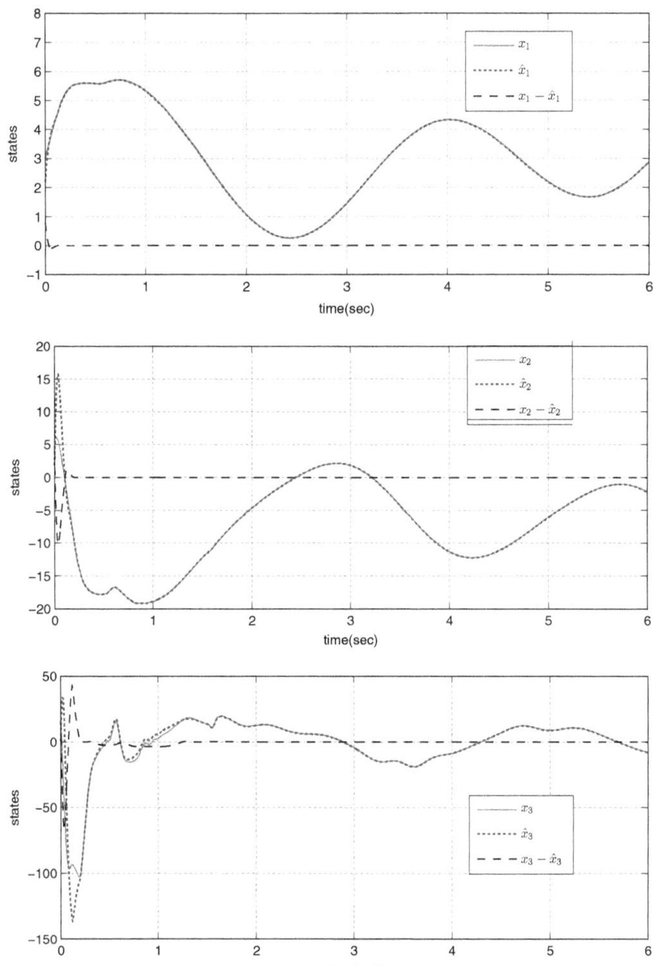

Figure 3.3 Performance of the state observer (Case 1).

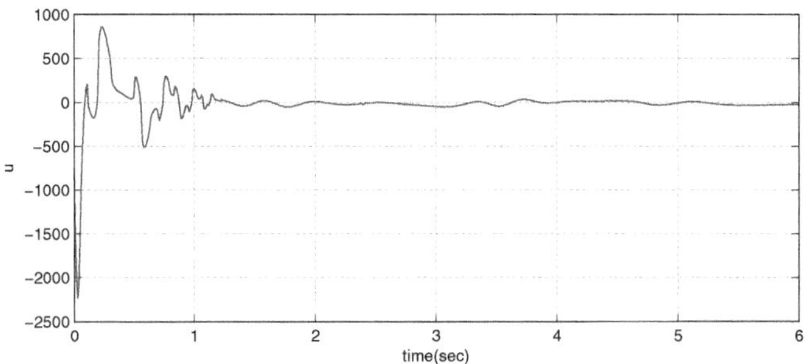

Figure 3.4 Time history of the control input (Case 1).

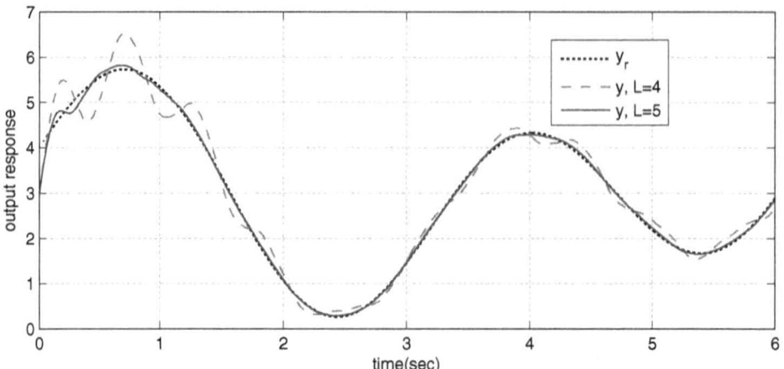

Figure 3.5 Practical output tracking performances (Case 2).

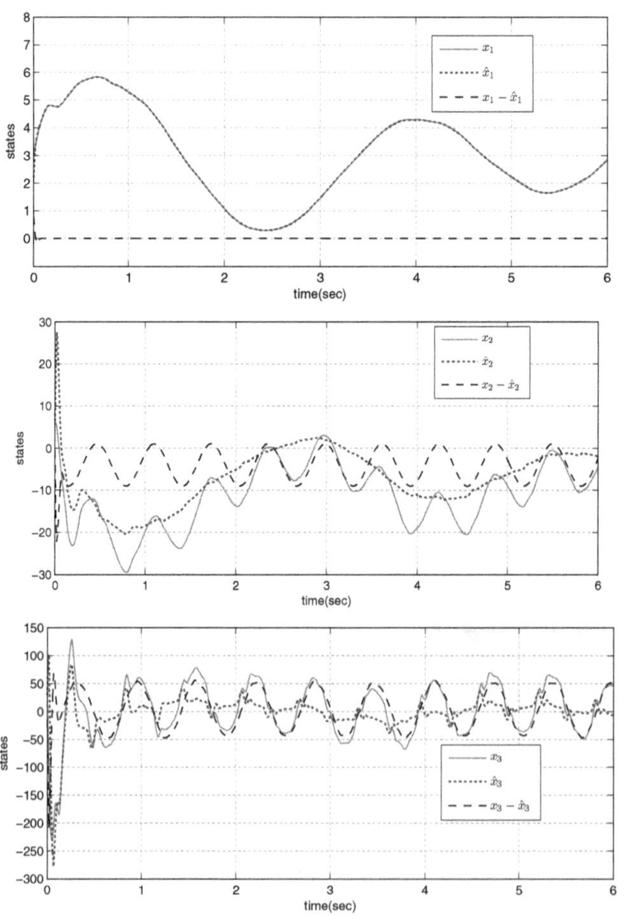

Figure 3.6 Response curves of x and their estimations $\hat{x}$ with $L = 5$ (Case 2).

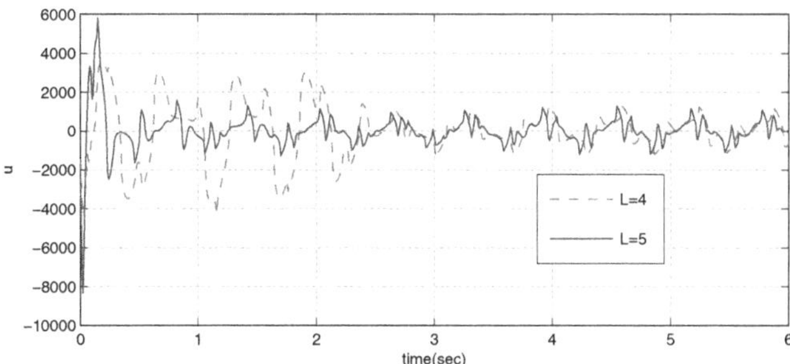

Figure 3.7 Time histories of the control inputs (Case 2).

Remark 3.10 As observed from the design procedure for the numerical example (3.69), there is no need to conquer various recursive calculations of virtual controllers, which are inevitable by utilizing all existing recursive approaches. Moreover, the proposed tracking law is much simpler and the control parameters are easier to be selected, hence will give much more convenience in practical implementations.

Chapter 4

Nonrecursive Adaptive Control for Nonlinear Systems with Non-Parametrized Uncertainties

In modern control practices, adaptive control method has been well established and popularly regarded as one of the most effective synthesis tools for systems involving unknown parameters with possible variations, unknown actuator nonlinearities, external disturbances, etc. Being an attractive research aspect in both control theory and control engineering, evidence can be regarded as the fact that various advanced nonlinear adaptive control approaches have been investigated in recent decades. Among them, adaptive backstepping control based on a feedback system form and its derivations have been reported in numerous references. By assuming that the uncertainties in the system should satisfy a parameterized structure, a systematic adaptive design procedure with a series of tuning functions is a popular handling strategy. However, most of those related methods strongly rely on the smooth continuity and certain nonlinearity requirements which are presented in a feedback structure. In addition, a notable drawback is the complexity of the controller expression derived by utilizing adaptive backstepping design procedures.

DOI: 10.1201/9781003399230-4

For nonlinear systems with the presence of non-parametric uncertainties, an alternative synthesis tool is of significance. In [33, 34, 36, 56], different kinds of time-varying scaling gain methods are presented, which could derive adaptive control laws by means of a dynamic high-gain domination technique. However, the Lipschitz continuous condition is required for the assurance of a global asymptotical stabilization result.

Regarding the finite-time adaptive stabilization design, under a polynomial hypothesis on the system structure, the work in [45] proposes an adaptive design framework based on feedback domination strategy while a later result in [22] extends the finite-time adaptive control issue to a class of p-normal uncertain nonlinear systems. Later on, Sun et al. [72] develop a finite-time stabilization result via the adding a power integrator technique. Sun and Yang [71] show the existence of a monotone dynamic scaling gain, whose explicit construction procedure for practice, however remains open.

It is worth pointing out here that most of the above-mentioned adaptive design approaches involve a recursive determination of virtual controllers. Direct drawback may exhibit on the expanding complexity when the system order is growing (also called as "explosion of terms"). Meanwhile, the gain selection guideline is made much conservative by going through a variety of necessary mathematical estimations. From a practical point of view, proposing a concise control scheme to simplify the gain tuning mechanism and ease their implementations is more imperatively demanded.

Inspired by the design facility of homogeneous system theory as illustrated in the existing literature, a nonrecursive adaptive controller design framework is proposed in this chapter, aiming to solve the dynamic stabilization problem for a class of nonlinear systems with the presence of general non-parametric uncertainties. As a direct benefit of the nonrecursive design approach, a simplest non-nested stabilizing controller can be constructed independently of the Lyapunov function-based stability analysis. Besides, the control gain vector can be easily determined while the dynamic scaling gain is online updated subjecting to an explicit self-tuning adaption law.

4.1 Nonrecursive $\mathbb{C}^1$ Adaptive Control Design

A nonrecursive $\mathbb{C}^1$ adaptive stabilizing control method, under a semi-global stability objective, is studied in this section for a class of non-parameterized (or, nonlinearly parameterized) uncertain nonlinear systems. By incorporating the homogeneous system theory and a nonrecursive synthesis approach, the resulting $\mathbb{C}^1$ adaptive stabilizing control scheme can now be constructed within a novel one-step design framework. Employing an adaption law to online determine the time-varying scaling gain, both the controller expression and gain tuning mechanism are much simpler than the related existing recursive design approaches.

4.1.1 Problem formulation

Consider a class of lower-triangular nonlinear systems with non-parametric uncertainties, depicted by the following form:

$$\begin{cases} \dot{x}_i = x_{i+1} + f_i(\theta, \bar{x}_i), \ i \in \mathbb{N}_{1:n-1}, \\ \dot{x}_n = u + f_n(\theta, x), \end{cases} \tag{4.1}$$

where $\bar{x}_i = \mathrm{col}(x_1, x_2, \cdots, x_i)$, $x = \bar{x}_n$, and u are the system partial state vector, full state vector, and control input, respectively. $\theta \in \mathbb{R}^r$ is an unknown bounded parameter vector which could be either time-varying or constant, $f_i(\cdot)$, $i \in \mathbb{N}_{1:n}$ is a continuous nonlinear function satisfying a vanishing condition, i.e., $f_i(\theta, 0) = 0$ for $i \in \mathbb{N}_{1:n}$.

The control objective is to find a dynamic stabilizing controller of the following non-nested form:

$$\begin{cases} u = u(L, x) \in \mathbb{C}^1, \ u(L, 0) = 0, \\ \dot{L} = \psi(L, x) \in \mathbb{C}^1\big([1, +\infty) \times \mathbb{R}^n \to [0, +\infty)\big), \\ L(t_0) = 1, \end{cases} \tag{4.2}$$

such that the following conclusions hold for the resulting closed-loop system (4.1)–(4.2).

- All the signals of the closed-loop system are globally uniformly bounded.
- The origin $x = 0$ is an asymptotically stable equilibrium.

4.1.2 System pre-treatment

First of all, define a x-z coordination transform by

$$z_i = x_i / L^i, \ i \in \mathbb{N}_{1:n}, \ v = u / L^{n+1}, \tag{4.3}$$

where $L : \mathbb{R}_{\geq 0} \to [1, \infty)$ is a time-varying scaling gain which will be online determined by a dual-layer update law. It is not difficult to verify that the z-system dynamics should be presented by

$$\begin{cases} \dot{z}_i = L z_{i+1} - i\dot{L}/L z_i + \dfrac{f_i(\theta, \bar{x}_i)}{L^i}, \ i \in \mathbb{N}_{1:n-1}, \\ \dot{z}_n = L v - n\dot{L}/L z_n + \dfrac{f_n(\theta, x)}{L^n}. \end{cases} \tag{4.4}$$

By denoting

$$A = \begin{bmatrix} 0 & 1 & \cdots & 0 \\ \vdots & \vdots & \cdots & \vdots \\ 0 & 0 & \cdots & 1 \\ 0 & 0 & \cdots & 0 \end{bmatrix}, \ B = [0, 0, \cdots, 1]^\top, \ D = [1, 2, \cdots, n],$$

$$F(\theta,L,x)=[f_1(\theta,\bar{x}_1)/L, f_2(\theta,\bar{x}_2)/L^2,\cdots,f_n(\theta,x)/L^n]^\top,$$

system (4.4) can also be rewritten as the following compact form:

$$\dot{z}=L(Az+Bv)-\dot{L}/LDz+F(\theta,L,x). \tag{4.5}$$

Motivated by [34,36], we know that there exist a dilation weight r defined by $r_1=1$, $r_i=r_{i-1}+\tau$, $i\in\mathbb{N}_{2:n}$ with $\tau\in[0,1)$, a constant κ satisfying $\kappa\geq r_n+\tau$, a gain vector $K=[k_1,k_2,\cdots,k_n]$, a positive constant a and a positive definite, symmetrical matrix P, such that the following relations hold:

$$\begin{aligned}&(A-BK)^\top P+P(A-BK)\leq -aI,\\ &\tilde{D}P+P\tilde{D}>0,\ \tilde{D}=\left[\frac{\kappa-\frac{\tau}{2}}{r_1},\frac{2\kappa-\tau}{r_1},\cdots,\frac{n\kappa-\frac{n\tau}{2}}{r_n}\right].\end{aligned} \tag{4.6}$$

4.1.3 Stabilizing control law construction

To present a stabilizing control result, the following assumption is required to restrain the system nonlinearities.

Assumption 4.1.1 *There exists a non-negative continuous function $\psi_i(\theta,\bar{x}_i)$, such that the following relation holds:*

$$|f_i(\theta,\bar{x}_i)|\leq\psi_i(\theta,\bar{x}_i)\sum_{j=1}^{i}|x_j|^{\frac{r_i+\tau}{r_j}},\ i\in\mathbb{N}_{1:n}.$$

Remark 4.1 In order to pave a way for carrying out homogeneous domination strategy, such an assumption is currently essential to restrict the nonlinearity growth rate for system (4.1). By noting that $\psi_i(\theta,\bar{x}_i)$ is allowed to be any non-negative continuous function, hence it is worth pointing out that Assumption 4.1.1 could include more general nonlinear functions than the common homogeneous growth conditions required in [55,73,91].

By using Lemma 1.10, we know that there must exist two smooth scalar functions $\delta_i(\theta)\geq 1$ and $\gamma_i(\bar{x}_i)\geq 1$, such that Assumption 4.1.1 can be further revised as

$$|f_i(\theta,\bar{x}_i)|\leq\delta_i(\theta)\gamma_i(\bar{x}_i)\sum_{j=1}^{i}|x_j|^{\frac{r_i+\tau}{r_j}},\ i\in\mathbb{N}_{1:n}.$$

Recalling that the uncertain parameter θ is assumed to be bounded, it is obvious that $\max_{i\in\mathbb{N}_{1:n}}\{\delta_i(\theta)\}\leq\sigma$ with $\sigma>1$ being an unknown constant to the designer.

By denoting $\hat{\sigma}$ as the estimation of σ, we could construct a continuously differentiable ($\mathbb{C}^1$) stabilizing control law, which is expressed by

$$\begin{cases} u = L^{n+1}v, \\ v = -\sum_{i=1}^{n} k_i \lfloor z_i \rceil^{\frac{r_n+\tau}{r_i}} \triangleq -K\lfloor z \rceil_{\Delta^r}^{r_n+\tau}, \\ \dot{\hat{\sigma}} = c_1 L^{\rho} \|z\|_{\Delta^r}^{2\kappa}, \\ \dot{L} = L^{1+\rho}\max\{0,\ c_2\hat{\sigma} - c_3 L^{1-\rho}\}\|z\|_{\Delta^r}^{\tau},\ L(0) \geq 1, \end{cases} \tag{4.7}$$

where $\rho = 1 - \frac{1-\tau}{r_n} \in [0,1)$, and c_1, c_2, c_3 are positive design constants which will be made precise in the stability analysis later on.

4.1.4 Explicit stability analysis

The main result of this section can be summarized by the following theorem.

Theorem 4.1

Consider the closed-loop system consisting of system (4.1) satisfying Assumption 4.1.1 and the adaptive stabilizing controller (4.7). For any initial states of the closed-loop system starting from a compact set, i.e., $(x(0)^\top, \hat{\sigma}(0))^\top \in \Gamma \triangleq [-\rho, \rho]^{n+1}$ *with* $\rho > 0$ *being a given constant which can be arbitrarily large, the following statements hold.*

- *All the signals in the closed-loop system are uniformly bounded.*
- *The equilibrium* $x = 0$ *is asymptotically stable.*

Proof: Construct a Lyapunov function of the form

$$U(z) = \left(\lfloor z \rceil_{\Delta^r}^{\kappa - \frac{\tau}{2}}\right)^\top P \lfloor z \rceil_{\Delta^r}^{\kappa - \frac{\tau}{2}}.$$

From the fact that $2\kappa - \tau \geq 2r_{n+1} - \tau \geq 2r_1 = 2$, we know that $U(z)$ is continuously differentiable and proper. With the dilation mapping $\Delta_\varepsilon^r z = (\varepsilon^{r_1} z_1, \cdots, \varepsilon^{r_n} z_n)$ where r is the weight vector, one can verify that $U(z) \circ \Delta_\varepsilon^r = \varepsilon^{2\kappa-\tau} U(z)$. It is also not difficult to obtain the following relations

$$\begin{cases} z_{i+1} \circ \Delta_\varepsilon^r = \varepsilon^{r_{i+1}} z_{i+1} = \varepsilon^{r_i+\tau} z_{i+1},\ i \in \mathbb{N}_{1:n-1}, \\ v \circ \Delta_\varepsilon^r = -\varepsilon^{r_n+\tau} K \lfloor z \rceil_{\Delta^r}^{r_n+\tau}, \end{cases}$$

which clearly conclude that the vector field $[z_1, z_2, \cdots, v]^\top$ is Δ^r-homogeneous of degree τ.

Hence by using Lemma 2.1, we know that there exists a constant $\varsigma \in (0,1)$, such that the system

$$\dot{z} = Az - BK\lfloor z \rceil_{\Delta^r}^{r_n+\tau}$$

is globally asymptotically stable for $\tau \in [0,\varsigma)$. Moreover, the following relation can be further obtained:

$$\frac{\partial U(z)}{\partial z^\top}(Az+Bv) \le -c\|z\|_{\Delta^r}^{2\kappa},\ \tau \in [0,\varsigma), \tag{4.8}$$

where $c > 0$ is a constant.

With the relation (4.6) in mind, it can be straightforwardly concluded from Lemma 1.1 that there exists a constant $\hat{c} > 0$ which is independent of L, such that the following relations hold:

$$\begin{aligned}\frac{\partial U(z)}{\partial z^\top}\dot{L}/LDz &= \dot{L}/L\left(\lfloor z\rceil_{\Delta^r}^{\kappa-\frac{\tau}{2}}\right)^\top(\tilde{D}P+P\tilde{D})\lfloor z\rceil_{\Delta^r}^{\kappa-\frac{\tau}{2}} \\ &\ge \hat{c}\dot{L}/L\|z\|_{\Delta^r}^{2\kappa-\tau}.\end{aligned} \tag{4.9}$$

Define the error $\tilde{\sigma} = \sigma - \hat{\sigma}$. Thereafter, we consider the new Lyapunov function

$$V(z,\tilde{\sigma}) = U(z) + \frac{1}{2\varepsilon}\tilde{\sigma}^2$$

where ε is a positive design parameter. With the help of (4.8) and (4.9), the time derivative of $V(z,\tilde{\sigma})$ along system (4.5)–(4.7) can be derived as

$$\begin{aligned}\dot{V}(z,\tilde{\sigma}) &= \frac{\partial U(z)}{\partial z^\top}L(Az+Bv) - \frac{\partial U(z)}{\partial z^\top}\dot{L}/LDz + \frac{\partial U(z)}{\partial z^\top}F(\theta,L,x) - \tilde{\sigma}\dot{\hat{\sigma}}/\varepsilon \\ &\le -cL\|z\|_{\Delta^r}^{2\kappa} - \hat{c}\dot{L}/L\|z\|_{\Delta^r}^{2\kappa-\tau} + \frac{\partial U(z)}{\partial z^\top}F(\theta,L,x) - \tilde{\sigma}\dot{\hat{\sigma}}/\varepsilon.\end{aligned} \tag{4.10}$$

Note that θ is bounded, which implies that there must exist two positive constants $\bar{\rho}_1$, $\bar{\rho}_2$ such that $1 \le \bar{\rho}_1 \le \sigma \le \bar{\rho}_2$. Hence for any initial states $(x(0)^\top,\hat{\sigma}(0))^\top \in \Gamma \triangleq [-\rho,\rho]^{n+1}$, we know that $\tilde{\sigma}(0) \in [\bar{\rho}_1-\rho,\bar{\rho}_2+\rho]$. Now define a level set

$$\Omega = \left\{(z,\tilde{\sigma}) \in \mathbb{R}^{n+1}\middle|V(z,\tilde{\sigma}) \le \sup_{(z^\top,\tilde{\sigma})^\top \in [-\wp,\wp]^{n+1}} V(z,\tilde{\sigma})\right\}$$

where $\wp = \max\{\rho,|\bar{\rho}_1-\rho|,|\bar{\rho}_2-\rho|\}$. Let $\Gamma_M = [-M,M]^n$, where M is a constant defined by $M = \max_{(z^\top,\tilde{\sigma})^\top \in \Omega}\|z\|_\infty$.

From Assumption 4.1.1, the following relations can be easily obtained

$$\begin{aligned}f_i(\theta,\bar{x}_i)/L^i\Big|_{\Gamma_M} &\le \frac{\gamma_i(\bar{x}_i)\sigma}{L^i}\left(|x_1|^{\frac{r_i+\tau}{r_1}} + |x_2|^{\frac{r_i+\tau}{r_2}} + \cdots + |x_i|^{\frac{r_i+\tau}{r_i}}\right) \\ &\le \frac{\bar{\gamma}_i\sigma}{L^i}\left(|L^1z_1|^{\frac{r_i+\tau}{r_1}} + |L^2z_2|^{\frac{r_i+\tau}{r_2}} + \cdots + |L^iz_i|^{\frac{r_i+\tau}{r_i}}\right) \\ &= \bar{\gamma}_i\sigma\left(L^{\frac{(1+i\tau)}{1}-i}|z_1|^{\frac{1+i\tau}{1}} + L^{2\frac{1+i\tau}{1+\tau}-i}|z_2|^{\frac{1+i\tau}{1+\tau}} + \cdots + L^{i\frac{1+i\tau}{1+(i-1)\tau}-i}|z_i|^{\frac{1+i\tau}{1+(i-1)\tau}}\right),\end{aligned} \tag{4.11}$$

where $\bar{\gamma}_i$ is a positive constant.

Noting that $L \geq 1$ and the restriction that $\tau \in [0,1)$, it can be concluded that for $j \in \mathbb{N}_{1:i}$

$$\begin{aligned}
j\frac{1+i\tau}{1+(j-1)\tau}-i &= \frac{j(1+(i-1)\tau)+j\tau-i-i(j-1)\tau}{1+(j-1)\tau} \\
&= \frac{(j-i)+j(i-1)\tau+j\tau-i(j-1)\tau}{1+(j-1)\tau} \\
&= 1-\frac{(1-\tau)(i-j+1)}{1+(j-1)\tau} \\
&\leq 1-\frac{1-\tau}{r_n} := \rho < 1.
\end{aligned} \tag{4.12}$$

Following (4.11), we further have

$$\begin{aligned}
f_i(\theta,\bar{x}_i)/L^i\Big|_{\Gamma_M} &\leq \bar{\gamma}_i \sigma L^\rho \left(|z_1|^{\frac{1+i\tau}{1}} + |z_2|^{\frac{1+i\tau}{1+\tau}} + \cdots + |z_i|^{\frac{1+i\tau}{1+(i-1)\tau}}\right) \\
&\leq \bar{c}_i \sigma L^\rho \|z\|_{\Delta^r}^{r_i+\tau},
\end{aligned} \tag{4.13}$$

where $\bar{c}_i \in \mathbb{R}_+$ is a constant independent of L.

Recalling that $U(z) \in \mathbb{H}_{\Delta^r}^{2\kappa-\tau}$ and utilizing Lemma 1.1, there exists a constant $\bar{c} \in \mathbb{R}_+$ which is independent of L, such that the following relations hold:

$$\begin{aligned}
\frac{\partial U(z)}{\partial z^\top} F(\theta,L,x) &\leq \sum_{i=1}^{n} \bar{c}_i \sigma \left|\frac{\partial U(z)}{\partial z_i}\right| |f_i(\theta,\bar{x}_i)|/L^i \\
&\leq L^\rho \sum_{i=1}^{n} \bar{c}_i \sigma \left|\frac{\partial U(z)}{\partial z_i}\right| \|z\|_{\Delta^r}^{r_i+\tau} \\
&\leq \bar{c}\sigma L^\rho \|z\|_{\Delta^r}^{2\kappa}.
\end{aligned} \tag{4.14}$$

With the assistance of (4.14), (4.10) can be further deduced to

$$\begin{aligned}
\dot{V}(z,\tilde{\sigma})\Big|_{\Gamma_M} \leq &- cL\|z\|_{\Delta^r}^{2\kappa} - \hat{c}\dot{L}/L\|z\|_{\Delta^r}^{2\kappa-\tau} + \bar{c}\sigma L^\rho \|z\|_{\Delta^r}^{2\kappa} - \tilde{\sigma}\dot{\hat{\sigma}}/\varepsilon \\
= &- (1-\mu)cL\|z\|_{\Delta^r}^{2\kappa} - \left(\hat{c}\dot{L}/L - (\bar{c}\hat{\sigma}L^\rho - \mu cL)\|z\|_{\Delta^r}^{\tau}\right)\|z\|_{\Delta^r}^{2\kappa-\tau} \\
&- \tilde{\sigma}\left(\dot{\hat{\sigma}}/\varepsilon - \bar{c}L^\rho \|z\|_{\Delta^r}^{2\kappa}\right),
\end{aligned} \tag{4.15}$$

where $\mu \in (0,1)$ is a constant.

Now we can utilize the adaptive stabilizing control law (4.7) with the design parameters set as $c_1 \triangleq \varepsilon\bar{c}$, $c_2 \triangleq \frac{\bar{c}}{\hat{c}}$, $c_3 = \frac{\mu c}{\hat{c}}$. It renders (4.15) to reduce to the following form

$$\dot{V}(z,\tilde{\sigma})\Big|_{\Gamma_M} \leq -(1-\mu)cL\|z\|_{\Delta^r}^{2\kappa}. \tag{4.16}$$

i) In what follows, we first prove that the following relation holds:

$$(z(0)^\top, \tilde{\sigma}(0))^\top \in [-\wp, \wp]^{n+1} \Rightarrow (z(t)^\top, \tilde{\sigma}(t))^\top \in \Omega,\ t \in [0, \infty). \tag{4.17}$$

If the claim of (4.17) is not true, there should exist a finite escape time, say, T_1, such that

$$(z(t)^\top, \tilde{\sigma}(t))^\top \notin \Omega,\ t \geq T_1.$$

With the fact that $[-\wp, \wp]^{n+1} \subset \Omega \subset \Gamma_M$ which yields

$$\dot{V}(z(0), \tilde{\sigma}(0)) \leq -(1-\mu)cL\|z(0)\|_{\Delta^r}^{2\kappa} < 0.$$

It is clear that there must exist two time instants $t^{'}$, $t^{''}$ satisfying $T_1 > t^{''} > t^{'} > 0$, such that the following relations hold:

$$\begin{aligned} &a)\ \dot{V}(z(t^{'}), \tilde{\sigma}(t^{'})) < 0;\\ &b)\ V(z(t^{''}), \tilde{\sigma}(t^{''})) = V(z(t^{'}), \tilde{\sigma}(t^{'}));\\ &c)\ \dot{V}(z(t^{''}), \tilde{\sigma}(t^{''})) > 0. \end{aligned} \tag{4.18}$$

Now we consider the trajectory within the time interval $[t^{'}, t^{''}]$. First, the estimation (4.14) is still valid for $t \in [t^{'}, t^{''}]$, hence one can obtain the following relations from (4.16):

$$\begin{aligned} V(z(t^{''}), \tilde{\sigma}(t^{''})) - V(z(t^{'}), \tilde{\sigma}(t^{'})) &= \int_{t^{'}}^{t^{''}} \dot{V}(z(s), \tilde{\sigma}(s))ds\\ &\leq -\int_{t^{'}}^{t^{''}} (1-\mu)cL\|z(s)\|_{\Delta^r}^{2\kappa} ds. \end{aligned} \tag{4.19}$$

With the relation b) of (4.18), (4.19) will clearly lead to a contradiction that

$$0 \geq \int_{t^{'}}^{t^{''}} (1-\mu)cL\|z(s)\|_{\Delta^r}^{2\kappa} ds > 0.$$

Hence, $(z(t)^\top, \tilde{\sigma}(t))^\top \in \Omega$, $t \in [0, \infty)$ is established, which implies that the trajectory of $(z(t), \tilde{\sigma}(t))$ is uniformly bounded on the time interval $[0, \infty)$. In addition, there exist positive constants $\bar{\sigma}$ and $\check{\sigma}$ such that

$$|\tilde{\sigma}| \leq \bar{\sigma},\ |\hat{\sigma}| \leq \check{\sigma},\ \forall t \in [0, \infty). \tag{4.20}$$

Now we proceed to prove that L is also well defined for $t \in [0, \infty)$. First, from the expression of $\dot{L}$ in (4.7), it is clear that $\dot{L} \geq 0$ for $t \in [0, \infty)$. By defining a monotone increasing time sequence t_ℓ, $L \in \mathbb{N}_{1:\infty}$, two cases can be addressed below.

Case 1): If $L(t_\ell)$ satisfies $\mu cL^{1-\rho}(t_\ell) - \bar{c}\hat{\sigma}(t_\ell) \geq 0$, by the continuity of L and the fact that $\dot{\hat{\sigma}} \geq 0$, there exist two time instants $t_{\ell+1} < t_{\ell+2} < \infty$, such that $\dot{L}(t) = 0,\ t \in [t_\ell, t_{\ell+1}); \dot{L}(t) > 0,\ t \in [t_{\ell+1}, t_{\ell+2})$. Then we shall turn to the following alternative case.

Case 2): If $\mu cL^{1-\rho}(t_{\ell+1}) - \bar{c}\hat{\sigma}(t_{\ell+1}) < 0$ (i.e., $\dot{L}(t_{\ell+1}) > 0$), one can conclude that the following relations hold:

$$\begin{cases} a)\ \mu cL^{1-\rho}(t) - \bar{c}\hat{\sigma}(t) < 0,\ t \in [t_{\ell+1}, t_{\ell+2}), \\ b)\ \mu cL^{1-\rho}(t_{\ell+2}) - \bar{c}\hat{\sigma}(t_{\ell+2}) \geq 0. \end{cases} \tag{4.21}$$

If the above statements are not true, it implies that

$$\mu cL^{1-\rho}(t) - \bar{c}\hat{\sigma}(t) < 0,\ t \in [t_{\ell+1}, \infty).$$

From (4.7), we know $L(t)$ keeps increasing for $t \in [t_{\ell+1}, \infty)$. On the other hand, with the fact that $|\hat{\sigma}| \leq \breve{\sigma},\ t \in [0, \infty)$, the following relation draws an obvious contradiction:

$$L(t) < \left(\frac{\bar{c}\breve{\sigma}}{\mu c}\right)^{\frac{1}{1-\rho}},\ t \in [t_{\ell+1}, \infty).$$

Hence, (4.21) indicates that for $t = t_{\ell+2}$, we are back to *Case 1)*. Then by repeatedly discussing the above two cases, and noticing the fact that $|\hat{\sigma}| \leq \breve{\sigma},\ t \in [0, \infty)$, one can draw a conclusion that $L(t)$ is bounded for the time interval $[0, \infty)$ and moreover, there exists a bounded positive constant $\bar{L}$ such that $\lim_{t\to\infty} L(t) = \bar{L}$.

ii) On the one hand, from (4.16), by noting that V is monotone decreasing and $L \geq 1$, we have

$$\begin{aligned} \int_0^\infty \|z(s)\|_{\Delta^r}^{2\kappa} ds &\leq -\frac{1}{(1-\mu)c}\int_0^\infty \dot{V}(z(s), \tilde{\sigma}(s))ds \\ &= \frac{1}{(1-\mu)c}\Big(V(z(0), \tilde{\sigma}(0)) - \lim_{t\to\infty} V(z(t), \tilde{\sigma}(t))\Big) < \infty. \end{aligned} \tag{4.22}$$

On the other hand, one can also conclude that $\dot{z}(t)$ and $z(t)$ are uniformly bounded for $t \in [0, \infty)$. It can be straightforwardly deduced from Barbalat's Lemma that $z(t) \to 0,\ t \to \infty$, which implies that the equilibrium $x = 0$ is asymptotically stable. This completes the proof of Theorem 4.1. ■

4.1.5 Extension to a linear control case

In practice, verifying Assumption 4.1.1 might present a non-easy task. In what follows, we show that by following Theorem 4.1, an interesting semi-global stabilizing control result can be obtained for lower-triangular systems of the form (4.1) with essentially relaxed nonlinearity constraints.

More specifically, for any initial states of system (4.1) starting from a compact set, we will show that the following adaptive control law:

$$\begin{cases} u = -L^{n+1}Kz, \dot{\hat{\sigma}} = c_1\|z\|_2^2, \\ \dot{L} = L\max\{0, c_2\hat{\sigma} - c_3L\},\ L(0) \geq 1, \end{cases} \tag{4.23}$$

is able to stabilize nonlinear system of the form (4.1) with only $\mathbb{C}^1$ nonlinearities. The result can be summarized by the following theorem.

Theorem 4.2

Assume the nonlinear functions of system (4.1) belong to $\mathbb{C}^1$. Consider the closed-loop system consisting of system (4.1) and the adaptive controller (4.23). If $(x(0)^\top, \hat{\sigma}(0))^\top \in \Gamma \triangleq [-\rho, \rho]^{n+1}$ with $\rho > 0$ being a given constant which can be arbitrary large, then the following statements hold.

- *All the signals in the closed-loop system are uniformly bounded.*
- *The system state will converge to the origin asymptotically.*

Proof: Similar to the handling procedures (4.11)–(4.13), using Mean-Value Theorem and Lemma 1.10, we know that the following relations hold for any function $f_i(\theta, \bar{x}_i) \in \mathbb{C}^1$:

$$\begin{aligned} \left|f_i(\theta, \bar{x}_i)\right|/L^i\big|_{\Gamma_M} &\leq \frac{\bar{\gamma}_i\sigma}{L^i}(|x_1| + |x_2| + \cdots + |x_i|) \\ &\leq \bar{\gamma}_i\sigma\frac{|Lz_1| + |L^2z_2| + \cdots + |L^iz_i|}{L^i} \\ &\leq \bar{\gamma}_i\sigma(|z_1| + |z_2| + \cdots + |z_i|) \leq \bar{c}_i\sigma\|z\|, \end{aligned} \tag{4.24}$$

where $\sigma > 0$ is an unknown constant dependent on θ, $\bar{\gamma}_i, \bar{c}_i$ are bounded positive constants which are dependent on the value of M. The remainder of the stability analysis could be almost identical to the proof of Theorem 4.1 by simply setting $\tau = 0$ and $\kappa = 1$, and hence is omitted here. ■

Remark 4.2 It can be observed from Theorems 4.1 and 4.2 that by setting the homogeneous degree $\tau = 0$, the smooth control result proposed in Theorem 4.1 can be specified to a linear adaptive control case. The main differences rely on the system nonlinearity pre-requirement and the employed homogeneous degree. In Theorem 4.1, by tuning the homogeneous degree to a larger value, a faster convergence rate can be achieved provided that the system state is far away from its equilibrium.

Remark 4.3 The result in Theorem 4.2 presents an essential enrichment in the area of adaptive control for strict-feedback systems. First, the requirement of sufficient smooth continuity of system nonlinearities for controller design with backstepping approaches is totally removed with a weaker $\mathbb{C}^1$ condition. Second, it is shown that a simple linear state feedback law will be sufficient to stabilize a nonlinear system via a time-varying scaling gain under a semi-global control objective. This feature greatly facilitates practical implementations.

In order to illustrate the simplicity of the proposed control design strategy, the following example is studied.

Example 4.1.1 *Consider a 3-D nonlinear system of the following form*

$$\begin{cases} \dot{x}_1 = x_2 + \theta x_1^{4/3} \\ \dot{x}_2 = x_3 + x_1 x_2 \\ \dot{x}_3 = u, \end{cases} \tag{4.25}$$

where $\theta = 0.5$ is an unknown parameter to the designer.

On one hand, system (4.25) is of interest here due to the fact that functions $x_1^{4/3}$, $x_1 x_2$ are not global Lipschitz continuous, which implies that this system does not satisfy the assumptions required in [33,34,36,56]. On the other hand, it is obvious that when carrying out the adaptive backstepping technique, the singularity problem appears when differentiating the virtual controller of α_2. However, with the proposed method in Theorem 4.2, one can obtain the following simple adaptive stabilizing control law:

$$\begin{cases} u = -L^4(k_1 z_1 + k_2 z_2 + k_3 z_3), \\ \dot{\hat{\sigma}} = 0.5(z_1^2 + z_2^2 + z_3^2), \\ \dot{L} = L\max\{0,\ 2/3\hat{\sigma} - 7/15L\}. \end{cases} \tag{4.26}$$

As shown by the simulation results in Figures 4.1 and 4.2, all the signals are bounded and the adaptive stabilization can be realized by setting $K = [27, 27, 9]$, $[x(0); \hat{\sigma}(0), L(0)] = [2, -2, 2; 0.1, 1.1]$.

Remark 4.4 Owing to the proposed systematic nonrecursive design approach, now the control gain vector can be determined simply by following a conventional pole placement manner, i.e., finding a proper vector K to render its companion matrix $A - BK$ Hurwitz. The time-varying scaling gain L will be updated online, which is dependent on the estimation of the unknown parameter σ.

4.1.6 Extension to non-triangular nonlinear systems

A feedback structure of nonlinear systems is currently considered as an essential precondition for adaptive controller design. Relaxing the restriction of lower-triangular structure is never a trivial task and few results can be referred to in

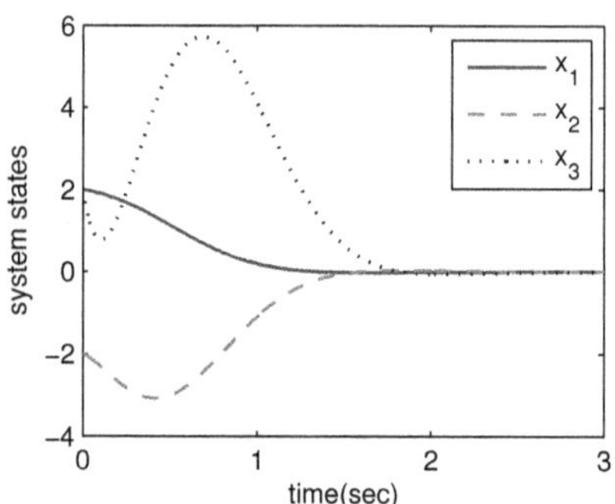

Figure 4.1 State response curves of the system (4.25) under the adaptive stabilizing control law (4.26).

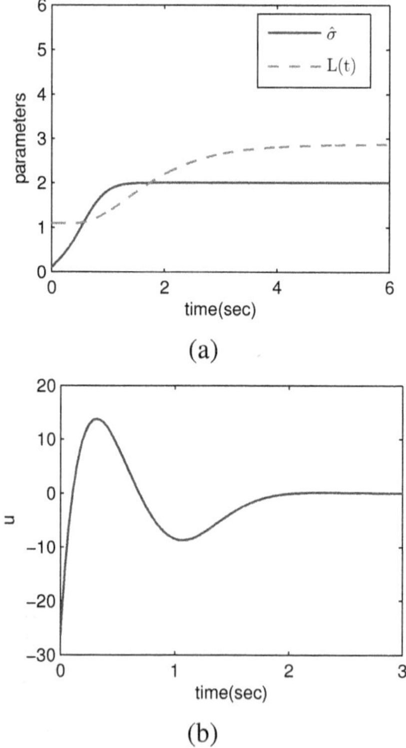

Figure 4.2 (a) The trajectories of the scaling gain and parameter; (b) Time history of the proposed control input (4.26).

the literature. However, owing to the design facility of a nonrecursive synthesis manner, the proposed result can also be applicable to a certain class of nonlinear systems which presents a non-triangular system structure, depicted by

$$\begin{cases} \dot{x}_i = x_{i+1} + f_i(\theta, x), \ i \in \mathbb{N}_{1:n-1}, \\ \dot{x}_n = u + f_n(\theta, x). \end{cases} \tag{4.27}$$

To proceed, the following assumption is presented in order to restrain the nonlinearity growth rate of system (4.27).

Assumption 4.1.2 *There exist an unknown constant $\sigma > 0$ dependent on θ, a constant $\rho < 1$, such that under the transform of coordinates in (4.3), the following relation holds:*

$$|f_i(\theta, x)|/L^i \le \sigma L^{\rho} \sum_{j=1}^{n} |z_j|^{\frac{r_i+\tau}{r_j}}, \ i \in \mathbb{N}_{1:n}.$$

By utilizing the same design procedure proposed in the above subsections, one can deduce the following theorem straightforwardly.

Theorem 4.3

Consider the closed-loop system consisting of system (4.27) satisfying Assumption 4.1.2 and the adaptive stabilizing controller (4.23). If $(x(0)^{\top}, \hat{\sigma}(0))^{\top} \in \Gamma \triangleq [-\rho, \rho]^{n+1}$ with $\rho > 0$ being a given constant which can be arbitrary large, then the following statements hold.

- *All the signals in the closed-loop system are uniformly bounded.*
- *The system state will converge to the origin asymptotically.*

Proof: By careful observation, Assumption 4.1.2 could directly render the relation (4.14) while the other stability analysis steps are exactly the same. The remainder proof can be carried out by an almost identical analysis procedure with the proof of Theorem 4.1, and hence is omitted here. ■

Remark 4.5 Note that the existing adaptive controllers are mainly built by recursive design procedures which strongly rely on the strict-feedback structure of the system. Assumption 4.1.2 allows a non-triangular structural hypothesis, a significant relaxation from the strict-feedback structure widely assumed in the existing literature. In a recent result [6], stabilizing systems in classical strict-feedback forms with sufficiently small non-triangular structural perturbations are investigated. However, the proposed controller design scheme is still based on the triangular structural nominal model of the system. Furthermore, a linear growth constraint is strongly depended on, which can be considered as a special case of Assumption 4.1.2 provided $\tau = 0$ and $f_i(\theta, 0) = 0$.

Remark 4.6 It is well acknowledged that the global adaptive control problem for nonlinear systems with non-triangular structure is a very challenging issue. Thus very few existing methods can be found in the literature. To some extent, Assumption 4.1.2 is still not straightforward to be pre-verified. The proposed result provides us a pioneer nonrecursive synthesis method for a class of complex nonlinear systems without structural restrictions and it possibly opens a new interesting topic for further investigation by researchers in the community.

The following example provides an illustrative demonstration of the effectiveness of the result proposed in Theorem 4.3.

Example 4.1.2 *Consider the following uncertain nonlinear system*

$$\begin{cases} \dot{x}_1 = x_2 + x_1 \ln(1 + (\theta_1 x_3)^2) \\ \dot{x}_2 = x_3 + \theta_2 x_1^{7/5} \\ \dot{x}_3 = u, \end{cases} \tag{4.28}$$

where $\theta_1 = 0.2$ *and* $\theta_2 = 0.5$ *are unknown constants to the designer.*

Denote $\sigma = \max\{2|\theta_1|^{1/7}, |\theta_2|\}$ and the x to z coordination transform $z_1 = \frac{x_1}{L}$, $z_2 = \frac{x_2}{L^2}$, $z_3 = \frac{x_3}{L^3}$, $v = \frac{u}{L^4}$. We have the following relations

$$\begin{aligned} |f_1| &= |x_1 \ln(1 + (\theta_1 x_3)^2)| \le 2|\theta_1|^{1/7} |Lz_1| \cdot |L^3 z_3|^{1/7} \le \sigma L^{1+3/7}(|z_1|^{6/5} + |z_3|^{6/7}), \\ |f_2| &= |\theta_2 x_1^{7/5}| \le \sigma L^{2-3/5} |z_1|^{7/5}. \end{aligned} \tag{4.29}$$

The above relations conclude that Assumption 4.1.2 holds for system (4.28) with $\rho = 3/7, \tau = 1/5, r_1 = 1, r_2 = 6/5, r_3 = 7/5$, and $\kappa = 8/5$. Hence following the proof of Theorem 4.3, after a series of calculations, we can design the following adaptive stabilizing controller:

$$\begin{cases} u(t) = -L^4 K \left(\lfloor z_1 \rceil^{8/5}, \lfloor z_2 \rceil^{4/3}, \lfloor z_3 \rceil^{8/7} \right)^{\top}, \\ \dot{\hat{\sigma}} = 0.09 L^{3/7} \|z\|_{\Delta^r}^{16/5}, \\ \dot{L} = L^{7/10} \max\{0,\ 9/15\hat{\sigma} - 7/15 L^{4/7}\} \|z\|_{\Delta^r}^{1/5}, \end{cases} \tag{4.30}$$

where $\|z\|_{\Delta^r} = \sqrt{|z_1|^2 + |z_2|^{5/3} + |z_3|^{7/10}}$.

In the simulation, the control gain vector K is selected as $[27, 27, 9]$. To conduct the simulation, the initial values are set as $[x(0), \hat{\sigma}(0), L(0)] = [2, -2, 2; 0.6, 2]$.

It can be observed from Figures 4.3–4.4 that all the signals in the closed-loop system (4.28) and (4.30) are bounded and the trajectories of the states converge to

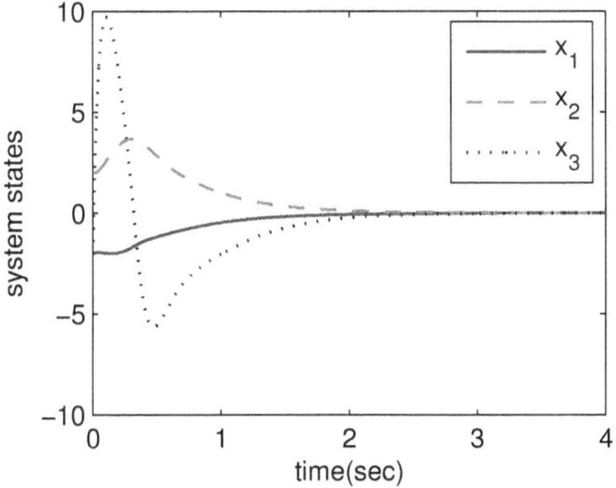

Figure 4.3 Response curves of the system states.

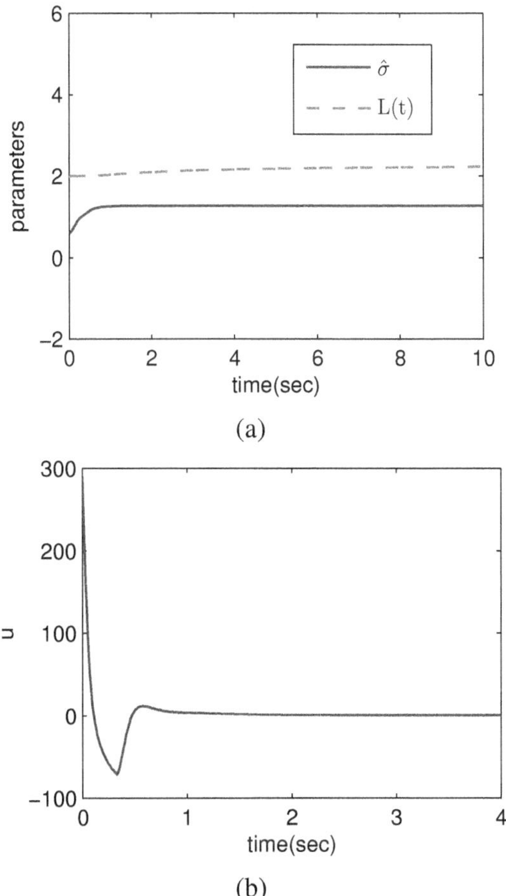

Figure 4.4 (a) The trajectories of the time-varying scaling gain and the estimated parameter; (b) Time history of the control input.

the origin in a satisfactory control performance. One distinct feature here is that the control gain vector K and the update laws can be selected in a very simple way. The time history of the control signal is shown in Figure 4.4 (b).

4.2 Nonrecursive $\mathbb{C}^0$ Adaptive Stabilization via State Feedback

4.2.1 Problem formulation

Let's revisit a class of lower-triangular nonlinear systems with non-parametric uncertainties, depicted as the form (4.1). The following assumption is made in prior.

Assumption 4.2.1 *There exists an unknown constant $\sigma \in \mathbb{R}_+$ which is dependent on θ, such that the following relation holds:*

$$|f_i(\theta,\bar{x}_i)| \le \sigma\left(|x_1|^{\frac{r_i+\tau}{r_1}} + |x_2|^{\frac{r_i+\tau}{r_2}} + \cdots + |x_i|^{\frac{r_i+\tau}{r_i}}\right),\ i \in \mathbb{N}_{1:n}.$$

Under the above assumption, the control objective is to find a dynamic stabilizing controller of the following non-nested form:

$$\begin{cases} u = u(L,x) \in \mathbb{C}^0,\ u(L,0) = 0, \\ \dot{L} = \psi(L,x) \in \mathbb{C}^1\left([1,+\infty) \times \mathbb{R}^n \to [0,+\infty)\right),\ L(t_0) = 1, \end{cases} \tag{4.31}$$

such that the following conclusions hold for the resulting closed-loop system (4.1)–(4.31),

- All the signals are globally uniformly bounded.
- The origin $x = 0$ is a finite-time stable equilibrium.

Subsequently, three handling procedures are presented to better illustrate the nonsmooth adaptive stabilization design.

4.2.2 Parameter configuration

With the denotation that A, B are matrices of controllable canonical form, $F(\theta,x) = (f_1(\cdot), f_2(\cdot), \cdots, f_n(\cdot))^\top$, system (4.1) can be rewritten as the following compact form:

$$\dot{x} = Ax + Bu + F(\theta,x). \tag{4.32}$$

The homogeneous dilation weight r is set as $r_1 = 1,\ r_i = r_{i-1} + \tau,\ i \in \mathbb{N}_{2:n}$ with a degree $\tau \in (-\frac{1}{n}, 0)$. Let $\kappa > 1$ be a pre-set design constant. Denote $\kappa - \tau/2 = \gamma$, and

$$\Xi_1 \triangleq \text{diag}\left\{\frac{\gamma}{r_1}, \frac{\gamma}{r_2}, \cdots, \frac{\gamma}{r_n}\right\},\ \Xi_2 \triangleq \text{diag}\left\{0, \frac{\gamma}{r_2}, \cdots, \frac{(n-1)\gamma}{r_n}\right\}.$$

It is obvious that $\gamma > 0$ and $r_i > 0$. Hence by following Lemma 1.8, there exist a positive definite, symmetrical matrix $P \in \mathbb{R}^{n\times n}$ and a gain vector K, such that

$$\begin{aligned} &(A-BK)^\top P + P(A-BK) \leq -I, \\ &\Xi_1 P + P\Xi_1 > 0. \end{aligned} \tag{4.33}$$

Based on the relations (4.33), we could define a sufficiently large constant design parameter ρ satisfying

$$\rho > \max\left\{0, -\frac{\lambda_{\min}(\Xi_2 P + P\Xi_2)}{\lambda_{\min}(\Xi_1 P + P\Xi_1)}\right\}.$$

4.2.3 Change of coordinates

Define an $x - \xi$ coordination transform by the following form

$$\xi_i = x_i / L^{\rho+i-1},\ i \in \mathbb{N}_{1:n},\ v = u/L^{\rho+n},$$

where L is a scaling gain function to be made precise by an update law later on. Then one have

$$\left\{\begin{array}{l} \dot{\xi}_i = L\xi_{i+1} - (\rho+i-1)\dot{L}/L\xi_i + f_i(\theta,\bar{x}_i)/L^{\rho+i-1},\ i \in \mathbb{N}_{1:n-1}, \\ \dot{\xi}_n = Lv - (\rho+n-1)\dot{L}/L\xi_n + f_n(\theta,x)/L^{\rho+n-1}. \end{array}\right. \tag{4.34}$$

By denoting

$$\begin{aligned} &\xi = (\xi_1, \xi_2, \cdots, \xi_n)^\top,\ D = \text{diag}\{0, 1, \cdots, n-1\}, \\ &\tilde{F}(L,\theta,x) = (f_1(\cdot)/L^\rho, f_2(\cdot)/L^{\rho+1}, \cdots, f_n(\cdot)/L^{\rho+n-1})^\top, \end{aligned}$$

and I_n as an identity matrix, it is not difficult to derive that the ξ-system dynamics should be expressed as the following compact form

$$\dot{\xi} = L(A\xi + Bv) - (\rho I_n + D)\xi\dot{L}/L + \tilde{F}(L,\theta,x). \tag{4.35}$$

4.2.4 Nonsmooth control law construction

At this stage, corresponding to (4.31), we are already able to design a nonsmooth dynamic state feedback stabilizing controller of the following simplest form

$$\left\{\begin{array}{l} u = L^{\rho+n}v,\ v = -K\lfloor\xi\rceil_{\Delta^r}^{r_n+\tau}, \\ \dot{L} = \mu\|\xi\|_{\Delta^r}^{2\kappa},\ L(t_0) = 1, \end{array}\right. \tag{4.36}$$

where $\mu \in \mathbb{R}_+$ is a design constant.

Remark 4.7 Compared to related existing results on universal adaptive design problem for non-parametric nonlinear systems, such as [24, 44, 71], etc., a distinguishable feature of the proposed adaptive controller is that both the proposed control law and design procedure are much simpler by totally neglecting the costly recursive calculations. More significantly, the controller design procedure is essentially detached from a recursive determination process of virtual controllers so that the step-by-step calculations of the tuning functions are also avoided. This characteristic could endow the practitioners much more convenience to implement the controller in real-life plants.

Remark 4.8 Inspired by the existing dynamic high gain techniques and the facility of homogeneous system theory, one could find that if an indirect adaption law for the dynamic gain is settled, rather than directly estimating the uncertain parameters via classical tuning functions, a nonrecursive adaptive synthesis approach is promising, which could facilitate the practitioners. Moreover, the proposed method can also be regarded as an essential nonsmooth extension of the result in [36, 56]. Indeed, in a special case by setting the homogeneous degree $\tau = 0$, the control law (4.36) could reduce to a linear universal controller of the following form:

$$\begin{cases} u = L^{\rho+n}v, \ v = -K\xi, \\ \dot{L} = \mu\|\xi\|^2, \ L(t_0) = 1. \end{cases} \tag{4.37}$$

4.2.5 Stability analysis

We will show that global stabilization can be realized by the proposed nonsmooth adaptive controller of the form (4.36).

The main result of this section can be stated as follows.

Theorem 4.4

Consider the closed-loop system consisting of (4.1) satisfying Assumption 4.2.1 and the nonsmooth adaptive control scheme (4.36). The following conclusions hold.

- *All the signals in the closed-loop system are globally uniformly bounded.*
- *The system state will converge to the origin within a finite time.*

Proof: Define a Lyapunov function candidate as

$$V(\xi) = \left(\lfloor \xi \rceil^{\gamma}_{\Delta^r}\right)^{\top} P \lfloor \xi \rceil^{\gamma}_{\Delta^r}.$$

From the fact that the homogeneous degree $\tau < 0$ and $\kappa > 1$, we know that $V(\xi) \in \mathbb{C}^1$. With the dilation mapping $\Delta_\varepsilon^r \xi = (\varepsilon^{r_1}\xi_1, \cdots, \varepsilon^{r_n}\xi_n)$, one can easily verify that $V(\Delta_\varepsilon^r \xi) = \varepsilon^{2\gamma} V(\xi)$.

Calculating the time derivative of $V(\xi)$ along system (4.35) gives

$$\dot{V}(\xi) = \frac{\partial V(\xi)}{\partial \xi^\top} L(A\xi + Bv) - \frac{\partial V(\xi)}{\partial \xi^\top}(\rho I_n + D)\xi \dot{L}/L + \frac{\partial V(\xi)}{\partial \xi^\top} \tilde{F}(L, \theta, x). \tag{4.38}$$

To proceed, a series of estimations on the terms in (4.38) will be given.

Firstly, it is not difficult to obtain the following relations:

$$\begin{cases} \xi_{i+1} \circ \Delta_\varepsilon^r = \varepsilon^{r_{i+1}} \xi_{i+1} = \varepsilon^{r_i + \tau} \xi_{i+1}, \ i \in \mathbb{N}_{1:n-1}, \\ v \circ \Delta_\varepsilon^r = -\varepsilon^{r_n + \tau} K \lfloor \xi \rceil_{\Delta^r}^{r_n + \tau}, \end{cases}$$

which conclude that the vector field $A\xi + Bv$ is Δ^r-homogeneous of degree τ. One can follow Lemma 2.1 directly that the closed-loop system is globally finite-time stable.

In addition, noting that $\frac{\partial V(\xi)}{\partial \xi^\top}(A\xi + Bv) \in \mathbb{H}_{\Delta^r}^{2\kappa}$ and $V(\xi) \in \mathbb{H}_{\Delta^r}^{2\gamma}$, by applying Lemma 1.1, there exists a constant $c_1 \in \mathbb{R}_+$, such that the following relation holds for $\tau \in (-1/n, 0)$

$$\frac{\partial V(\xi)}{\partial \xi^\top}(A\xi + Bv) \le -c_1 \|\xi\|_{\Delta^r}^{2\kappa}. \tag{4.39}$$

Second, with Assumption 4.2.1 in mind, we have

$$\begin{aligned} \frac{f_i(\cdot)}{L^{\rho+i-1}} &\le \frac{\sigma}{L^{\rho+i-1}} \left(|x_1|^{\frac{1+i\tau}{1}} + |x_2|^{\frac{1+i\tau}{1+\tau}} + \cdots + |x_i|^{\frac{1+i\tau}{1+(i-1)\tau}} \right) \\ &= \frac{\sigma}{L^{\rho+i-1}} \left(|L^\rho \xi_1|^{\frac{1+i\tau}{1}} + |L^{\rho+1}\xi_2|^{\frac{1+i\tau}{1+\tau}} + \cdots + |L^{\rho+i-1}\xi_i|^{\frac{1+i\tau}{1+(i-1)\tau}} \right) \\ &= \sigma \Big(L^{\frac{\rho(1+i\tau)}{1} - (\rho+i-1)} |\xi_1|^{\frac{1+i\tau}{1}} + L^{(\rho+1)\frac{1+i\tau}{1+\tau} - (\rho+i-1)} |\xi_2|^{\frac{1+i\tau}{1+\tau}} \\ &\quad + \cdots + L^{(\rho+i-1)\frac{1+i\tau}{1+(i-1)\tau} - (\rho+i-1)} |\xi_i|^{\frac{1+i\tau}{1+(i-1)\tau}} \Big). \end{aligned} \tag{4.40}$$

Noting that $L \ge 1$, $0 < \frac{1+i\tau}{1+(j-1)\tau} < 1$ and $\rho > 0$, it is concluded that

$$L^{(\rho+j-1)\frac{1+i\tau}{1+(j-1)\tau} - (\rho+i-1)} < L^0 = 1, \ j \in \mathbb{N}_{1:i}.$$

Hence, following (4.40), we have

$$\frac{f_i(\theta, \bar{x}_i)}{L^{\rho+i-1}} \le \sigma \left(|\xi_1|^{\frac{1+i\tau}{1}} + |\xi_2|^{\frac{1+i\tau}{1+\tau}} + \cdots + |\xi_i|^{\frac{1+i\tau}{1+(i-1)\tau}} \right) \le \bar{c}_{2,i} \sigma \|\xi\|_{\Delta^r}^{r_i + \tau}, \tag{4.41}$$

where $\bar{c}_{2,i} \in \mathbb{R}_+$ is a constant independent of L.

It is clear that $V(\xi) \in \mathbb{H}_{\Delta^r}^{2\gamma}$. By utilizing Lemma 1.1, there exists a constant $c_2 \in \mathbb{R}_+$ which is independent of L, such that the following relations hold:

$$\begin{aligned}\frac{\partial V(\xi)}{\partial \xi^\top}\tilde{F}(L,\theta,x) &\leq \sum_{i=1}^{n} \bar{c}_{2,i}\sigma \left|\frac{\partial V(\xi)}{\partial \xi_i}\right| \|\xi\|_{\Delta^r}^{r_i+\tau} \\ &\leq c_2\sigma\|\xi\|_{\Delta^r}^{2\kappa}. \end{aligned} \tag{4.42}$$

Third, from the definition of ρ and $\dot{L}/L \geq 0$, it is clear that

$$(\Xi_1 P + P\Xi_1)\rho + \Xi_2 P + P\Xi_2 > 0.$$

Then with Lemma 1.1, there must exist a constant $c_3 \in \mathbb{R}_+$ such that the following relations hold:

$$\begin{aligned}\frac{\partial V(\xi)}{\partial \xi^\top}(\rho I_n + D)\xi\dot{L}/L &= \left(\lfloor\xi\rceil_{\Delta^r}^{\gamma}\right)^\top \left((\Xi_1 P + P\Xi_1)\rho + \Xi_2 P + P\Xi_2\right)\lfloor\xi\rceil_{\Delta^r}^{\gamma}\dot{L}/L \\ &\geq c_3\dot{L}/L\|\xi\|_{\Delta^r}^{2\gamma}. \end{aligned} \tag{4.43}$$

Substituting (4.39), (4.42) and (4.43) into (4.38), the time derivative of $V(\xi)$ along system (4.35)–(4.36) can be reduced to the following form:

$$\begin{aligned}\dot{V}(\xi) &\leq -c_1 L\|\xi\|_{\Delta^r}^{2\kappa} - c_3\dot{L}/L\|\xi\|_{\Delta^r}^{2\gamma} + c_2\sigma\|\xi\|_{\Delta^r}^{2\kappa} \\ &\leq -(c_1 L - c_2\sigma)\|\xi\|_{\Delta^r}^{2\kappa}. \end{aligned} \tag{4.44}$$

i) In what follows, we will use contradiction arguments to prove that there will be no finite escape time of the trajectory $(\xi^\top, L)^\top$ starting from t_0. If the above statement is not true, there must exist a maximal time interval $[t_0, t_f)$ where $t_f > t_0$, such that $(\xi^\top, L)^\top$ are well defined and $\lim_{t\to t_f}\|(\xi^\top, L)^\top\| = +\infty$.

Firstly, we show that L cannot escape at $t = t_f$. Otherwise, the relation $\lim_{t\to t_f} L = +\infty$ holds. Noting that c_1, c_2, σ are all bounded, hence there exists a time instant t_1 satisfying $t_0 < t_1 < t_f$, such that

$$c_1 L - c_2\sigma \geq 1,\ t \in [t_1, t_f].$$

This relation could lead to a fact that

$$\dot{V}(\xi) \leq -\|\xi\|_{\Delta^r}^{2\kappa},\ t \in [t_1, t_f].$$

Therefore, with the update law (4.36), it is straightforward that

$$\begin{aligned}+\infty &= L(t_f) - L(t_1) \\ &= \int_{t_1}^{t_f} \dot{L}(s)ds \\ &= \int_{t_1}^{t_f} \mu\|\xi(s)\|_{\Delta^r}^{2\kappa}ds \\ &\leq \mu V(\xi(t_1)) < +\infty, \end{aligned} \tag{4.45}$$

which is clearly a contradiction. Then one could claim that L is well defined in the time interval $[t_0, t_f]$.

Secondly, we show that ξ cannot escape at $t = t_f$. Integrating both sides of (4.44) over the time interval $[t_0, t]$ yields

$$
\begin{aligned}
V(\xi(t)) - V(\xi(t_0)) &\leq -\int_{t_0}^{t} c_1 L(s)\|\xi(s)\|_{\Delta^r}^{2\kappa} ds + \int_{t_0}^{t} c_2\sigma\|\xi(s)\|_{\Delta^r}^{2\kappa} ds \\
&\leq -\int_{t_0}^{t} \frac{c_1}{\mu} L(s)\dot{L}(s) ds + \int_{t_0}^{t} \frac{c_2\sigma}{\mu} \dot{L}(s) ds \\
&\leq -\frac{c_1}{2\mu}(L^2(t) - 1) + \frac{c_2\sigma}{\mu}(L(t) - 1).
\end{aligned}
$$

Hence one could deduce the relation that

$$
V(\xi(t_f)) \leq V(\xi(t_0)) - \frac{c_1}{2\mu}(L^2(t_f) - 1) + \frac{c_2\sigma}{\mu}(L(t_f) - 1).
$$

From the fact that $L(t_f)$ is bounded, it is obvious that $V(\xi(t_f))$ is also bounded. With Lemma 1.1 in mind, it is clear that there exists a constant $c_4 \in \mathbb{R}_+$ such that

$$
\|\xi(t_f)\|_{\Delta^r} \leq c_4 V^{\frac{1}{2\gamma}}(\xi(t_f)),
$$

which implies that $\xi(t_f)$ is bounded. This fact clearly contradicts the finite escape time claim.

In conclusion, we have already proved that $(\xi^\top, L)^\top$ is well defined and globally uniformly bounded for $t \in [t_0, \infty)$.

ii) Using the boundness of $\dot{\xi}$ and ξ for $t \in [t_0, \infty)$, it can be straightforwardly deduced from Barbalat's Lemma that $\xi \to 0,\ t \to \infty$.

In addition, by recalling from the expression (4.36) that L is monotone increasing and c_1, c_2, σ are bounded, we know there must exist a time instant $T_1 \geq t_0$ such that

$$
c_1 L - c_2\sigma > 0,\ \forall t \geq T_1.
$$

Note that with Lemma 1.1, the following relation holds:

$$
\|\xi\|_{\Delta^r}^{2\kappa} \geq c_5 V^{\frac{2\kappa}{2\kappa-\tau}}(\xi),
$$

with c_5 being a positive constant. Then with (4.44) in mind, the following relation holds

$$
\dot{V}(\xi) + c_5(c_1 L - c_2\sigma)V(\xi)^{\frac{2\kappa}{2\kappa-\tau}} \leq 0,\ t \geq T_1. \tag{4.46}
$$

Based on (4.46), it can be deduced straightforwardly from Lemma 1.4 that there exists a finite time $T_2 \geq T_1$, such that $\xi(t) = 0,\ \forall t \geq T_2$.

This completes the proof of Theorem 4.4. ■

Remark 4.9 As implied by the proof, the closed-loop system stability is directly dependent on a sufficiently large scaling gain L which is subject to a simple self-tuning update law in (4.36). Indeed, it can be concluded from the initial value $L(t_0) = 1$ that the closed-loop system could possibly be unstable at the initial stage. However, L is made monotone increasing as $\dot{L}(t) \geq 0$, which is illustrated by Figure 4.6 in Example 4.2.1. After a short period, $L(t)$ can be sufficiently large to render the closed-loop system stable.

Remark 4.10 Owing to the proposed systematic nonrecursive synthesis approach, now the control gain parameter determination can be simplified as preliminary works. Similar to the method in [94], one can find a constant gain vector K to render $A - BK$ Hurwitz first, and then obtain P by solving LMI in (4.33). Therefore the value ρ can be selected. Noting that the value of μ is critical to the increasing rate of L, hence, a proper choice of the parameter μ can be determined via "trial and error" method. Subsequently, the scaling gain L will be self-tuned online from its initial value $L(0) = 1$. A detailed illustration will be shown in Example 4.2.1 later. Such a gain tuning mechanism can be regarded as a mediate role between adaptive and robust methods. It provides the controlled system with certain robustness, but the conservative selection guideline of L in the existing related robust methods, such as [55, 91], is no longer required.

4.2.6 An illustrative example

An example as an illustrative demonstration of the simplicity and effectiveness of the proposed control scheme is given below.

Example 4.2.1 *Consider the following uncertain nonlinear system*

$$\begin{cases} \dot{x}_1 = x_2 + \ln(1 + (\theta_1 x_1)^2) \\ \dot{x}_2 = x_3 + \sin(\theta_2 x_2) \\ \dot{x}_3 = u, \end{cases} \tag{4.47}$$

where $\theta_1 = 0.2\sin(t)$ *and* $\theta_2 = 0.1$ *are unknown parameters to the designer.*

Denoting $\sigma = \max\{2|\theta_1|^{6/7}, |\theta_2|^{5/6}\}$, one can verify the following relations:

$$|f_1(\theta, x_1)| = |\ln(1 + (\theta_1 x_1)^2)| \leq 2|\theta_1|^{6/7} \cdot |x_1|^{6/7} \leq \sigma |x_1|^{6/7},$$
$$|f_2(\theta, x_2)| = |\sin(\theta_2 x_2)| \leq \sigma |x_2|^{5/6}.$$

Thus system (4.47) satisfies Assumption 4.2.1 with $\tau = -1/7$, $r_1 = 1$, $r_2 = 6/7$, $r_3 = 5/7$. Hence without going through a series of recursive determination of virtual controllers, one can design directly from system (4.47) of the following dynamic stabilizing controller:

$$\begin{cases} u(t) = -L^{\rho+3}K\left(\lfloor\xi_1\rceil^{4/7}, \lfloor\xi_2\rceil^{2/3}, \lfloor\xi_3\rceil^{4/5}\right)^\top, \\ \dot{L} = \mu\|\xi\|_{\Delta^r}^{2\kappa},\ L(0) = 1, \end{cases} \tag{4.48}$$

where $\|\xi\|_{\Delta^r} = \sqrt{|\xi_1|^2 + |\xi_2|^{7/3} + |\xi_3|^{14/5}}$.

Before doing simulation, K is first selected as in $[1,3,3]$ to render its companion matrix $A - BK$ Hurwitz, which is quite easy, and by solving $(A-BK)^\top P + P(A-BK) \leq -I$, one can determine a proper matrix of the form:

$$P = \begin{bmatrix} 2.31 & 1.94 & 0.50 \\ 1.94 & 3.25 & 0.81 \\ 0.50 & 0.81 & 0.44 \end{bmatrix}.$$

Second, set $\kappa = 1.2$. Calculating the smallest eigenvalue of $\Xi_1 P + P\Xi_1$ gives 0.76, then the design parameter ρ can be set as 2. Aiming to show the impact on the control performance from the design parameter μ, we select two different values in the simulation, i.e., $\mu = 2$ and 20, respectively. To conduct the simulation, the initial values are set as $x(0) = [-2,3,4]$.

It can be observed from Figures 4.5–4.6 that all the signals in the closed-loop system (4.47)–(4.48) are uniformly bounded and the trajectories of the states converge to the origin in a satisfactory control performance. The time history of the dynamic control signal is shown in Figure 4.7. Note that a larger μ could require a larger control effort, but lead to a faster response.

4.3 Nonrecursive $\mathbb{C}^0$ Adaptive Stabilization via Output Feedback

In this section, we will present the procedure of how to design a nonsmooth nonrecursive adaptive stabilizing controller for a class of nonlinear systems with non-parameterized uncertainties via output feedback strategy.

4.3.1 Parameter configuration

Define two matrices satisfying $Q = Q^\top > 0,\ P = P^\top > 0$ and the homogeneous degree as $\tau \in (-\frac{1}{n}, 0)$. Hence, we know that $2 - \tau > 0$ holds. Subsequently, with the definition of two auxiliary matrices as

$$E_1 = \text{diag}\left\{\frac{2-\tau}{2r_1}, \frac{2-\tau}{2r_2}, \cdots, \frac{2-\tau}{2r_n}\right\},$$

$$E_2 = \text{diag}\left\{0, \frac{2-\tau}{2r_2}, \cdots, \frac{(n-1)(2-\tau)}{2r_n}\right\},$$

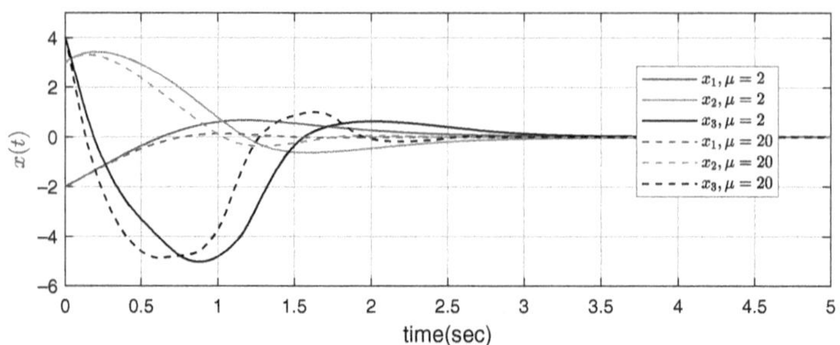

Figure 4.5 State response curves.

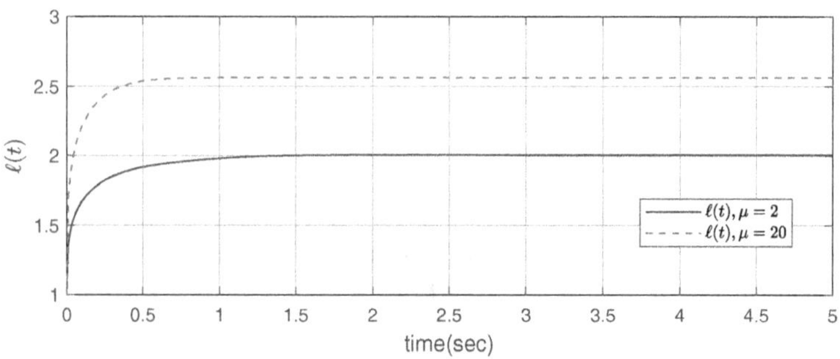

Figure 4.6 The trajectory of the dynamic scaling gain L.

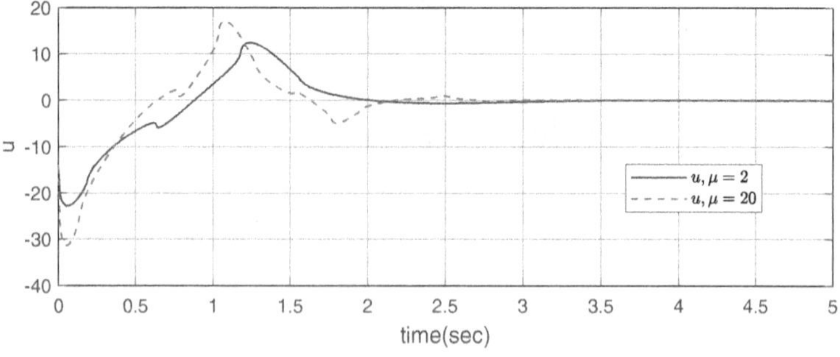

Figure 4.7 Time history of the proposed control input.

the following relationships exist according to Lemma 1.8:

$$(A-BK)^\top P+P(A-BK)\leq -I_n,\ E_1P+PE_1>0,$$
$$\tilde{A}^\top Q+Q\tilde{A}\leq -I_n,\ E_2Q+QE_2>0,$$

where $K=[k_1,k_2,\cdots,k_n]$ is the gain vector, $\tilde{A}=\begin{bmatrix}-\tilde{\alpha} & I_{n-1}\\ -\alpha_n & 0\end{bmatrix}$ with $\tilde{\alpha}=[\alpha_1,\alpha_2,\cdots,\alpha_{n-1}]^\top$. Furthermore, the parameters $\alpha_i>0, i=1,2,\cdots,n$ and $k_i, i=1,2,\cdots,n$ are the coefficients of the Hurwitz polynomial $P(s)=s^n+p_1s^{n-1}+\cdots+p_{n-1}s+p_n$. In what follows, a key design parameter ρ can be determined by the following guideline:

$$\rho>\max\left\{0,-\frac{\lambda_{\min}(E_2P+PE_2)}{\lambda_{\min}(E_1P+PE_1)},-\frac{\lambda_{\min}(E_2Q+QE_2)}{\lambda_{\min}(E_1Q+QE_1)}\right\}. \tag{4.49}$$

4.3.2 Controller construction

Without going through any complex determination procedures of the virtual controllers, the proposed nonrecursive output feedback controller can be depicted as the following nonsmooth form:

$$\begin{cases}\dot{\hat{x}}_1=\hat{x}_2+\alpha_1L\lfloor y-\hat{x}_1\rceil^{1+\tau},\\ \dot{\hat{x}}_2=\hat{x}_3+\alpha_2L^2\lfloor y-\hat{x}_1\rceil^{1+2\tau},\\ \quad\vdots\\ \dot{\hat{x}}_n=u+\alpha_nL^n\lfloor y-\hat{x}_1\rceil^{1+n\tau};\end{cases} \tag{4.50}$$

$$\begin{cases}\dot{L}=\lambda\left(\dfrac{y-\hat{x}_1}{L^\rho}\right)^2,\ L_0=1,\\ u=L^{\rho+n}\upsilon,\ \upsilon=-K\lfloor z\rceil_\Delta^{r_n+\tau}.\end{cases} \tag{4.51}$$

where $z=(z_1,z_2,\cdots,z_n)^\top$, $z_i=\hat{x}_i/L^{\rho+i-1}$, λ is a positive constant related to the convergence of scaling gain, and $\hat{x}_i$ represents the state estimation.

4.3.3 Stability analysis

The main result of this section can be summarized by the following theorem.

Theorem 4.5
Under Assumption 4.2.1, consider the closed-loop system consisting of (4.1)–(4.50)–(4.51). The following statements hold:

- *All the signals in the closed-loop system are globally uniformly bounded;*
- *The system state will converge to the origin within a finite time.*

Proof: Firstly, define $e_i = (x_i - \hat{x}_i)/L^{\rho+i-1}, i = \mathbb{N}_{1:n}$, and $e = (e_1, e_2, \cdots, e_n)^\top$. Then, the compact form of z-dynamic and e-dynamic can be expressed as the following equations:

$$\begin{aligned}
\dot{z} &= L(Az + Bv) - (\rho I_n + \Theta)\frac{\dot{L}}{L}z + \tilde{F}(x, \theta, L),\\
\dot{e} &= LAe - (\rho I_n + \Theta)\frac{\dot{L}}{L}e + \tilde{F}(x, \theta, L) - \Phi(L, e_1, \tau),
\end{aligned} \tag{4.52}$$

where

$$\begin{aligned}
\Theta &= \mathrm{diag}\{0, 1, \cdots, n-1\},\\
\Phi &= \begin{bmatrix} \alpha_1 L^{\rho\tau+1}\lfloor e_1 \rceil^{1+\tau} \\ \alpha_2 L^{2\rho\tau+1}\lfloor e_1 \rceil^{1+2\tau} \\ \vdots \\ \alpha_n L^{n\rho\tau+1}\lfloor e_1 \rceil^{1+n\tau} \end{bmatrix},\\
\tilde{F} &= \left[\frac{f_1(\theta, x_1)}{L^\rho}, \frac{f_2(\theta, x_1, x_2)}{L^{\rho+1}} \cdots \frac{f_n(\theta, x)}{L^{\rho+n-1}}\right].
\end{aligned}$$

The candidate Lyapunov function can be constructed as $W(e,z) : \mathbb{R}^n \times \mathbb{R}^n \to \mathbb{R}^+$, and specifically, is depicted as:

$$\begin{aligned}
W(e,z) &= W_1(e) + W_2(z),\\
W_1(e) &= \left(\lfloor e \rceil^{1-\tau/2}\right)^\top Q \lfloor e \rceil^{1-\tau/2},\\
W_2(z) &= \left(\lfloor z \rceil^{1-\tau/2}\right)^\top P \lfloor z \rceil^{1-\tau/2}.
\end{aligned} \tag{4.53}$$

Taking the derivative of W along (4.52) will yield

$$\begin{aligned}
\dot{W} =& \frac{\partial W_1}{\partial e^\top} L(Ae - \bar{\Phi}(L, e_1, \tau)) - \frac{\partial W_1}{\partial e^\top}(\rho I_n + \Theta)\frac{\dot{L}}{L}e\\
&+ \frac{\partial W_1}{\partial e^\top}\tilde{F}(x, \theta, L) + \frac{\partial W_2}{\partial z^\top}L(Az + Bv) + \frac{\partial W_2}{\partial z^\top}\Phi(L, e_1, \tau)\\
&- \frac{\partial W_2}{\partial z^\top}(\rho E_n + \Theta)\frac{\dot{L}}{L}z.
\end{aligned} \tag{4.54}$$

In what follows, we shall analyze the right-side of (4.54). First of all, the following relations can be verified:

$$\begin{aligned}
(e_{i+1} - \alpha_i L^{i\rho\tau}\lfloor e_1 \rceil^{1+\tau}) \circ \Delta^\varepsilon &= \varepsilon^{r_i+\tau}(e_{i+1} - \alpha_i L^{i\rho\tau}\lfloor e_1 \rceil^{1+\tau}),\ i \in \mathbb{N}_{1:n},\\
-\alpha_n L^{n\rho\tau}\lfloor e_1 \rceil^{1+n\tau} \circ \Delta^\varepsilon &= -\varepsilon^{r_n+\tau}\alpha_n L^{n\rho\tau}\lfloor e_1 \rceil^{1+n\tau}.
\end{aligned} \tag{4.55}$$

Then, it is obvious that $(Ae - \bar{\Phi}(L, e_1, \tau)) \in \mathbb{H}^{\tau}_{\Delta^r}$. Hence, we get that $\frac{\partial W_1}{\partial e^\top}(Ae - \bar{\Phi}(L, e_1)) \in \mathbb{H}^2_{\Delta^r}$, and further, the following expressions can be obtained for $\tau \in [(-1/n, 0)$

$$\frac{\partial W_1}{\partial e^\top} L(Ae - \bar{\Phi}(L, e_1)) \le -La_1 \|e\|^2_{\Delta^r}, \tag{4.56}$$

$$\frac{\partial W_2}{\partial z^\top} L(Az + Bv) \le -Lc_1 \|z\|^2_{\Delta^r}, \tag{4.57}$$

where $a_1 \in \mathbb{R}_+, c_1 \in \mathbb{R}_+$ are two constants.

Subsequently, combing with the selection guideline of ρ, the following relations can be obtained:

$$\begin{aligned}\frac{\partial W_1}{\partial e^\top}(\rho I_n + \Theta)\frac{\dot{L}}{L}e &= \frac{\dot{L}}{L}(\lfloor e \rceil^{1-\frac{\tau}{2}}_{\Delta^r})^\top ((PE_1 + E_1 P)\rho + PE_2 + E_2 P) \lfloor e \rceil^{1-\frac{\tau}{2}}_{\Delta^r} \\ &\ge a_2 \frac{\dot{L}}{L} \|e\|^{2-\tau}_{\Delta^r};\end{aligned} \tag{4.58}$$

$$\frac{\partial W_2}{\partial z^\top}(\rho I_n + \Theta)\frac{\dot{L}}{L}z \ge c_2 \frac{\dot{L}}{L} \|z\|^{2-\tau}_{\Delta^r}, \tag{4.59}$$

where $a_2 \in \mathbb{R}_+, c_2 \in \mathbb{R}_+$ are two constants.

Then, it is clear that $\Phi(L, e_1, \tau) \in \mathbb{H}^{\tau}_{\Delta^r}$ by verifying $\Phi_i \circ \Delta^{\varepsilon} = \alpha_i L^{i\tau+1} \lfloor \varepsilon^{r_1} e_1 \rceil^{1+\tau} = \varepsilon^{r_i+\tau}\Phi_i$, and we get that $\frac{\partial W_2}{\partial z^\top}\Phi(L, e_1, \tau) \in \mathbb{H}^2_{\Delta^r}$. Furthermore, the following expression can be obtained with a positive constant c_3:

$$\frac{\partial W_2}{\partial z^\top}\Phi(L, e_1, \tau) \le c_3 \|z\|^2_{\Delta^r}. \tag{4.60}$$

Finally, considering Assumption 4.2.1, one can get that

$$\begin{aligned}&\frac{f_i(\theta, x_1, x_2, \cdots, x_i)}{L^{\rho+i-1}} \le \frac{\sigma\left(|x_1|^{\frac{1+i\tau}{r_1}} + \cdots + |x_i|^{\frac{1+i\tau}{r_i}}\right)}{L^{\rho+i-1}} \\ &= \frac{\sigma\left(|L^\rho e_1 + \hat{x}_1|^{\frac{1+i\tau}{r_1}} + \cdots + |L^{\rho+i-1}e_i + \hat{x}_i|^{\frac{1+i\tau}{1+(i-1)\tau}}\right)}{L^{\rho+i-1}} \\ &= \sigma\left(L^{\frac{\rho(1+i\tau)}{r_1} - (\rho+i-1)}(|e_1 + z_1|)^{\frac{1+i\tau}{r_1}} + \cdots + L^{\frac{(\rho+j-1)(1+i\tau)}{1+(i-1)\tau} - (\rho+i-1)}(|e_i + z_i|)^{\frac{1+i\tau}{1+(i-1)\tau}}\right).\end{aligned} \tag{4.61}$$

Noting that $L \ge 1, 0 < \frac{1+i\tau}{1+(j-1)\tau} < 1$ with $j = 1, 2, \cdots, i$ and $\rho > 0$, we have

$$L^{(\rho+j-1)\frac{1+i\tau}{1+(j-1)\tau} - (\rho+i-1)} < L^0 = 1.$$

Hence, one can conclude that

$$\begin{aligned}\frac{f_i(\theta, x_1, \cdots, x_i)}{L^{\rho+i-1}} &\le \sigma\left(|e_1 + z_1|^{\frac{1+i\tau}{r_1}} + \cdots + |e_i + z_i|^{\frac{1+i\tau}{1+(i-1)\tau}}\right) \\ &\le \sigma\left(|e_1|^{\frac{1+i\tau}{r_1}} + |z_1|^{\frac{1+i\tau}{r_1}} + \cdots + |e_i|^{\frac{1+i\tau}{1+(i-1)\tau}} + |z_i|^{\frac{1+\tau}{1+(i-1)\tau}}\right) \\ &\le a_{3,i}\sigma(\|z\|^{1+i\tau}_{\Delta^r} + \|e\|^{1+i\tau}_{\Delta^r}),\end{aligned}$$

where $a_{3,i}$ is a positive constant independent of L.

Then, integrating with $\frac{\partial W_1}{\partial e_i} \in \mathbb{H}_{\Delta^r}^{2-\tau-r_i}$ and $\|z\|_{\Delta^r}^{1+i\tau} + \|e\|_{\Delta^r}^{1+i\tau} \in \mathbb{H}_{\Delta^r}^{r_i+\tau}$, the following relations can be derived:

$$\begin{aligned}\frac{\partial W_1}{\partial e^\top}\tilde{F}(x,L,\theta) &\leq \sum_{i=1}^{n} a_{3,i}\sigma\left|\frac{\partial W_1}{\partial e_i}\right|(\|z\|_{\Delta^r}^{1+i\tau} + \|e\|_{\Delta^r}^{1+i\tau})\\ &\leq a_3\sigma\|e\|_{\Delta^r}^2 + a_3\sigma\|z\|_{\Delta^r}^2,\end{aligned} \tag{4.62}$$

where $a_3 = \max_{i\in\mathbb{N}_{1:n}}\{a_{3,i}\}$.

Substituting the relationships (4.56) and (4.62) into (4.54), the time derivative of $W(e,z)$ along (4.52) can be clearly rewritten as

$$\begin{aligned}\dot{W} \leq &-L(a_1\|e\|_\Delta^2 + c_1\|z\|_\Delta^2) - a_2\frac{\dot{L}}{L}\|e\|_\Delta^{2-\tau} - c_2\frac{\dot{L}}{L}\|z\|_\Delta^{2-\tau}\\ &+ c_3\|z\|_\Delta^2 + a_3\sigma(\|e\|_\Delta^2 + \|z\|_\Delta^2)\\ \leq &-(\check{c}L - c_3 - a_3\sigma)(\|e\|_\Delta^2 + \|z\|_\Delta^2),\end{aligned} \tag{4.63}$$

where $\check{c} = \min\{c_1, a_1\}$.

Part 1: The Boundness of L

In what follows, the boundness analysis of dynamic gain L will be specifically presented.

Taking the fact that $L(t)$ is a non-decreasing function and $\check{c}, c_3, a_3, \sigma$ are all bounded constants into consideration, there should exist a time instant $t_1 > t_0$, such that $\check{c}L - c_3 - a_3\sigma \geq 1, \forall t \geq t_1$. In this situation, (4.63) can be simplified as

$$\dot{W} \leq -(\|e\|_\Delta^2 + \|z\|_\Delta^2),\ \forall t \geq t_1.$$

For the convenience of statement, suppose that L will escape at $t = t_f$, i.e., $\lim_{t\to t_f} L = +\infty$, and clearly, $t_1 < t_f$ holds. Combined with the mentioned relationships above, the following contradiction holds:

$$\begin{aligned}+\infty = &\int_{t_1}^{t_f}\dot{L}(s)ds = \int_{t_1}^{t_f}\lambda e_1^2 ds\\ \leq &\lambda\int_{t_1}^{t_f}\|e\|_{\Delta^r}^2 + \|z\|_{\Delta^r}^2 ds\\ \leq &\lambda\int_{t_1}^{t_f} -\dot{W}ds = \lambda\left(W(t_1) - W(t_f)\right) < +\infty.\end{aligned} \tag{4.64}$$

Hence one can claim that L is well defined in the time interval $[t_0, t_f]$ by employing this contradiction.

Part 2: Uniformly Boundness and $\lim_{t=t_f}(z,e)^\top = 0$

At this stage, we are able to declare that z is well defined and bounded in $t \in [t_0, t_f]$. To see why, noticing that the Lyapunov function linking to the z-dynamic system is depicted as $W_2(z) = \left(\lfloor z \rceil^{1-\frac{\tau}{2}}\right)^\top P \lfloor z \rceil^{1-\frac{\tau}{2}}$, and for ease of illustrating the philosophy, combined with the dynamic scaling gain configuration and the boundness of L, we know that the signal of z cannot escape within a finite-time instant.

Subsequently, since L and z are all bounded in $t \in [t_0, t_f]$, we shall prove that e is also well defined. Similar to the proof procedure for the boundness of z, for e-dynamic system, the Lyapunov function is $W_1(e) = \left(\lfloor e \rceil^{1-\frac{\tau}{2}}\right)^\top Q \lfloor e \rceil^{1-\frac{\tau}{2}}$, then the following relations hold with ι_2 being a positive constant

$$\begin{aligned}
&\iota_2 \|e\|_\Delta^{2-\tau} - \left(\lfloor e(t_1))^{1-\frac{\tau}{2}}\right)^\top Q \lfloor e(t_1))\rceil^{1-\frac{\tau}{2}} \leq W_1(e(t)) - W_1(e(t_1)) \\
&\leq \int_{t_1}^{t} -(a_1 L - a_3\sigma)\|e\|_{\Delta^r}^2 ds + \int_{t_0}^{t} a_3\sigma \|z\|_{\Delta^r}^2 ds \\
&\leq \int_{t_1}^{t} -\|e\|_{\Delta^r}^2 ds + \int_{t_1}^{t} a_3\sigma \|z\|_{\Delta^r}^2 ds,
\end{aligned} \tag{4.65}$$

that is,

$$\begin{aligned}
&\|e\|_\Delta^{2-\tau} \leq \frac{1}{\iota_2}\left(W_1(e(t_1)) + \int_{t_1}^{t} a_3\sigma \|z\|_\Delta^2 ds\right), \\
&\int_{t_1}^{t} \|e\|_\Delta^2 ds \leq W_1(e(t_1)) + \int_{t_1}^{t} a_3\sigma \|z\|_\Delta^2 ds.
\end{aligned} \tag{4.66}$$

Since $\int_{t_1}^{t} a_3\sigma \|z\|_\Delta^2 ds$ is bounded on $[t_0, t_f]$, it is concluded from (4.66) that e and $\int_{t_1}^{t_f} \|e\|_\Delta^2 ds$ are uniformly bounded on $[t_1, t_f]$.

Summarily, the boundness analysis of z and e have been completed in $t \in [0, t_f]$.

Part 3: Finite-Time Convergence

From Lemma 1.9, integrating with the boundness of (L, z, e) on $[t_0, \infty)$, it is straightforward to deduce that $\dot{z} \in L_\infty$ and $\dot{e} \in L_\infty$. Besides, we know that $\|e\|_\Delta^2 + \|z\|_\Delta^2 \geq \iota_3 W^{\frac{2}{2-\tau}}(e, z)$ with ι_3 being a positive constant from Lemma 1.1. Then, the following relation holds:

$$\dot{W} + \iota_3(\check{c}L - c_3 - a_3\sigma) W^{\frac{2}{2-\tau}} \leq 0,\ t \geq t_1. \tag{4.67}$$

According to (4.67), there should exist a finite time $T \geq t_1$, such that $\big(z(t), e(t)\big) = 0, \forall t \geq T$ from Lemma 1.4.

This completes the proof of Theorem 4.5. ■

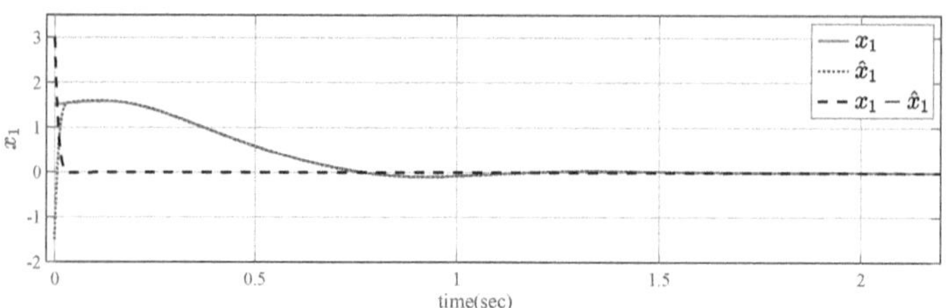

Figure 4.8 Response curves of the output.

4.3.4 Numerical simulation

Consider the system in Example 4.2.1. According to the proposed output feedback control design framework, the observer can be designed as:

$$\begin{cases} \dot{\hat{x}}_1 = \hat{x}_2 + \alpha_1 L \lfloor y - \hat{x}_1 \rceil^{1+\tau} \\ \dot{\hat{x}}_2 = \hat{x}_3 + \alpha_2 L^2 \lfloor y - \hat{x}_1 \rceil^{1+2\tau} \\ \dot{\hat{x}}_3 = u + \alpha_3 L^3 \lfloor y - \hat{x}_1 \rceil^{1+3\tau}. \end{cases} \tag{4.68}$$

Subsequently, the design procedure can be implemented as:

$$z_1 = \hat{x}_1 / L^{\rho},\ z_2 = \hat{x}_2 / L^{\rho+1},\ z_3 = \hat{x}_3 / L^{\rho+2}.$$

Furthermore, the controller can be depicted as:

$$\begin{cases} \dot{L} = c \left(\dfrac{y - \hat{x}_1}{L^{\rho}} \right)^2, \ L_0 = 1, \\ u = -L^{\rho+3} \left(k_1 \lfloor z_1 \rceil^{1+3\rho} + k_2 \lfloor z_2 \rceil^{\frac{1+3\rho}{1+\rho}} + k_3 \lfloor z_3 \rceil^{\frac{1+3\rho}{1+2\rho}} \right). \end{cases} \tag{4.69}$$

In the simulation, the non-parametric uncertainties are set as: $\theta_1 = 0.2\sin(t), \theta_2 = 0.1$, meanwhile, the controller gains are chosen as: $\alpha_1 = 27, \alpha_2 = 27, \alpha_3 = 9; c = 300, \rho = 0.2; k_1 = 9, k_2 = 27, k_3 = 27, \tau = -0.15$. The initial value of the state is $x_1(0) = 0.5, x_2(0) = 1.5, x_3(0) = 0$, and $\hat{x}_1 = -1, \hat{x}_2 = 0, \hat{x}_3(0) = 1$.

It can be observed from Figures 4.8–4.10 that all the signals are uniformly bounded and the trajectories of the states converge to the origin in a satisfactory control performance. Meanwhile, the state estimation can converge in a short time. The time history of the nonsmooth dynamic control signal is shown in Figure 4.11.

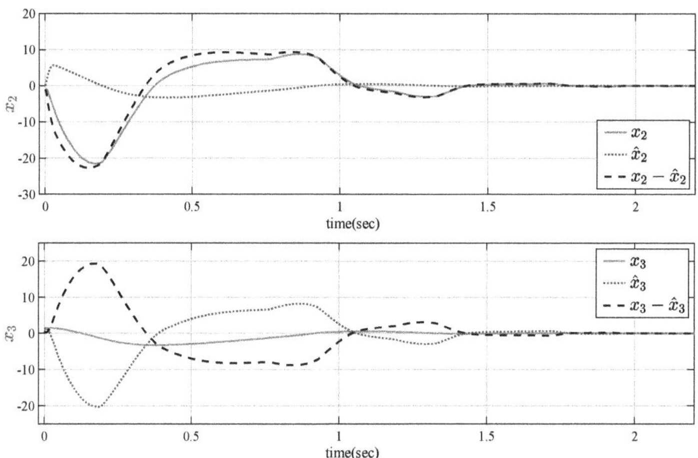

Figure 4.9 Response curves of system states.

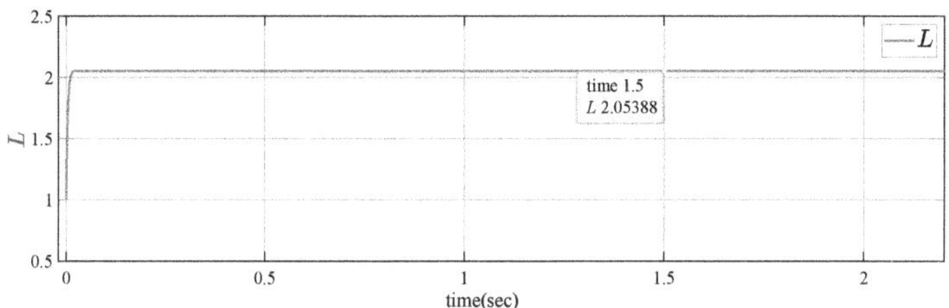

Figure 4.10 The trajectory of the dynamic scaling gain L.

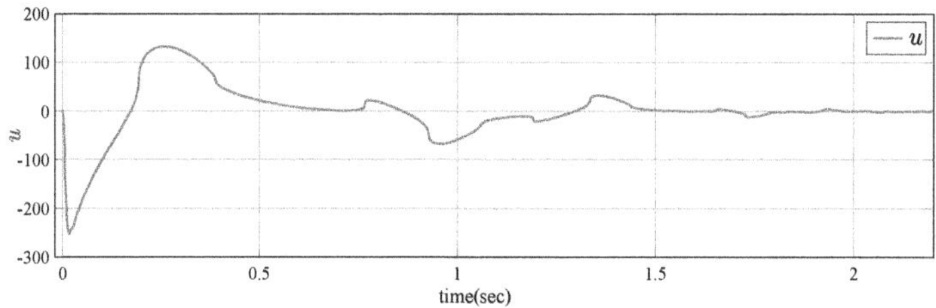

Figure 4.11 Time history of the control input.

Chapter 5

Nonrecursive Robust Output Regulation for Nonlinear Systems with Mismatched Disturbances

Developing high performance output regulation schemes for nonlinear systems with the presence of mismatched uncertainties/disturbances is always a challenging but imperative issue in the control community. Owing to the blossom of uncertainty/disturbance measurement and identification tools over the past two decades, the issue of exact output tracking realization for nonlinear systems in the presence of mismatched uncertainties/disturbances has aroused great efforts in the control community. Output regulation theory [23, 28, 82, 85], sliding mode techniques [7, 11, 19], and other control Lyapunov function based approaches [40, 97] can be extensively found as three main effective branches.

In this chapter, we are considering a more challenging theoretical issue of designing a nonsmooth output regulation law to achieve a finite-time tracking realization even for systems subject to mismatched disturbances. Compared to its asymptotical exact tracking control counterpart, very few existing tools can be found to solve the nonsmooth tracking control problem. To be specific, without imposing any nonlinear growth conditions, most of the existing related results are non-applicable.

Note that when the system is disturbed by mismatched disturbances, rather than popularly assumed matched disturbance, the control performance will

DOI: 10.1201/9781003399230-5

inevitably be affected and the objective of achieving an exact output tracking result is well acknowledged as a very challenging issue. In the existing literature, under different assumptions, various disturbance observation tools can be found. One can refer to [40] for more details. Specifically, as a main concern, finite-time disturbance observation will yield a finite-time system performance recovery by adopting a feedforward compensation design, is clearly of significance in the exact tracking problems. An existing result can be referred to the famous higher order sliding mode (HOSM) disturbance observer. However considering the parameter configuration for the HOSM observer is a complicated process, motivated by the homogeneous observer design tools in the literature, an alternative continuous finite-time disturbance observer is also presented. Under a hypothesis that the mismatched disturbances are guided by certain nonlinear exogenous systems, it is shown that a saturated disturbance observer will also realize a finite-time system performance recovery while the observer gain tuning mechanism follows a simple tuning mechanism. Further, considering a practical case that even the exosystem cannot be exactly identified, a generalized practically effective disturbance estimation result can also be achieved, which has been illustrated by a real-life application.

In this chapter, we shall propose, by developing a novel nonrecursive homogeneous domination strategy, an alternative practically oriented design which can directly address finite-time composite controllers with a simpler expression. Moreover, by proposing a less demanding but more practical control objective, namely, semi-global instead of global control, the restrictive nonlinearity growth constraints in most of the existing related continuous finite-time control works can be fully removed. As a matter of fact, an essential smooth condition for the system nonlinearities will be sufficient to derive an exact tracking control law, even with the presence of mismatched disturbances, covering both smooth and nonsmooth control cases. Firstly, a delicate coordinates transformation is given by virtue of exactly calculating the steady-state equations. Secondly, the conventional virtual controllers in recursive design steps are no longer required in essence, while a concise nonsmooth composite control law can be explicitly constructed. Thirdly, it is shown, by rigorous analysis of the semi-global attractivity, any trajectory from an arbitrary large compact set will be captured by an arbitrary small set and then converge to its equilibrium within a finite-time.

5.1 Problem Formulation

In this chapter, we specify the control objective as to realize the finite-time exact tracking task for the following nonlinear system with the presence of mismatched disturbances

$$\begin{cases} \dot{x}_i(t) = x_{i+1}(t) + f_i(\bar{x}_i(t)) + d_i(t), \ i \in \mathbb{N}_{1:n-1}, \\ \dot{x}_n(t) = u(t) + f_n(x(t)) + d_n(t), \\ y(t) = x_1(t), \end{cases} \tag{5.1}$$

where $\bar{x}_i = (x_1, x_2, \cdots, x_i)^\top \in \mathbb{R}^i$ is the system partial state vector with $i \in \mathbb{N}_{1:n}$, $x = \bar{x}_n$ is the full state vector, y is the system output, u is the system input. $f_i(\cdot)$, $i \in \mathbb{N}_{1:n}$ is a known smooth nonlinear function (or, at least $\mathbb{C}^{n-i}$), $d_i(t)$, $i \in \mathbb{N}_{1:n}$ represents a nonvanishing mismatched disturbance item. The output reference signal, denoted by y_r satisfies Assumption 3.1.1. Without loss of generality, the initial time is set as zero.

5.2 Disturbance Observer Design

Aiming to realize an accurate tracking result for the nonlinear systems in the presence of mismatched disturbances, a promising handling procedure is to identify or reconstruct the disturbance terms. In what follows, we will present a few disturbance observer design methods to enable the subsequent feedforward compensation processes.

5.2.1 A nonrecursive finite-time disturbance observer design

First, a nonrecursive finite-time disturbance observer is designed based on a certain assumption for the mismatched disturbances as shown below.

Assumption 5.2.1 *The mismatched disturbance $d_i(t)$ is generated by a nonlinear exosystem*

$$d_i^{(n-i+2)} = \phi_i\left(d_i, d_i^{(1)}, \cdots, d_i^{(n-i+1)}\right), \ \phi_i \in \mathbb{C}^1(\mathbb{R}^{n-i+2}, \mathbb{R}),$$

and satisfies $\max\limits_{i \in \mathbb{N}_{1:n}, \ j \in \mathbb{N}_{0:n-i+1}} \left\{ \sup\limits_{t \geq 0} \left| \frac{\partial d_i^j(t)}{\partial t^j} \right| \right\} \leq D$ *with $D \in \mathbb{R}_+$ being a constant.*

Consider the i-th subsystem

$$\dot{x}_i = x_{i+1} + f_i(\bar{x}_i) + d_i.$$

Assume that all the states are physically measurable. By setting $x_i = z_{i,1}$, $d_i^{(j)} = z_{i,j+2}$, $j \in \mathbb{N}_{0:n-i+1}$, and with Assumption 5.2.1 in mind, the following extended system can be achieved

$$\begin{cases} \dot{z}_{i,1} = x_{i+1} + f_i(\bar{x}_i) + z_{i,2}, \\ \dot{z}_{i,j} = z_{i,j+1}, \ j \in \mathbb{N}_{2:n-i+1}, \\ \dot{z}_{i,n-i+2} = \phi_i(z_{i,2}, z_{i,3} \cdots, z_{i,n-i+1}). \end{cases} \tag{5.2}$$

Then we employ a saturated high-gain finite-time observer of the following form:

$$\begin{cases} \dot{\hat{z}}_{i,1} = x_{i+1} + f_i(\bar{x}_i) + \hat{z}_{i,2} + \ell_{i,1}\sigma_i \lfloor z_{i,1} - \hat{z}_{i,1} \rceil^{1+\tau}, \\ \dot{\hat{z}}_{i,j} = \hat{z}_{i,j+1} + \ell_{i,j}\sigma_i^j \lfloor z_{i,1} - \hat{z}_{i,1} \rceil^{1+j\tau}, \ j \in \mathbb{N}_{2:n-i+1}, \\ \dot{\hat{z}}_{i,n-i+2} = \phi_i(\langle \hat{z}_{i,2} \rangle_M, \langle \hat{z}_{i,3} \rangle_M, \cdots, \langle \hat{z}_{i,n-i+1} \rangle_M) \\ \qquad + \ell_{i,n-i+2}\sigma_i^{n-i+2} \lfloor z_{i,1} - \hat{z}_{i,1} \rceil^{1+(n-i+2)\tau}, \end{cases} \tag{5.3}$$

where $\sigma_i \geq 1$ is a design parameter to be determined later, $M \geq D$ is a saturation threshold, $H_i = \text{col}(\ell_{i,1}, \ell_{i,2}, \cdots, \ell_{i,n-i+2})$ is the observer gain vector with $\ell_{i,j}$ being the corresponding coefficient of a Hurwitz polynomial

$$p(s) = s^{n-i+2} + \ell_{i,1}s^{n-i+1} + \cdots + \ell_{i,n-i+1}s + \ell_{i,n-i+2}.$$

By combing (5.2) and (5.3) together with the denotation of $e_{i,j} = (z_{i,j} - \hat{z}_{i,j})/\sigma_i^{j-1}$, $j \in \mathbb{N}_{1:n-i+2}$ and $e_i = \text{col}(e_{i,1}, e_{i,2}, \cdots, e_{i,n-i+2})$, the error dynamics can be expressed in the following form:

$$\begin{cases} \dot{e}_{i,j} = \sigma_i e_{i,j+1} - \ell_{i,j}\sigma_i \lfloor e_{i,1} \rceil^{1+j\tau}, \ j \in \mathbb{N}_{1:n-i+1}, \\ \dot{e}_{i,n-i+2} = -\ell_{i,n-i+2}\sigma_i \lfloor e_{i,1} \rceil^{1+(n-i+2)\tau} \\ \qquad + \left(\phi_i(z_{i,2}, \cdots, z_{i,n-i+1}) - \phi_i(\langle \hat{z}_{i,2} \rangle_M, \cdots, \langle \hat{z}_{i,n-i+1} \rangle_M)\right)/\sigma_i^{n-i+1}, \end{cases}$$

which can also be rewritten as the following compact form:

$$\begin{aligned} \dot{e}_i = A\sigma_i e_i - &\begin{bmatrix} \ell_{i,1}\sigma_i \lfloor e_{i,1} \rceil^{1+\tau} \\ \vdots \\ \ell_{i,n-i+2}\sigma_i \lfloor e_{i,1} \rceil^{1+(n-i+2)\tau} \end{bmatrix} \\ &+ E(\phi_i(z_{i,2}, \cdots, z_{i,n-i+1}) - \phi_i(\langle \hat{z}_{i,2} \rangle_M, \cdots, \langle \hat{z}_{i,n-i+1} \rangle_M))/\sigma_i^{n-i+1} \\ &\triangleq \sigma_i \Psi_i(H_i, e_i, \tau) + E(\phi_i - \hat{\phi}_{iM})/\sigma_i^{n-i+1}, \end{aligned} \tag{5.4}$$

where $A = \begin{bmatrix} 0 & I_{n-i+1} \\ 0 & 0 \end{bmatrix}$ and $E = [0, \cdots 0, 1]^\top$.

Denote a homogeneous dilation $r_i = (1, 1+\tau, \cdots, 1+(n-i+2)\tau)$ for the vector field $\Psi_i(H_i, e_i, \tau)$. Construct the following Lyapunov candidate function:

$$W_i(e_i) = \left(\lfloor e_i \rceil_{\Delta^{r_i}}^{1-\frac{\tau}{2}} \right)^\top Q_i \lfloor e_i \rceil_{\Delta^{r_i}}^{1-\frac{\tau}{2}}, \tag{5.5}$$

where Q_i is a positive definite and symmetrical matrix satisfying $\tilde{A}_i^\top Q_i + Q_i\tilde{A}_i = -I$, $\tilde{A}_i$ is the companion matrix of H_i.

Now we can present the following theorem.

Theorem 5.1

Under Assumption 5.2.1, for any well defined trajectory $x(t)$ in system (5.1), there exist proper parameters σ_i, $i \in \mathbb{N}_{1:n}$, such that all the signals in system (5.4) are uniformly bounded and the origin is a finite-time stable equilibrium, that is, there exists a time instant T_1, such that

$$\widehat{d_i^{(j)}} = d_i^{(j)},\ i \in \mathbb{N}_{1:n},\ j \in \mathbb{N}_{0:n-i+1},\ \forall t \geq T_1.$$

Proof: By Assumption 5.2.1, we know that $z_{i,j}$, $j \in \mathbb{N}_{2:n-i+1}$ is bounded and $\phi_i \in \mathbb{C}^1$. Hence by using Mean-Value Theorem, it is clear that the following relation can be satisfied:

$$\phi_i - \phi_{iM} \leq \hat{\gamma}_i \Big(|z_{i,2} - \langle \hat{z}_{i,2} \rangle_M| + |z_{i,3} - \langle \hat{z}_{i,3} \rangle_M| + \cdots + |z_{i,n-i+1} - \langle \hat{z}_{i,n-i+1} \rangle_M| \Big),$$

where $\hat{\gamma}_i$ is a bounded constant.

In the case when $|z_{i,j} - \langle \hat{z}_{i,j} \rangle_M| \geq 1$, $\forall j \in \mathbb{N}_{2:n-i+1}$, by noting that $|z_{i,j}| \leq D$ and $|\langle \hat{z}_{i,j} \rangle_M| \leq M$, the following relation holds:

$$|z_{i,j} - \langle \hat{z}_{i,j} \rangle_M| \leq M + D \leq (M+D) |z_{i,j} - \langle \hat{z}_{i,j} \rangle_M|^{\frac{1+(n-i+2)\tau}{1+(j-1)\tau}}.$$

In the case when $|z_{i,j} - \langle \hat{z}_{i,j} \rangle_M| < 1$, $\forall j \in \mathbb{N}_{2:n-i+1}$, with the fact $\frac{1+(n-i+2)\tau}{1+(j-1)\tau} \leq 1$, one can also have

$$|z_{i,j} - \langle \hat{z}_{i,j} \rangle_M| \leq |z_{i,j} - \langle \hat{z}_{i,j} \rangle_M|^{\frac{1+(n-i+2)\tau}{1+(j-1)\tau}}.$$

With $\sigma_i \geq 1$ in mind, summarizing the above two cases and using Lemma 1.13, the following inequalities can be obtained with a constant $\gamma_i = \hat{\gamma}_i \max\{M + D, 1\}$ which is independent of σ_i:

$$\begin{aligned}
&\big(\phi_i - \phi_{iM}\big)/\sigma_i^{n-i+1} \\
\leq &\frac{\gamma_i}{\sigma_i^{n-i+1}} \Big(|z_{i,2} - \langle \hat{z}_{i,2} \rangle_M|^{\frac{1+(n-i+2)\tau}{1+\tau}} + |z_{i,3} - \langle \hat{z}_{i,3} \rangle_M|^{\frac{1+(n-i+2)\tau}{1+2\tau}} + \cdots \\
&+ |z_{i,n-i+1} - \langle \hat{z}_{i,n-i+1} \rangle_M|^{\frac{1+(n-i+2)\tau}{1+(n-i+1)\tau}} \Big) \\
\leq &\frac{\gamma_i}{\sigma_i^{n-i+1}} \Big(|\sigma_i^1 e_{i,2}|^{\frac{1+(n-i+2)\tau}{1+\tau}} + |\sigma_i^2 e_{i,3}|^{\frac{1+(n-i+2)\tau}{1+2\tau}} + \cdots + |\sigma_i^{n-i} e_{i,n-i+1}|^{\frac{1+(n-i+2)\tau}{1+(n-i+1)\tau}} \Big) \\
\leq &\gamma_i \Big(|e_{i,2}|^{\frac{1+(n-i+2)\tau}{1+\tau}} + |e_{i,3}|^{\frac{1+(n-i+2)\tau}{1+2\tau}} + \cdots + |e_{i,n-i+1}|^{\frac{1+(n-i+2)\tau}{1+(n-i+1)\tau}} \Big).
\end{aligned} \tag{5.6}$$

It is easy to verify that

$$\Big(|e_{i,2}|^{\frac{1+(n-i+2)\tau}{1+\tau}} + |e_{i,3}|^{\frac{1+(n-i+2)\tau}{1+2\tau}} + \cdots + |e_{i,n-i+1}|^{\frac{1+(n-i+2)\tau}{1+(n-i+1)\tau}} \Big) \in \mathbb{H}_{\Delta^r}^{1+(n-i+2)\tau}.$$

Further, using Lemma 1.1 and recalling that $W_i(e_i) \in \mathbb{H}_{\Delta^r}^{2-\tau}$, the following relations hold for a constant $\bar{c}_i \in \mathbb{R}_+$:

$$\begin{aligned}
&\frac{\partial W_i(e_i)}{\partial e_{i,n-i+2}}(\phi_i - \hat{\phi}_{iM})/\sigma_i^{n-i+1} \\
&\leq \gamma_i \left|\frac{\partial W_i(e_i)}{\partial e_{i,n-i+2}}\right| \left(|e_{i,2}|^{\frac{1+(n-i+2)\tau}{1+\tau}} + |e_{i,3}|^{\frac{1+(n-i+2)\tau}{1+2\tau}} + \cdots + |e_{i,n-i+1}|^{\frac{1+(n-i+2)\tau}{1+(n-i+1)\tau}}\right) \\
&\leq \bar{c}_i W_i^{\frac{2}{2-\tau}}(e_i).
\end{aligned} \tag{5.7}$$

Note that the vector field $\Psi_i(H_i, e_i, \tau)$ is homogeneous with a degree τ w.r.t. the dilation weight Δ^r. By following Lemma 2.1, we can arrive at the conclusion that the system $\dot{e}_i = \sigma_i \Psi_i(H_i, e_i, \tau)$ is globally finite-time stable for $\tau \in (-\kappa, 0)$ where $\kappa \in (0, 1/(n+1))$ is a constant.

In addition, noting that $\frac{\partial W_i(e_i)}{\partial e_i^{\top}}\Psi_i(\cdot) \in \mathbb{H}_{\Delta^r}^{2}$ and $W(e_i) \in \mathbb{H}_{\Delta^r}^{2-\tau}$, then by using Lemma 1.1, the following relation can be easily obtained for a constant $c \in \mathbb{R}_+$:

$$\frac{\partial W_i(e_i)}{\partial e_i^{\top}}\Psi_i(H_i, e_i, \tau) \leq -c_i W_i^{\frac{2}{2-\tau}}(e_i), \ \tau \in (-\kappa, 0). \tag{5.8}$$

With (5.7) and (5.8) in mind, the time derivative of $W_i(e_i)$ gives

$$\begin{aligned}
\dot{W}_i(e_i) &= \frac{\partial W_i(e_i)}{\partial e_i^{\top}}\sigma_i \Psi_i(H_i, e_i, \tau) + \frac{\partial W_i(e_i)}{\partial e_{i,n-i+2}}(\phi_i - \hat{\phi}_{iM})/\sigma_i^{n-i+1} \\
&\leq -(c_i\sigma_i - \bar{c}_i) W_i^{\frac{2}{2-\tau}}(e_i).
\end{aligned} \tag{5.9}$$

With the following simple tuning mechanism to determine σ_i

$$c_i\sigma_i - \bar{c}_i \geq 1, \tag{5.10}$$

(5.9) reduces to

$$\dot{W}_i(e_i) + W_i^{\frac{2}{2-\tau}}(e_i) \leq 0.$$

Therefore, for any well defined $x(t)$, all the signals in the error dynamics (5.4) are uniformly bounded. Then with $\tau < 0$ in mind, it follows straightforwardly from Lemma 1.4 that the error dynamics (5.4) is finite-time stable.

This completes the proof of Theorem 5.1. ■

5.2.2 A practical nonrecursive disturbance observer

Under the assumption that the disturbances are subject to known external exosystems, Theorem 5.1 has presented a rigorous theoretical result on exact disturbance estimation. However, by noticing that the disturbance models are usually a non-trivial task to be identified, the direct application of the finite-time disturbance observer (5.3) might be somehow difficult. In what follows, we show that an alternative practical nonsmooth composite control result can also be established based on the following weaker and more practical assumption.

Assumption 5.2.2 *The mismatched disturbance $d_i(t) \in \mathbb{C}^{n-i+1}$ satisfies*

$$\max_{i\in\mathbb{N}_{1:n},\ j\in\mathbb{N}_{0:n-i+1},\ t\in\mathbb{R}_+}\left\{\left|\frac{\partial d_i^j(t)}{\partial t^j}\right|\right\}\le D,$$

where $D \in \mathbb{R}_+$ is a constant.

First, similar to (5.2), we have

$$\begin{cases}\dot{z}_{i,1} = x_{i+1} + f_i(\bar{x}_i) + z_{i,2},\\ \dot{z}_{i,j} = z_{i,j+1},\ j\in\mathbb{N}_{2:n-i+1},\\ \dot{z}_{i,n-i+2} = h(t),\end{cases} \tag{5.11}$$

where $h(t) = d_i^{(n-i+1)}(t)$.

Under Assumption 5.2.2, one is able to directly modify the nonsmooth observer (5.3) to the following simpler form:

$$\begin{cases}\dot{\hat{z}}_{i,1} = x_{i+1} + f_i(\bar{x}_i) + \hat{z}_{i,2} + \ell_{i,1}\sigma_i\lfloor z_{i,1}-\hat{z}_{i,1}\rceil^{1+\tau},\\ \dot{\hat{z}}_{i,j} = \hat{z}_{i,j+1} + \ell_{i,j}\sigma_i^j\lfloor z_{i,1}-\hat{z}_{i,1}\rceil^{1+j\tau},\ j\in\mathbb{N}_{2:n-i+1},\\ \dot{\hat{z}}_{i,n-i+2} = \ell_{i,n-i+2}\sigma_i^{n-i+2}\lfloor z_{i,1}-\hat{z}_{i,1}\rceil^{1+(n-i+2)\tau}.\end{cases} \tag{5.12}$$

Similarly, the following theorem can be presented.

Theorem 5.2

Consider the disturbance observer (5.12) under Assumption 5.2.2. For any well defined trajectory $x(t)$ in system (5.1) and an arbitrarily small positive constant ε_i, there exist sufficiently large parameters σ_i, $i\in\mathbb{N}_{1:n}$, and a finite time instant T_2, such that

$$|\widehat{d_i^{(j)}} - d_i^{(j)}| \le \varepsilon_i,\ i\in\mathbb{N}_{1:n},\ j\in\mathbb{N}_{0:n-i+1},\ \forall t\ge T_2.$$

Proof: With (5.11) and (5.12), the error dynamics gives

$$\begin{cases}\dot{e}_{i,j} = \sigma_i e_{i,j+1} - \ell_{i,j}\sigma_i\lfloor e_{i,1}\rceil^{1+j\tau},\ j\in\mathbb{N}_{1:n-i+1},\\ \dot{e}_{i,n-i+2} = -\ell_{i,n-i+2}\sigma_i\lfloor e_{i,1}\rceil^{1+(n-i+2)\tau} + h(t)/\sigma^{n-i+1}.\end{cases} \tag{5.13}$$

With Assumption 5.2.2 in mind, we know that $|h(t)|\le D$. Hence with a similar stability analysis to the proof of Theorem 5.1 and using Lemma 1.12, we can also have the following relation:

$$\begin{aligned}\dot{W}_i(e_i) &= \frac{\partial W_i(e_i)}{\partial e_i^T}\sigma_i\Psi_i(H_i, e_i,\tau) + \frac{\partial W_i(e_i)}{\partial e_{i,n-i+2}}h(t)/\sigma_i^{n-i+1}\\ &\le -c_i\sigma_i W_i^{\frac{2}{2-\tau}}(e_i) + \left|\frac{\partial W_i(e_i)}{\partial e_{i,n-i+2}}\right|(D/\sigma_i^{n-i+1})^{\frac{1+(n-i+2)\tau}{1+(n-i+2)\tau}}\\ &\le -(c_i\sigma_i - \tilde{c}_i)W_i^{\frac{2}{2-\tau}}(e_i) + D^*,\end{aligned} \tag{5.14}$$

where $\tilde{c}_i\in\mathbb{R}_+$ is a constant and $D^* = \tilde{c}_i(D/\sigma_i^{n-i+1})^{\frac{2}{1+(n-i+2)\tau}}$.

By selecting σ_i to be sufficiently large such that $c_i\sigma_i - \tilde{c}_i > 0$, we know that $e_{i,j}$, $j \in \mathbb{N}_{1:n-i+2}$ in system (5.13) is uniformly bounded and can be made arbitrarily approach to zero by tuning σ_i to be a larger value. Hence, there exists a finite time instant T_2, such that

$$|\widehat{d_i^{(j)}} - d_i^{(j)}| \leq \varepsilon_i,\ i \in \mathbb{N}_{1:n},\ j \in \mathbb{N}_{0:n-i+1},\ \forall t \geq T_2.$$

This completes the proof of Theorem 5.2. ■

Remark 5.1 The disturbance observer (5.12) can also be seen as a nonsmooth extension of existing GPI disturbance observers, which could handle a variety of commonly seen disturbances.

5.2.3 Higher-order sliding mode observer design

Under Assumption 5.2.2, the following higher-order sliding mode observer can also be built to realize a finite-time accurate estimation of the mismatched disturbances [38]:

$$\begin{aligned}
&\dot{z}_{i,0} = \hbar_{i,0} + x_{i+1} + \phi_i(\bar{x}_i),\\
&\hbar_{i,0} = z_{i,1} - l_{i,0}\lambda_i^{\alpha_{i,0}} \lfloor z_{i,0} - x_i \rceil^{1-\alpha_{i,0}},\\
&\dot{z}_{i,1} = \hbar_{i,1},\ \cdots\ \dot{z}_{i,k} = \hbar_{i,k},\ k \in \mathbb{N}_{1:n-i+1},\\
&\hbar_{i,j} = z_{i,j+1} - l_{i,j}\lambda_i^{\alpha_{i,j}} \lfloor z_{i,j} - \hbar_{i,j-1} \rceil^{1-\alpha_{i,j}}, j \in \mathbb{N}_{1:n-i},\\
&\hbar_{i,n-i+1} = -l_{i,n-i+1}\lambda_i^{\alpha_{i,n-i+1}} \lfloor z_{i,n-i+1} - \hbar_{i,n-i} \rceil^{1-\alpha_{i,n-i+1}},\ i \in \mathbb{N}_{1:n-1},\\
&\dot{z}_{n,0} = \hbar_{n,0} + u + \phi_n(x),\ \dot{z}_{n,1} = \hbar_{n,1},
\end{aligned} \tag{5.15}$$

where $\alpha_{i,j} = \frac{1}{n+2-i-j}$, $l_{i,j} \in \mathbb{R}_+$, $\lambda_i \in \mathbb{R}_+$ are design parameters, and $z_{i,0} = \hat{x}_i$, $z_{i,1} = \hat{d}_i$, and $z_{i,j} = \widehat{d_i^{(j-1)}}$ represent the estimates of x_i, d_i, and $d_i^{(j-1)}$, respectively.

Denote $e_{i,0} = \hat{x}_i - x_i$ and $e_{i,j} = z_{i,j} - d_i^{(j-1)}$. Combining (5.1) and (5.15), the error dynamics gives

$$\begin{aligned}
&\dot{e}_{i,0} = e_{i,1} - l_{i,0}\lambda_i^{\alpha_{i,0}} \lfloor e_{i,0} \rceil^{1-\alpha_{i,0}},\\
&\dot{e}_{i,j} = e_{i,j+1} - l_{i,j}\lambda_i^{\alpha_{i,j}} \lfloor e_{i,j} - \dot{e}_{i,j-1} \rceil^{1-\alpha_{i,j}}, j \in \mathbb{N}_{1:n-i},\\
&\dot{e}_{i,n-i+1} \in [-D, D] - l_{i,n-i+1}\lambda_i^{\alpha_{i,n-i+1}} \lfloor e_{i,n-i+1} - \dot{e}_{i,n-i} \rceil^{1-\alpha_{i,n-i+1}}.
\end{aligned} \tag{5.16}$$

Lemma 5.1

([38]) Assume the observer gain λ_i satisfies $\lambda_i > D$, $i \in \mathbb{N}_{1:n}$. For all possible well defined trajectories $x(t)$, all signals in (5.16) are uniformly bounded and there exists a finite-time $T_1 \in \mathbb{R}_+$ such that $e_{i,j}(t) = 0$, $t \in [T_1, \infty)$.

5.3 Robust Output Regulation—A Linear Case Study

Based on the disturbance observation technologies presented above, this section presents a simple linear composite control design procedure to illustrate the non-recursive design framework.

5.3.1 System pre-treatment

Provided that all the disturbance terms $d_i's$ are exactly known, we are thereafter able to define an auxiliary variable $\bar{\chi}_i = (\chi_1, \chi_2, \cdots, \chi_i)^\top$, $i \in \mathbb{N}_{1:n+1}$, where each element χ_i is determined by the following output regulation equations:

$$\begin{cases} \chi_1 = y_r, \\ \chi_i = \dfrac{d\chi_{i-1}}{dt} - f_{i-1}(\bar{\chi}_{i-1}) - d_{i-1},\ i \in \mathbb{N}_{2:n+1}. \end{cases} \tag{5.17}$$

Note that (5.17) is clearly inaccessible in practice. However, with the corresponding estimates from the disturbance observers proposed in the above section (e.g., (5.15)), replacing $\frac{\partial d_i^j}{\partial t^j}$ by $z_{i,j+1}$ for $i \in \mathbb{N}_{1:n}$, $j \in \mathbb{N}_{0:n-i+1}$, one can therefore obtain the following implementable state trajectory reference function:

$$x_i^* = \chi_i(z, \bar{y}_r),\ i \in \mathbb{N}_{1:n+1} \tag{5.18}$$

where $z = (z_{1,0}, z_{1,1}, \cdots, z_{1,n}, \cdots, z_{n,1})^\top$, $\bar{y}_r = (y_r, y_r^{(1)}, \cdots, y_r^{(n)})^\top$.

Further, let $\eta_i := x_i - x_i^*$, $i \in \mathbb{N}_{1:n}$ and $\eta = (\eta_1, \eta_2, \cdots, \eta_n)^\top$. By defining a change of coordinates of the following form:

$$\xi_i = \eta_i / L^{i-1},\ i \in \mathbb{N}_{1:n},\ v = (u - x_{n+1}^*)/L^n \tag{5.19}$$

where $L \geq 1$ is a scaling gain to be made precise in the stability analysis later on, system (5.1) can be transformed to the following stabilizable form:

$$\begin{cases} \dot{\xi}_i = L\xi_{i+1} + (f_i(\bar{x}_i) - f_i(\bar{x}_i^*) + \varepsilon_i)/L^{i-1},\ i \in \mathbb{N}_{1:n-1}, \\ \dot{\xi}_n = Lv + (f_n(x) - f_n(x^*) + \varepsilon_n)/L^{n-1}, \end{cases} \tag{5.20}$$

where $\bar{x}_i^* = (x_1^*, \cdots, x_i^*)^\top$, $i \in \mathbb{N}_{1:n}$, $x^* = \bar{x}_n^*$, $\varepsilon_i = \phi_i(\bar{x}_i^*) - \phi_i(\bar{\chi}_i) + x_{i+1}^* - \chi_{i+1} + \dot{\chi}_i - \dot{x}_i^*$.

Keeping (5.20) in mind, the compact form of the stabilizable system can be written as:

$$\dot{\xi} = L(A\xi + Bv) + F(\cdot) + \varepsilon, \tag{5.21}$$

where (A, B) is the n-th order controllable canonical matrix pair, $F(\cdot) = (\frac{f_1(x_1) - f_1(x_1^*)}{L^0}, \frac{f_2(\bar{x}_2) - f_2(\bar{x}_2^*)}{L^1}, \cdots, \frac{f_n(x) - f_n(\bar{x}^*)}{L^{n-1}})$ and $\varepsilon = (\varepsilon_1, \varepsilon_2/L, \cdots, \varepsilon_n/L^{n-1})$.

5.3.2 Composite output regulation law design

At this stage, without any recursive determination of virtual controllers and tuning functions, a linear composite output regulation law can be straightfowardly achieved:

$$u = L^n v + x_{n+1}^*, \; v = -K\xi, \tag{5.22}$$

where $L \geq 1$ is a sufficiently large constant, K is the control gain vector that determined by a Hurwitz polynomial $H(s) = s^n + k_n s^{n-1} + \cdots + k_1$.

5.3.3 Stability analysis

The main result of this section is presented by the following theorem.

Theorem 5.3
Under Assumptions 3.1.1 and 5.2.2. For the closed-loop system consisting of (5.1)–(5.15)–(5.22) *and any state satisfying that* $x_0 \in \mho \triangleq [-\rho, \rho]^n$, *with* $\rho \in \mathbb{R}_+$ *being an arbitrarily large constant, the following statements hold:*

- *All signals in the closed-loop system are uniformly bounded;*
- $\lim\limits_{t\to\infty} y = y_r$.

Proof: In what follows, the detailed stability analysis of the proposed controller is given step by step. For system (5.21), the candidate Lyapunov function regarding ξ is constructed as $V(\xi) = \xi^\top P\xi$ in which P satisfies $P = P^\top$ and $(A-BK)^\top P + P(A-BK) \leq -I_n$.

Taking the derivative of $V(\xi)$ along (5.21), we will arrive at

$$\dot{V}(\xi) = \frac{\partial V(\xi)}{\partial \xi^\top} L(A-BK)\xi + \frac{\partial V(\xi)}{\partial \xi^\top} F(\cdot) + \frac{\partial V(\xi)}{\partial \xi^\top}\varepsilon. \tag{5.23}$$

According to the guideline of the parameter matrix P, one can get that

$$\frac{\partial V(\xi)}{\partial \xi^\top} L(A-BK)\xi \leq -L\|\xi\|^2. \tag{5.24}$$

For the nonlinear function $f_i(\cdot)$, from (5.18) and Lemma 5.1, we know that for any well-defined $x(t)$, x_i^* is uniformly bounded, i.e., there is a positive constant $\bar{\rho}$ satisfying $\max\limits_{i\in\mathbb{N}_{1:n}}\{\sup\limits_{t\geq 0}\{x_i^*(t)\}\} \leq \bar{\rho}$. Hence a compact set $\mho \triangleq [-\rho, \rho]^n$ is given. Define a level set

$$\Omega = \left\{ \eta \in \mathbb{R}^n | \eta^\top P\eta \leq c_0 \triangleq \sup_{\eta\in\Omega_\eta = [-(\rho+\bar{\rho}),(\rho+\bar{\rho})]} \{\eta^\top P\eta\} \right\}.$$

Furthermore, one can get that $\forall \eta(t) \in \Omega \Rightarrow \xi(t) \in \Omega$.

For a smooth nonlinear function, the relationship

$$f_i(\bar{x}_i) - f_i(\bar{x}_i^*) \leq \bar{\mu}_i(\bar{x}_i, \bar{x}_i^*)(|x_1 - x_1^*| + |x_2 - x_2^*| + \cdots + |x_i - x_i^*|)$$

derived from the Mean-Value Theorem holds, where $\bar{\mu}_i(\bar{x}_i, \bar{x}_i^*) \in \mathbb{C}^1$ is a non-negative function.

If $\eta \in \Omega$, there is a positive constant $\bar{N}$ satisfying $|x_j| \leq \bar{N}, j \in \mathbb{N}_{1:i}$. Then there exists a positive constant μ_i independent to scaling gain L, satisfying $\bar{\mu}_i(\bar{x}_i, \bar{x}_i^*) \leq \mu_i$. Hence, the following relationship holds:

$$\begin{aligned}
\frac{f_i(\bar{x}_i) - f_i(\bar{x}_i^*)}{L^{i-1}} &\leq \frac{\mu_i}{L^{i-1}}(|x_1 - x_1^*| + |x_2 - x_2^*| + \cdots + |x_i - x_i^*|) \\
&\leq \frac{\mu_i}{L^{i-1}}(L^0|\xi_1| + L|\xi_2| \cdots + L^{i-1}|\xi_i|) \leq \mu_i \cdot \sqrt{n}\|\xi\|, \ \eta \in \Omega. \qquad (5.25)
\end{aligned}$$

Further, one could get

$$\begin{aligned}
\frac{\partial V(\xi)}{\partial \xi^\top} F(L, x, x^*)|_{\eta \in \Omega} &\leq \sum_{i=1}^{n} \frac{\partial V(\xi)}{\partial \xi_i} \cdot \frac{f_i(\bar{x}_i) - f_i(\bar{x}_i^*)}{L^{i-1}} \\
&\leq c_1 \|\xi\|^2, \qquad (5.26)
\end{aligned}$$

where $c_1 > 0$ is a constant.

Finally, from Assumption 5.2.2 and Lemma 5.1, ε is related to the convergence time instant T_1, hence, the mathematical analysis of ε is categorized into two parts, i.e., $t \in [0, T_1)$ and $t \in [T_1, \infty)$.

For the case when $t \in [0, T_1)$, from Assumption 5.2.2, one can get that there must exist a bounded constant $E \in \mathbb{R}_+$ rendering to $\sup_{i \in \mathbb{N}_{1:n}} \{|\varepsilon_i|\} \leq E$. Keeping Lemma 1.12 in mind, we have

$$\begin{aligned}
\frac{\partial V(\xi)}{\partial \xi^\top}\varepsilon &\leq 2\lambda_{\max}(P)\|\xi\|\|\varepsilon\| \leq 2\sqrt{n}\lambda_{\max}(P)\|\xi\|E \\
&\leq c_2\|\xi\|^2 + E^2, \qquad (5.27)
\end{aligned}$$

where $c_2 > 0$ is a constant.

At this point, the selection guideline of the fixed scaling gain L is able to be presented. For ease of description, define a suitable positive constant $\delta \in (0, c_0/2)$. Then, we can choose a sufficiently large L satisfying

$$L \geq c_1 + c_2 + 1; \ E^2 \leq \frac{L - c_1 - c_2}{2\lambda_{\max}(P)}\delta. \qquad (5.28)$$

In what follows, the boundness of ξ should be claimed, i.e., for any non-zero initial state $\eta(0) \in \Omega_\eta$, any trajectory of $\xi(t)$ will stay in Ω.

Suppose that ξ will escape in the convergence process. Considering the escape phenomenon, there exists a time point $t_1 > 0$, such that

$$\eta^\top(t_1)P\eta(t_1) = c_0,\ \dot{V}(\xi(t_1)) > 0. \tag{5.29}$$

It is clear that $\forall t \in [0,t_1], \eta(t) \in \Omega$. Substituting (5.24), (5.26), and (5.27) into (5.23), it yields

$$\dot{V}(\xi) \leq -(L-c_1-c_2)\|\xi\|^2 + E^2. \tag{5.30}$$

Integrating the fact that $\|\xi\|^2 \geq \lambda_{\max}^{-1}(P)V(\xi)$, (5.30) can be rewritten as

$$\dot{V}(\xi(t_1)) \leq -(L-c_1-c_2)\lambda_{\max}^{-1}(P)\big(V(\xi(t_1)) - 0.5\delta\big) \leq 0,$$

which clearly leads to a contradiction to the hypothesis, then we can arrive at the conclusion that

$$\forall x(0) \in \mho \Rightarrow \eta \in \Omega \Rightarrow \xi \in \Omega, \forall t \geq 0.$$

For the case when $t \in [T_1,\infty)$, the disturbance information has been estimated precisely, then we have $\varepsilon = 0$.

With the above discussions, (5.23) can be rewritten as

$$\begin{aligned} &\dot{V}(\xi) \leq -(L-c_1-c_2)\|\xi\|^2 + E^2,\ t \in [0,T_1), \\ &\dot{V}(\xi) \leq -(L-c_1)\|\xi\|^2,\ t \in [T_1,\infty). \end{aligned} \tag{5.31}$$

Combining with the selection guideline of L, i.e., (5.28), the closed-loop system (5.1)–(5.15)–(5.22) is asymptotically stable. Hence, $\lim\limits_{t\to\infty} \xi(t) = 0$ can be achieved, i.e., $\lim\limits_{t\to\infty} x(t) = x^*(t)$.

It means that the nonrecursive robust controller proposed in this section can achieve the control objective of exact tracking, i.e., $\lim\limits_{t\to\infty} y = y_r$. This completes the proof of Theorem 5.3. ■

5.3.4 An illustrative example

Consider the offset-free tracking control problem of a single-link manipulator to illustrate the effectiveness of the proposed linear robust controller.

Example 5.3.1 *The system dynamics of a single-link manipulator can be expressed as [13]:*

$$\begin{cases} D\ddot{q} + B\dot{q} + N\sin(q) = \tau + \tau_{d_1}, \\ M\dot{\tau} + H_m\tau = u - K_m\dot{q} + \tau_{d_2}, \end{cases} \tag{5.32}$$

where $q, \dot{q}, \ddot{q}$ represent the connection position, speed, and acceleration, respectively. τ is the torque generated by the electrical system, τ_{d_1} corresponds to the *torque disturbance, τ_{d_2} is the measurement error of the system measurable data, u is the control input of the system, usually referring to the electro-mechanical torque. $D = 1\,\mathrm{kg}\cdot\mathrm{m}^2$ is the mechanical inertia. $B = 1\,\mathrm{N}\cdot\mathrm{m}\cdot\mathrm{s}\cdot\mathrm{rad}^{-1}$ is the coefficient of viscous friction. $N = 10$ is a constant whose value is related to the mass of the load and the coefficient of gravity. $M = 1\,\mathrm{H}$ is armature reactance. $H_m = 1.0\,\Omega$ is armature resistance. $K_m = 0.2\,\mathrm{N}\cdot\mathrm{m}\cdot\mathrm{A}^{-1}$ is the back EMF coefficient.*

Letting $x_1 = q$, $x_2 = \dot{q}$, $x_3 = \tau$, then system (5.32) can be rewritten in the following form:

$$\begin{cases} \dot{x}_1 = x_2, \\ \dot{x}_2 = x_3 - x_2 - 10\sin(x_1) + \tau_{d_1}, \\ \dot{x}_3 = u - 0.2x_2 - x_3 + \tau_{d_2}. \end{cases}$$

In the subsequent simulation, set $y_r = \sin(t) + \cos(t + \pi/3)$ and the disturbances as $\tau_{d_1} = 2 + 0.5\cos(2t)$, $\tau_{d_2} = 0.5e^{-t}$.

Then, according to the above-mentioned controller design process, one can utilize the high-order sliding mode disturbance observer to estimate τ_{d_1}, τ_{d_2}, by formula (5.15), the observer can be specifically expressed as:

$$\tau_{d_1} \begin{cases} \dot{z}_{2,0} = x_3 - x_2 - 10\sin(x_1) + \hbar_{2,1},\ \dot{z}_{2,1} = \hbar_{2,2}, \\ \dot{z}_{2,2} = \hbar_{2,3},\ \zeta_{1,i} = z_{1,i} - \hbar_{1,i}, i = 0,1,2, \\ \hbar_{2,1} = -l_{1,0}\lambda_1^{1/3}\lfloor\zeta_{2,0}\rceil^{2/3} + z_{2,1}, \\ \hbar_{2,2} = -l_{1,1}\lambda_1^{1/2}\lfloor\zeta_{2,1}\rceil^{1/2} + z_{2,2}, \\ \hbar_{2,3} = -l_{1,2}\lambda_1\operatorname{sign}(\zeta_{2,2}); \end{cases} \tag{5.33}$$

$$\tau_{d_2} \begin{cases} \dot{z}_{3,0} = u - 0.2x_2 - x_3 + \hbar_{3,1}, \\ \dot{z}_{3,1} = \hbar_{3,2},\ \zeta_{3,i} = z_{3,i} - \hbar_{3,i},\ i = 0,1, \\ \hbar_{3,1} = -l_{2,0}\lambda_2^{1/2}\lfloor\zeta_{3,0}\rceil^{1/2} + z_{3,1}, \\ \hbar_{3,2} = -l_{2,1}\lambda_2\operatorname{sign}(\zeta_{3,1}). \end{cases} \tag{5.34}$$

Next, calculate the steady state reference functions as:

$$\begin{aligned} x_1^* &= y_r, \\ x_2^* &= \dot{y}_r, \\ x_3^* &= \ddot{y}_r + \dot{y}_r + 10\sin(y_r) - z_{2,1}, \\ u^* &= y_r^{(3)} + \ddot{y}_r + 10\cos(y_r)\dot{y}_r - z_{2,2} + 0.2x_2^* + x_3^* - z_{3,1}. \end{aligned} \tag{5.35}$$

Through the following coordinate transformation:

$$\begin{aligned} \eta_1 &= x_1 - x_1^*, \\ \eta_2 &= (x_2 - x_2^*)/L, \\ \eta_3 &= (x_3 - x_3^*)/L^2, \end{aligned} \tag{5.36}$$

one can build the following simple controller:

$$u = -L^3(k_1\eta_1 + k_2\eta_2 + k_3\eta_3) + u^*. \tag{5.37}$$

The parameters are set as: $x(0)^\top = [0, 1.5, 2]$, $[l_{1,0}, l_{1,1}, l_{1,2}, \lambda_1; l_{2,0}, l_{2,1}, \lambda_2] = [15, 23, 32, 1; 12, 18, 1]$, $[L, K^\top] = [3, 15, 18, 22]$.

According to Figure 5.1, one can clearly observe that the proposed controller is able to ensure that the output regulation objective can be realized. Besides, the disturbance estimation performance is shown in Figure 5.2.

5.4 Nonrecursive $\mathbb{C}^0$ Robust Output Regulation

5.4.1 Controller construction

Illustrated by the controller design procedure in the above section, similarly, we will show that, for system (5.1), without going through a series of recursive design steps, a simple nonsmooth controller can be explicitly pre-built of the following form:

$$v = -K\lfloor \xi \rceil_{\Delta^r}^{r_n+\tau}, \; u = L^n v + x_{n+1}^*. \tag{5.38}$$

5.4.2 Stability analysis

The main result of this section can be summarized by the following theorem.

Theorem 5.4

Consider the closed-loop system consisting of (5.1) under Assumption 5.2.2 and the dynamic compensator (5.15)–(5.38) with a sufficiently large scaling gain L. Then for any given constant $\rho \in \mathbb{R}_+$ which could be arbitrarily large, all trajectories of $x(t)$ starting from the compact set $\mho \triangleq [-\rho, \rho]^n$ will converge to the equilibrium point within a finite-time. Moreover, there exists a finite time instant $T_2 > T_1 > 0$ such that $\forall t \geq T_2$, $y(t) = y_r$.

Proof: Construct a candidate Lyapunov function $U(\xi) \in \mathbb{C}^1 \cap \mathbb{H}_{\Delta^r}^{2-\tau}$ of the form

$$U(\xi) = \left(\lfloor \xi \rceil_{\Delta^r}^{1-\frac{\tau}{2}}\right)^\top P \lfloor \xi \rceil_{\Delta^r}^{1-\frac{\tau}{2}}, \tag{5.39}$$

with P being a positive definite and symmetrical matrix satisfying $\Lambda^\top P + P\Lambda = -I$ and Λ being a companion matrix of K. The time derivative of $U(\xi)$ along the

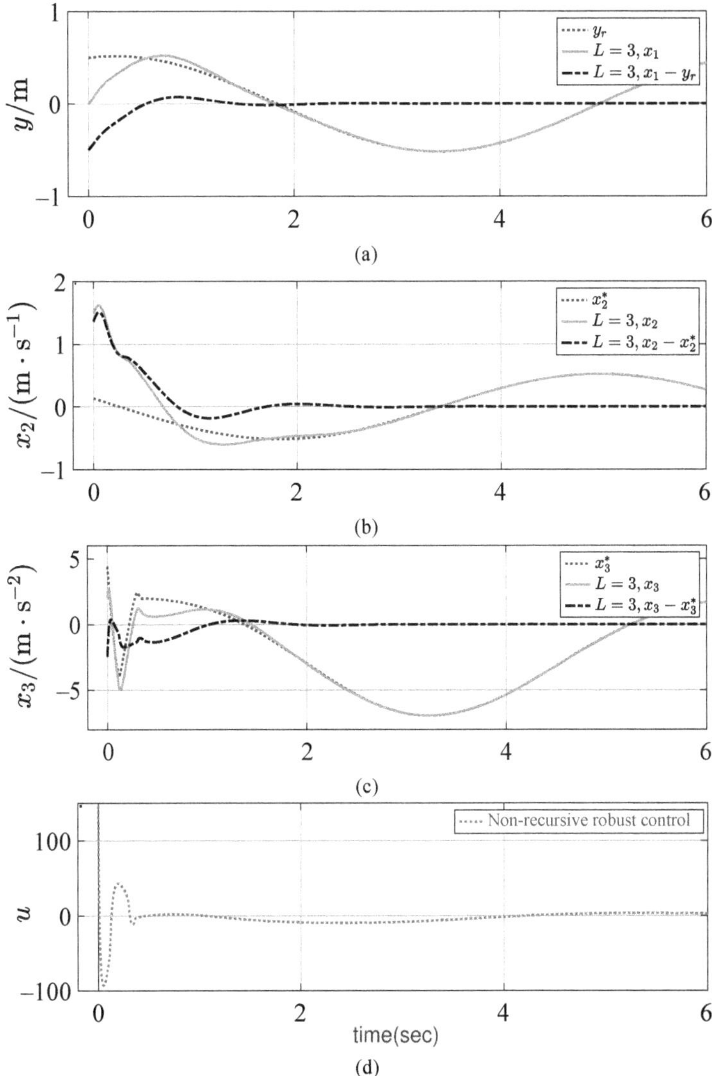

Figure 5.1 The output response curves, state curves x_2, x_3, and the control effort of one-link manipulator with time-varying disturbance. (a) Output response; (b) State x_2; (c) State x_3; (d) The control effort.

closed-loop system (5.20)–(5.38) is given by

$$\dot{U}(\xi) = \frac{\partial U(\xi)}{\partial \xi^\top} L\left(\xi_2, \cdots, \xi_n, -K\lfloor \xi \rceil_{\Delta^r}^{r_n+\tau}\right)^\top + \sum_{i=1}^{n} \frac{\partial U(\xi)}{\partial \xi_i}\left(\frac{f_i(\bar{x}_i) - f_i(\bar{x}_i^*)}{L^{i-1}} + \frac{\varepsilon_i}{L^{i-1}}\right). \tag{5.40}$$

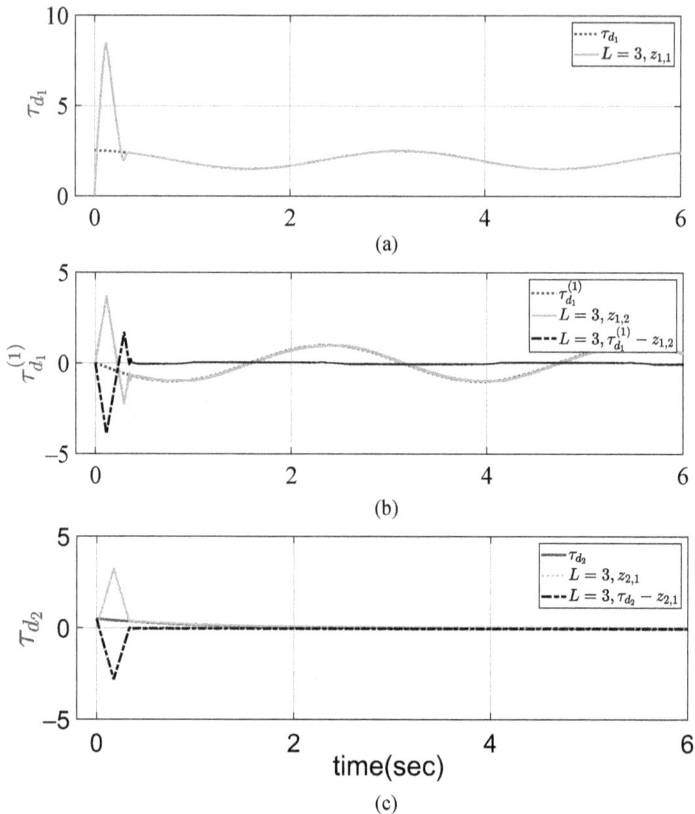

Figure 5.2 The time-varying disturbance curves of one-link manipulator. (a) τ_{d_1}; (b) $\tau_{d_1}^{(1)}$; (c) τ_{d_2}

By the definition of x_i^* in (5.18) and Lemma 5.1, we know that for any well defined $x(t)$, the signal x_i^* is uniformly bounded, that is, there exists a constant $\bar{\rho} > 0$ such that $\max_{i \in \mathbb{N}_{1:n}}\{\sup_{t \geq 0}\{x_i^*(t)\}\} \leq \bar{\rho}$.

Thereafter, for a given compact set $\mho \triangleq [-\rho, \rho]^n$, define a level set

$$\Omega = \left\{\eta \in \mathbb{R}^n | (\lfloor \eta \rceil_{\Delta^r}^{1-\frac{\tau}{2}})^\top P \lfloor \eta \rceil_{\Delta^r}^{1-\frac{\tau}{2}} \leq c_0 \right\},$$

where

$$c_0 \triangleq \sup_{\eta \in [-(\rho+\bar{\rho}),(\rho+\bar{\rho})]^n} \{(\lfloor \eta \rceil_{\Delta^r}^{1-\frac{\tau}{2}})^\top P \lfloor \eta \rceil_{\Delta^r}^{1-\frac{\tau}{2}}\}.$$

On the other hand, with $L \geq 1$ and the relation (5.19) in mind, we know that $\forall \eta(t) \in \Omega \Rightarrow \xi(t) \in \Omega$.

In order to proceed, this section will present the detailed stability analysis step by step within a homogeneous framework.

First, by following a similar proof as Lemma 2.1, we know that $\frac{\partial U(\xi)}{\partial \xi^\top}(\xi_2,\cdots,\xi_n,-K\lfloor\xi\rceil_{\Delta^r}^{r_n+\tau})^\top$ is negative definite for $\tau \in (-\varsigma,0)$. With $U(\xi) \in \mathbb{H}_{\Delta^r}^{2-\tau}$ and $\left(\xi_2,\cdots,\xi_n,-K\lfloor\xi\rceil_{\Delta^r}^{r_n+\tau}\right)^\top \in \mathbb{H}_{\Delta^r}^{\tau}$ in mind, using Lemma 1.1, one will arrive at the prior conclusion that there exist a constant $\alpha \in \mathbb{R}_+$ and a constant $\varsigma \in (0,\frac{1}{n})$, such that

$$\frac{\partial U(\xi)}{\partial \xi^\top}\left(\xi_2,\cdots,\xi_n,-K\lfloor\xi\rceil_{\Delta^r}^{r_n+\tau}\right)^\top \leq -\alpha\|\xi\|_{\Delta^r}^2$$

holds for $\tau \in (-\varsigma,0)$.

Second, noting that f_i, $i \in \mathbb{N}_{1:n}$ is a smooth function, by Mean-Value Theorem, one can obtain that

$$\phi_i(\bar{x}_i)-\phi_i(\bar{x}_i^*) \leq \bar{c}_i(\bar{x}_i,\bar{x}_i^*)\Big(|x_1-x_1^*|+\cdots+|x_i-x_i^*|\Big)$$

where $\bar{c}_i(\bar{x}_i,\bar{x}_i^*)$ is a $\mathbb{C}^1$ nonnegative function.

If $\eta \in \Omega$, we know there must exist a constant $N>0$ such that $|x_j| \leq N$ for $j \in \mathbb{N}_{1:i}$. Then subsequently, there exists a constant $\tilde{c}_i$ such that $\bar{c}_i(\bar{x}_i,\bar{x}_i^*) \leq \tilde{c}_i$. In what follows, the following two cases will be studied.

1) In the case when $|x_j-x_j^*| \geq 1$, $\forall j \in \mathbb{N}_{1:i}$, by noting that $|x_j| \leq N$ and $|x_j^*| \leq \bar{\rho}$, the following relation holds:

$$|x_j-x_j^*| \leq N+\bar{\rho} \leq (N+\bar{\rho})|x_j-x_j^*|^{\frac{1+i\tau}{1+(j-1)\tau}}.$$

2) In the case when $|x_j-x_j^*|<1$, $\forall j \in \mathbb{N}_{1:i}$, by noting $\frac{1+i\tau}{1+(j-1)\tau} \leq 1$, we know that

$$|x_j-x_j^*| \leq |x_j-x_j^*|^{\frac{1+i\tau}{1+(j-1)\tau}}.$$

By summarizing the above two cases and noting that $(j-1)\frac{1+i\tau}{1+(j-1)\tau}-(i-1) \leq 0$, $\forall j \in \mathbb{N}_{1:i}$, the following relation holds with constants $\check{c}_i = \tilde{c}_i\max\{N+\bar{\rho},1\}$ and $c_i \in \mathbb{R}_+$ which are independent of L:

$$\begin{aligned}(\phi_i(\bar{x}_i)-\phi_i(\bar{x}_i^*))/L^{i-1} \leq& \check{c}_i\frac{\left(|L^0\xi_1|^{\frac{1+i\tau}{1}}+|L^1\xi_2|^{\frac{1+i\tau}{1+\tau}}+\cdots+|L^{i-1}\xi_i|^{\frac{1+i\tau}{1+(i-1)\tau}}\right)}{L^{i-1}}\\ \leq& c_i\|\xi\|_{\Delta^r}^{1+i\tau},\ \eta \in \Omega.\end{aligned}$$

It is clear that $U(\xi) \in \mathbb{H}_{\Delta^r}^{2-\tau}$ and $\bar{f}_i \in \mathbb{H}_{\Delta^r}^{1+i\tau}$. With Lemma 1.1 in mind, it is straightforward to conclude that there exists a constant $\tilde{\alpha} \in \mathbb{R}_+$ which is independent of L, such that

$$\sum_{i=1}^n \frac{\partial U(\xi)}{\partial \xi_i}\frac{\phi_i(\bar{x}_i)-\phi_i(\bar{x}_i^*)}{L^{i-1}} \leq \tilde{\alpha}\|\xi\|_{\Delta^r}^2,\ \forall \eta \in \Omega.$$

With Lemma 5.1 and Assumption 5.2.2 in mind, we know that for any well defined $x(t)$, $\forall 0 \leq t < T_1$, there exists a bounded constant $\Gamma \in \mathbb{R}_+$, such that $\max_{i\in\mathbb{N}_{1:n}} \{|\varepsilon_i|\} \leq \Gamma$. Utilizing Lemmas 1.1 and 1.12, the following relation can be obtained for constants $\hat{\alpha} \in \mathbb{R}_+$ and $\bar{\alpha} \in \mathbb{R}_+$:

$$\begin{aligned}\sum_{i=1}^{n} \frac{\partial U(\xi)}{\partial \xi_i} \frac{\varepsilon_i}{L^{i-1}} &\leq \hat{\alpha} \sum_{i=1}^{n} \|\xi\|_{\Delta^r}^{1-i\tau} \left(\frac{\Gamma}{L^{i-1}}\right)^{\frac{1+i\tau}{1+i\tau}} \\ &\leq \bar{\alpha} \left(\|\xi\|_{\Delta^r}^2 + \sum_{i=1}^{n} \frac{\Gamma^{\frac{2}{1+i\tau}}}{L^{\frac{2(i-1)}{1+i\tau}}} \right) \\ &\triangleq \bar{\alpha} \|\xi\|_{\Delta^r}^2 + \Gamma^*,\ t \in [0, T_1).\end{aligned} \tag{5.41}$$

Using Lemma 1.1 again, the following relation holds with a constant $\mu \in \mathbb{R}_+$:

$$\|\xi\|_{\Delta^r}^2 \geq \mu U^{\frac{2}{2-\tau}}(\xi). \tag{5.42}$$

Then for any arbitrarily small tolerance $\delta \in (0, c_0/2)$, now we are able to choose the scaling gain $L \geq 1$ under the following guideline:

$$\alpha L - \tilde{\alpha} - \bar{\alpha} \geq 1/\mu,\ \Gamma^* \leq \frac{(\alpha L - \tilde{\alpha} - \bar{\alpha})\mu}{2} \delta^{\frac{2}{2-\tau}}. \tag{5.43}$$

In what follows, we will first show the uniform boundness of trajectory ξ, and then prove that a local finite-time convergence can be achieved.

1) *Uniform boundness:* In this regard, we shall prove that for any non-zero initial states satisfying $\eta(0) \in [-(\rho+\bar{\rho}), (\rho+\bar{\rho})]^n$, all the trajectories of $\eta(t)$ and $\xi(t)$ will stay in Ω forever.

If the above statement is not true, that is, at least one trajectory of $\eta(t)$ will escape Ω within a finite-time. Regarding the finite-time escaping phenomenon, two cases described in Figure 5.3 will be discussed as follows.

Case 1: There exist two time instants $t_2 > t_1 > 0$, such that

$$\begin{aligned}&i)\ \dot{U}(\xi(t_1)) < 0, \\ &ii)\ U(\xi(t_2)) = U(\xi(t_1)) > \delta, \\ &iii)\ \dot{U}(\xi(t_2)) > 0.\end{aligned} \tag{5.44}$$

It is clear that for $t \in [t_1, t_2]$, $\eta(t) \in \Omega$. Hence with the propositions analyzed above, (5.41), (5.42), and (5.43), the following relation can be obtained from (5.40):

$$\begin{aligned}\dot{U}(\xi) &\leq -\alpha L \|\xi\|_{\Delta^r}^2 + \tilde{\alpha}\|\xi\|_{\Delta^r}^2 + \bar{\alpha}\|\xi\|_{\Delta^r}^2 + \Gamma^* \\ &\leq -(\alpha L - \tilde{\alpha} - \bar{\alpha})\|\xi\|_{\Delta^r}^2 + \Gamma^* \\ &\leq -(\alpha L - \tilde{\alpha} - \bar{\alpha})\mu \left(U^{\frac{2}{2-\tau}}(\xi) - \frac{1}{2}\delta^{\frac{2}{2-\tau}} \right) < 0,\ t \in [t_1, t_2).\end{aligned} \tag{5.45}$$

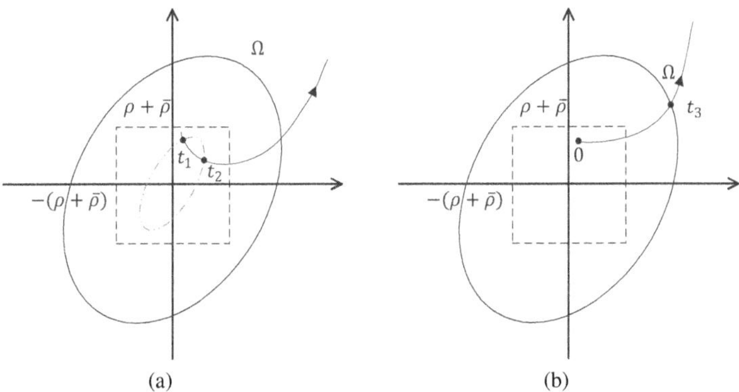

Figure 5.3 Sketch figure of finite-time escaping phenomenon. (a) Case 1; (b) case 2.

It is noted that (5.45) implies

$$U(\xi(t_2)) - U(\xi(t_1)) = \int_{t_1}^{t_2} \dot{U}(\xi(s))ds < 0.$$

Recalling the relation ii) of (5.44), it will lead to an obvious contradiction, i.e.,

$$0 = \int_{t_1}^{t_2} \dot{U}(\xi(s))ds < 0.$$

Case 2: There exists one time instant $t_3 \geq 0$, such that

$$\begin{aligned} &i)\ \left(\lfloor \eta(t_3) \rceil_{\Delta^r}^{1-\frac{\tau}{2}}\right)^\top P \lfloor \eta(t_3) \rceil_{\Delta^r}^{1-\frac{\tau}{2}} = c_0, \\ &ii)\ \dot{U}(\xi(t_3)) > 0, \\ &iii)\ U(\xi(t_3)) > \delta. \end{aligned} \tag{5.46}$$

In this case, we know that $\forall t \in [0, t_3]$, $\eta(t) \in \Omega$ and $\xi(t) \in \Omega$. Similarly with (5.45), we have

$$\dot{U}(\xi(t_3)) \leq -(\alpha L - \tilde{\alpha} - \bar{\alpha})\mu\left(U^{\frac{2}{2-\tau}}(\xi(t_3)) - \frac{1}{2}\delta^{\frac{2}{2-\tau}}\right) < 0$$

which clearly contradicts the claim ii) in (5.46).

In a summary of the above two cases, we can arrive at the conclusion that

$$\forall x(0) \in \mho \Rightarrow \eta(t) \in \Omega \Rightarrow \xi(t) \in \Omega,\ \forall t \geq 0.$$

2) *Local finite-time convergence:* Now it is true that the relation (5.45) also holds for $t \in [0, T_1]$. With this relation in mind, we know that any trajectory of the

closed-loop system (5.20)–(5.38) will be well defined. In the case when $t \geq T_1$, it concludes from Lemma 5.1 that $\varepsilon_i = 0,\ t \geq T_1,\ i \in \mathbb{N}_{1:n}$. Based on (5.43) and (5.45), one can also have

$$\dot{U}(\xi) + U^{\frac{2}{2-\tau}}(\xi)\Big|_{\Omega} \leq 0,\ t \in [T_1, \infty). \tag{5.47}$$

By Lemma 1.4 and with the fact that $0 < \frac{2}{2-\tau} < 1$ in mind, the relation (5.47) leads to a straightforward conclusion that there exists another time instant $T_2 > T_1 > 0$, such that $y(t) - y_r = 0,\ t \in [T_2, \infty)$.

This completes the proof of Theorem 5.4. ■

Remark 5.2 In the proof of Theorem 5.4, a delicate contradiction argument is employed to guarantee the avoidance of finite-time escaping phenomenon. Under the framework of nonrecursive homogeneous domination approach, we first show that under the guideline (5.43), the semi-global attractivity of the level set Ω can be ensured via a contradiction argument, and then all signals in the closed-loop system will be uniformly bounded. Moreover, by Lyapunov function based analysis, the finite-time convergence property of the system states is eventually guaranteed.

5.4.3 Extensions

Consider the case when the mismatched disturbances in the system (5.1) satisfy Assumption 5.2.1, one could also directly obtain the following theorem.

Theorem 5.5

Under Assumptions 3.1.1 and 5.2.1, consider the closed-loop system consisting of system (5.1), finite-time disturbance observer (5.3) and the controller (5.38) with a sufficiently large scaling gain L which is dependent on the compact set $\mho$. The following statements hold:

- *Any trajectory $x(t)$ starting from $\mho$ will be uniformly bounded.*
- *The closed-loop system is locally finite-time stable.*

Proof: By recalling Theorem 5.1, it is noted that besides the convergence analysis of the disturbance observer, the remainder of the proof of Theorem 5.5 is almost identical to the proof of Theorem 5.4 and hence is omitted here. ■

In addition, if the exosystems of the disturbances are very difficult to be identified, a practical tracking control result under the proposed nonrecursive synthesis framework is easier to be implemented in practice.

Corollary 5.1

Under Assumptions 3.1.1 and 5.2.2, consider the closed-loop system consisting of system (5.1) and the composite controller (5.12)–(5.38). For any trajectory $x(t)$ starting from $\mho$, there exists a sufficiently large scaling gain L such that the closed-loop system can be rendered semi-globally practically stable.

Proof: The proof is very similar to the proof of Theorem 5.5, thus is omitted here and left for the readers. ■

5.4.4 Illustrative examples and numerical simulations

Aiming to clarify the difference of nonsmooth and smooth control framework, consider a simple second-order system whose expression is presented as follows:

Example 5.4.1 *Consider the following 2-D nonlinear system:*

$$\begin{cases} \dot{x}_1 = x_2 + \sin(x_1) + d_1, \\ \dot{x}_2 = u + x_1 x_2 + d_2, \\ y = x_1, \end{cases} \tag{5.48}$$

where d_1 and d_2 are the matched and mismatched disturbances of the system, respectively.

In order to achieve the high-precision tracking control objective of the system, based on the method in [69], its composite controller can be expressed as the following form:

$$\begin{aligned} &d_1 : \begin{cases} \hat{d}_1 = q_1(x_1 - p_1), \\ \dot{p}_1 = x_2 + \sin(x_1) + \hat{d}_1; \end{cases} \\ &d_2 : \begin{cases} \hat{d}_2 = q_2(x_2 - p_2), \\ \dot{p}_2 = u + x_1 x_2 + \hat{d}_2; \end{cases} \\ &\begin{cases} \chi_1 = y_r,\ z_i = x_i - \chi_i,\ i = 1,2, \\ \chi_2 = -k_1 z_1 - \sin(x_1) + \dot{\chi}_1 - \hat{d}_1, \\ u = -k_2 z_2 - x_1 x_2 + k_1 \dot{\chi}_1 + \ddot{\chi}_1 - \hat{d}_2 - \\ \qquad \left(k_1 + \cos(x_1)\right)\left(x_2 + \sin(x_1) + \hat{d}_1\right). \end{cases} \end{aligned} \tag{5.49}$$

Among them, $q_i > 0$, p_i, $i = 1,2$ are the design parameters and auxiliary states respectively, $\hat{d}_1, \hat{d}_2$ are the estimated value of system disturbances. $k_1 > 0, k_2 > 0$. χ_1, χ_2 are the corresponding expected reference trajectories of state x_1, x_2.

Then, according to the proposed nonrecursive integrated composite control design framework, the disturbance is first estimated by HOSM observer, which can be described as:

$$d_1\begin{cases}\dot{z}_{1,0}=x_2+\sin(x_1)+\hbar_{1,1},\ \dot{z}_{1,1}=\hbar_{1,2},\\ \dot{z}_{1,2}=\hbar_{1,3},\ \zeta_{1,i}=z_{1,i}-\hbar_{1,i}, i=0,1,2,\\ \hbar_{1,1}=-l_{1,0}\lambda_1^{1/3}\lfloor\zeta_{1,0}\rceil^{2/3}+z_{1,1},\\ \hbar_{1,2}=-l_{1,1}\lambda_1^{1/2}\lfloor\zeta_{1,1}\rceil^{1/2}+z_{1,2},\\ \hbar_{1,3}=-l_{1,2}\lambda_1\operatorname{sign}(\zeta_{1,2});\end{cases}$$

$$d_2\begin{cases}\dot{z}_{2,0}=u+x_1x_2+\hbar_{2,1},\\ \dot{z}_{2,1}=\hbar_{2,2},\ \zeta_{2,i}=z_{2,i}-\hbar_{2,i}, i=0,1,\\ \hbar_{2,1}=-l_{2,0}\lambda_2^{1/2}\lfloor\zeta_{2,0}\rceil^{1/2}+w_{2,1},\\ \hbar_{2,2}=-l_{2,1}\lambda_2\operatorname{sign}(\zeta_{2,1}).\end{cases}$$

Subsequently, the steady-state reference signals can be calculated as

$$\begin{aligned}x_1^*&=y_r,\\ x_2^*&=\dot{y}_r-\sin(y_r)-z_{1,1},\\ u^*&=\ddot{y}_r-\dot{y}_r\cos(y_r)-z_{1,2}-y_rx_2^*-z_{2,1},\end{aligned}$$

and the coordinate transformation can be introduced as

$$\begin{aligned}&\eta_1=x_1-x_1^*,\ \eta_2=(x_2-x_2^*)/L,\\ &\upsilon=(u-u^*)/L^2\end{aligned}$$

Then the smooth/nonsmooth nonrecursive robust controller is designed as

$$\text{smooth case: } u=-L^{\rho+2}(k_1\eta_1+k_2\eta_2)+u^*,\ L\in\mathbb{R}_+.$$

$$\text{nonsmooth case: }\begin{cases}u=L^2v+u^*,\\ v=-k_1\lfloor\eta_1\rceil^{1+2\tau}-k_2\lfloor\eta_2\rceil^{\frac{1+2\tau}{1+\tau}},\ \tau\in(-0.5,0).\end{cases}\tag{5.50}$$

In subsequent simulations, by setting $y_r=1+0.5\cos(t+\pi/4)$ and selecting the disturbance dynamics as $d_1=\sin(0.5t+\pi/3)$, $d_2=\cos(t)+\sin(0.5t)$, using two different control strategies (recursive and nonrecursive algorithm), the following simulation results can be obtained, as shown in Figure 5.4.

The recursive controller parameters are set as: $[k_1,k_2,q_1,q_2]=[6,10,28,32]$. The nonrecursive controller parameters are: $[l_{1,0},l_{1,1},l_{1,2},\lambda_1]=[12,13,14,1]$, $[l_{2,0},l_{2,1},\lambda_2]=[12,13,1]$, and $[k_1,k_2,L,\tau]=[25,18,3,-0.1]$.

It can be seen from Figure 5.4(a) that both control strategies can achieve the precise tracking control target of the system (5.48). What is worth paying attention to is that by comparing (5.49) and (5.50), it can be clearly observed that the control law (5.50) has a more concise form, and the nonrecursive controller

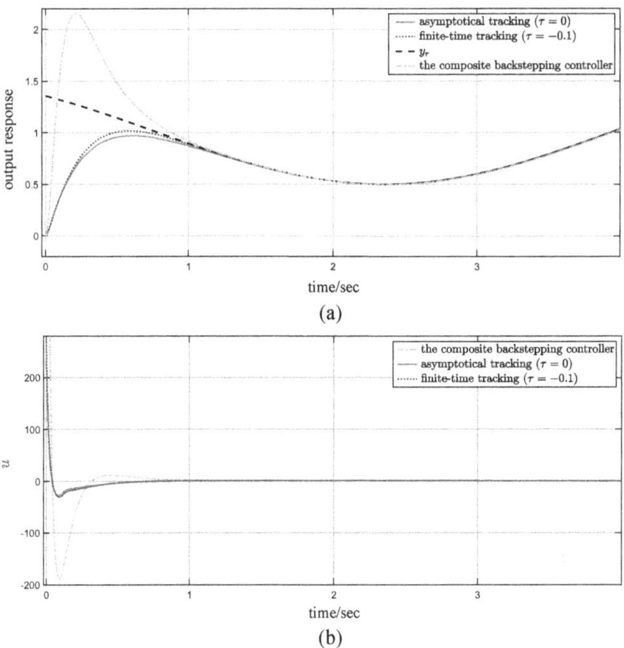

Figure 5.4 Control performance comparisons. (a) Tracking performance; (b) Control effort.

mainly depends on the nominal model of the system, while the nonlinear term and its partial differential term are not directly reflected in the control law, which is essentially different with the recursive control algorithm such as the backstepping method.

Example 5.4.2 *Consider the following 3-D disturbed nonlinear system:*

$$\begin{cases} \dot{x}_1 = x_2 + \sin(x_1) + d_1(t), \\ \dot{x}_2 = x_3 + \ln(1 + x_1^2), \\ \dot{x}_3 = u + x_1 x_3^{1/3} + d_2(t), \\ y = x_1, \end{cases} \tag{5.51}$$

where $d_1(t)$ and $d_2(t)$ are mismatched and matched disturbances, respectively. The control objective is to realize finite-time exact tracking of a given reference signal $y_r = 1 + \sqrt{2}\sin(t + \pi/4)$ while the disturbances are set as $d_1 = 0.1\sin(t)$, $d_2 = 1$.

In what follows, we show that by considering a semi-global control objective, the exact tracking control problem for Example 5.4.2 can now be solved by following a simple synthesis procedure presented in this section.

First, by skipping the pre-verifications of nonlinearity growth constraints, one can straightforwardly utilize the proposed control method with a series of pre-calculations as:

$$\begin{cases} y_r^{(1)} = \cos(t) - \sin(t), \\ y_r^{(2)} = -\sin(t) - \cos(t), \\ y_r^{(3)} = -\cos(t) + \sin(t); \end{cases} \tag{5.52}$$

$$\begin{cases} x_1^* = y_r, \\ x_2^* = y_r^{(1)} - \sin(y_r) - z_{1,1}, \\ x_3^* = y_r^{(2)} - \cos(y_r)y_r^{(1)} - z_{1,2} - \ln(1+y_r^2), \\ u^* = y_r^{(3)} + \sin(y_r)(y_r^{(1)})^2 - \cos(y_r)y_r^{(2)} - z_{1,3} - 2y_r y_r^{(1)}/(1+y_r^2) \\ \qquad - y_r(x_3^*)^{1/3} - z_{2,1}. \end{cases}$$

With the coordinates transformation

$$\begin{aligned} \xi_1 &= x_1 - x_1^*, \\ \xi_2 &= (x_2 - x_2^*)/L, \\ \xi_3 &= (x_3 - x_3^*)/L^2, \\ v &= (u - u^*)/L^3, \end{aligned} \tag{5.53}$$

the obtained exact tracking controller is depicted explicitly in the following expression:

$$d_1 : \begin{cases} \dot{z}_{1,0} = x_2 + \sin(x_1) + \hbar_{1,0} \\ \dot{z}_{1,i} = \hbar_{1,i}, \ i \in \mathbb{N}_{1:3} \\ \hbar_{1,0} = -l_{1,0}\lambda_1^{1/4} \lfloor z_{1,0} - x_1 \rceil^{3/4} + z_{1,1} \\ \hbar_{1,1} = -l_{1,1}\lambda_1^{1/3} \lfloor z_{1,1} - \hbar_{1,0} \rceil^{2/3} + z_{1,2} \\ \hbar_{1,2} = -l_{1,2}\lambda_1^{1/2} \lfloor z_{1,2} - \hbar_{1,1} \rceil^{1/2} + z_{1,3} \\ \hbar_{1,3} = -l_{1,3}\lambda_1 \lfloor z_{1,3} - \hbar_{1,2} \rceil^{0}; \end{cases} \tag{5.54}$$

$$d_2 : \begin{cases} \dot{z}_{2,0} = u + x_1 x_3^{1/3} + \hbar_{2,0} \\ \dot{z}_{2,1} = \hbar_{2,1} \\ \hbar_{2,0} = -l_{2,0}\lambda_2^{1/2} \lfloor z_{2,0} - x_3 \rceil^{1/2} + z_{2,1} \\ \hbar_{2,1} = -l_{2,1}\lambda_2 \lfloor z_{2,1} - \hbar_{2,0} \rceil^{0}; \end{cases} \tag{5.55}$$

$$\begin{cases} v = -K \left[\lfloor \xi_1 \rceil^{1+3\tau}, \lfloor \xi_2 \rceil^{\frac{1+3\tau}{1+\tau}}, \lfloor \xi_3 \rceil^{\frac{1+3\tau}{1+2\tau}} \right]^{\top}, \\ u = L^3 v + u^*. \end{cases} \tag{5.56}$$

In reference to existing backstepping-based approaches, a clear improvement is the design simplicity under the proposed nonrecursive design framework. For instance, following the backstepping based design in [69], a more complex control scheme with nested virtual controllers can be carried out via recursive design steps, as depicted sketchily as

$$d_1:\begin{cases}\dot{\hat{d}}_1=\lambda_1(x_1-p_1)\\ \dot{p}_1=x_2+\sin(x_1)+\hat{d}_1;\end{cases}$$

$$d_2:\begin{cases}\dot{\hat{d}}_2=\lambda_2(x_3-p_2)\\ \dot{p}_2=u+x_1x_3^{1/3}+\hat{d}_2;\end{cases}$$

$$\begin{cases}x_1^*=y_r,\ \zeta_1=x_1-x_1^*,\\ x_2^*=-k_1\zeta_1-f_1-\hat{d}_1,\ \zeta_2=x_2-x_2^*,\\ x_3^*=-\zeta_1-\left(k_2+\left(\dfrac{\partial x_2^*}{\partial x_1}+\dfrac{\partial x_2^*}{\partial \hat{d}_1}\lambda_1\right)^2\right)\zeta_2-f_2\\ \qquad+\left(\dfrac{\partial x_2^*}{\partial x_1}(x_1+f_1+\hat{d}_1)+\dfrac{\partial x_2^*}{\partial x_1^*}\dot{x}_1^*+\dfrac{\partial x_2^*}{\partial \dot{x}_1^*}\ddot{x}_1^*\right),\\ \zeta_3=x_3-x_3^*,\\ u=-\zeta_2-\left(k_3+\left(\dfrac{\partial x_3^*}{\partial x_1}+\dfrac{\partial x_3^*}{\partial \hat{d}_1}\lambda_1\right)^2\right)\zeta_3-f_3-\hat{d}_2\\ \qquad+\dfrac{\partial x_3^*}{\partial x_1}(x_1+f_1+\hat{d}_1)+\dfrac{\partial x_3^*}{\partial x_2}(x_3+f_2)+\dfrac{\partial x_3^*}{\partial x_1^*}\dot{x}_1^*+\dfrac{\partial x_3^*}{\partial \dot{x}_1^*}\ddot{x}_1^*+\dfrac{\partial x_3^*}{\partial \ddot{x}_1^*}\dddot{x}_1^*,\\ k_1>1/2,k_2>0,k_3>1/2,\lambda_1>3/2,\lambda_2>1.\end{cases}$$

In the simulation, by following the proposed design procedure, the gain vector K can be selected following the classical pole placement manner. The gain parameter is set as $K=[27,27,9]$ to place the pole of the nominal system into $(-3,-3,-3)$. The scaling gain is selected as $L=1.5$ according to the guideline (5.43). The designed homogeneous degree is set as $\tau=-0.1$. The observer gains are chosen as $\lambda_1=1$, $l_{1,0}=5$, $l_{1,1}=4$, $l_{1,2}=2$, $l_{1,3}=1$, and $\lambda_2=2$, $l_{2,0}=2$, $l_{2,1}=1$. The initial values are given as $[x(0);z(0)]=[3,2,-4;3,-1,0,-1,-4,0]$. For the backstepping based controller, the parameters are set as $k_1=0.51$, $k_2=0.02$, $k_3=0.51$, $\lambda_1=2.3$, and $\lambda_2=5$ while the initial value for the disturbance observer is $[p_1(0),p_2(0)]=[3,-4]$.

As shown in Figure 5.5, the finite-time tracking objective is realized under the designed tracking scheme while the backstepping controller could only render a practical tracking result. Figure 5.7 shows that under the proposed method, the states x_2 and x_3 also approach to their desired reference signal x_2^*, x_3^* within a finite-time. The time histories of two control input signals are shown in Figure 5.6. In Figure 5.8, the performance of the finite-time disturbance observer is demonstrated.

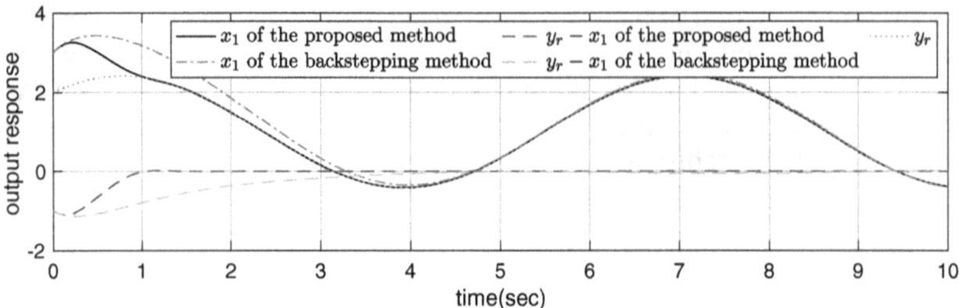

Figure 5.5 Output tracking performance.

Example 5.4.3 *Consider the following 3-D nonlinear system:*

$$\begin{cases} \dot{x}_1 = x_2 + x_1^2 + d(t) \\ \dot{x}_2 = x_3 + \cos(x_2) \\ \dot{x}_3 = u + x_1^{4/3} x_3 \\ y = x_1, \end{cases} \tag{5.57}$$

where $d(t)$ is a mismatched disturbance term and the control objective is to realize finite-time exact tracking of a given reference signal $y_r = 2\cos(2t)$.

Assume the disturbance as $d(t) = 2\sin(2t+1)$ which can be verified to be guided by an exosystem $d^{(4)}(t) = -4d^{(2)}(t)$. Without any pre-verifications, one can utilize the general semi-global exact tracking control method proposed in this section with a series of nonrecursive calculations and the coordinates transformation, the obtained exact tracking control law is depicted explicitly in the

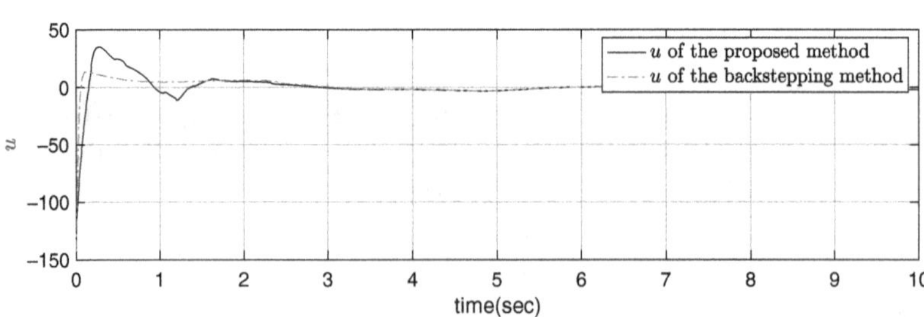

Figure 5.6 Time histories of the control inputs.

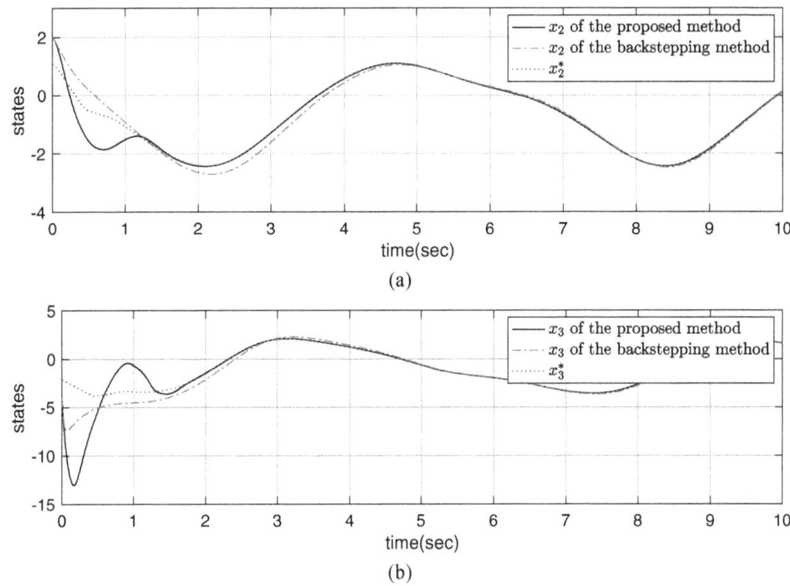

Figure 5.7 State response curves. (a) x_2; (b) x_3.

following form:

$$\begin{cases} \dot{\hat{z}}_1 = x_2 + x_1^2 + \hat{z}_2 + \ell_1 \sigma \lfloor z_1 - \hat{z}_1 \rceil^{1+\tau} \\ \dot{\hat{z}}_2 = \hat{z}_3 + \ell_2 \sigma^2 \lfloor z_1 - \hat{z}_1 \rceil^{1+2\tau} \\ \dot{\hat{z}}_3 = \hat{z}_4 + \ell_3 \sigma^3 \lfloor z_1 - \hat{z}_1 \rceil^{1+3\tau} \\ \dot{\hat{z}}_4 = -4\langle z_3 \rangle_M + \ell_4 \sigma^4 \lfloor z_1 - \hat{z}_1 \rceil^{1+4\tau}; \end{cases} \tag{5.58}$$

$$u = -L^3 K \left[\lfloor \zeta_1 \rceil^{1+3\omega}, \lfloor \zeta_2 \rceil^{\frac{1+3\omega}{1+\omega}}, \lfloor \zeta_3 \rceil^{\frac{1+3\omega}{1+2\omega}} \right]^\top + u^*. \tag{5.59}$$

To better demonstrate the efficacy of the disturbance attenuation ability enabled by the proposed composite method, a comparison result with a practical nonsmooth controller and a composite backstepping controller is given. The practical controller is achieved simply by replacing $\hat{z}_i$ in (5.59) to zero, i.e., the practical controller is designed without the disturbance estimation and feedforward compensation processes. Based on a recent result in [69], a composite backstepping controller can be derived in the following form:

$$\begin{aligned}
&\hat{d} = \lambda(x_1 - p_1),\ \dot{p}_1 = x_2 + x_1^2 + \hat{d},\\
&x_2^* = -(k_1 + \frac{1}{4\varepsilon})(x_1 - y_r) - x_1^2 - \hat{d} + y_r^{(1)},\\
&x_3^* = -(x_1 - y_r) - (k_2 + \pi_2)(x_2 - x_2^*) - \cos x_2 + \frac{x_2^*}{\partial x_1}(x_2 + x_1^2 + \hat{d})\\
&\quad + \frac{\partial x_2^*}{\partial y_r} y_r^{(1)} + \frac{\partial x_2^*}{\partial y_r^{(1)}} y_r^{(2)},
\end{aligned} \tag{5.60}$$

$$\begin{aligned}
u = &-(x_2 - x_2^*) - (k_3 + \pi_3)(x_3 - x_3^*) - \Big(x_1^{4/3} x_3 - (\frac{\partial x_3^*}{\partial x_1}(x_2 + x_1^2 + \hat{d})\\
&+ \frac{\partial x_3^*}{\partial x_2}\dot{x}_2 + \frac{\partial x_3^*}{\partial y_r} y_r^{(1)} + \frac{\partial x_3^*}{\partial y_r^{(1)}} y_r^{(2)} + \frac{\partial x_3^*}{\partial y_r^{(2)}} y_r^{(3)})\Big),
\end{aligned} \tag{5.61}$$

$$\pi_2 = \frac{1}{4\varepsilon}\left(\frac{\partial x_2^*}{\partial x_1} + \frac{\partial x_2^*}{\partial \hat{d}}\lambda\right)^2,\ \pi_3 = \frac{1}{4\varepsilon}\left(\frac{\partial x_3^*}{\partial x_1} + \frac{\partial x_3^*}{\partial \hat{d}}\lambda\right)^2, \tag{5.62}$$

where $\lambda, k_1, k_2, k_3, \varepsilon$ are positive design parameters.

In the simulation, we choose the saturation threshold $M = 16$, and two negative degrees $\tau = -0.1$, $\omega = -0.15$ respectively, the finite-time disturbance observer gain vector is set to be $H = [15, 75, 125, 90]^\top$ and the design parameter $\sigma = 2$. The control parameters for the proposed composite controller and practical nonsmooth controller are selected the same as $K = [27, 27, 9]$ to place the poles to $[-3, -3, -3]$, and the scaling gain is set to be $L = 5$ according to the guideline (5.43). The control gain parameters for the composite backstepping controller (5.62) are chosen as $\varepsilon = 8$, $\lambda = 10$, $k_1 = 2.9$, $k_2 = k_3 = 0.1$. To conduct the simulation, the initial values are given as $[x(0)|\hat{z}(0)] = [1, -1, 0|0, 0, 0, 0]$.

It is shown in Figure 5.9 that the finite-time exact tracking objective is realized under the proposed controller (5.59) while the practical controller can only render a practical tracking result. Even the composite backstepping controller can achieve a satisfactory practical tracking control result, however the steady state error cannot be fully eliminated. Moreover, it is easy to discover that the control expression (5.62) is much more complex than the proposed controller (5.59). In Figure 5.10, the performance of the designed disturbance observer (5.58) is demonstrated.

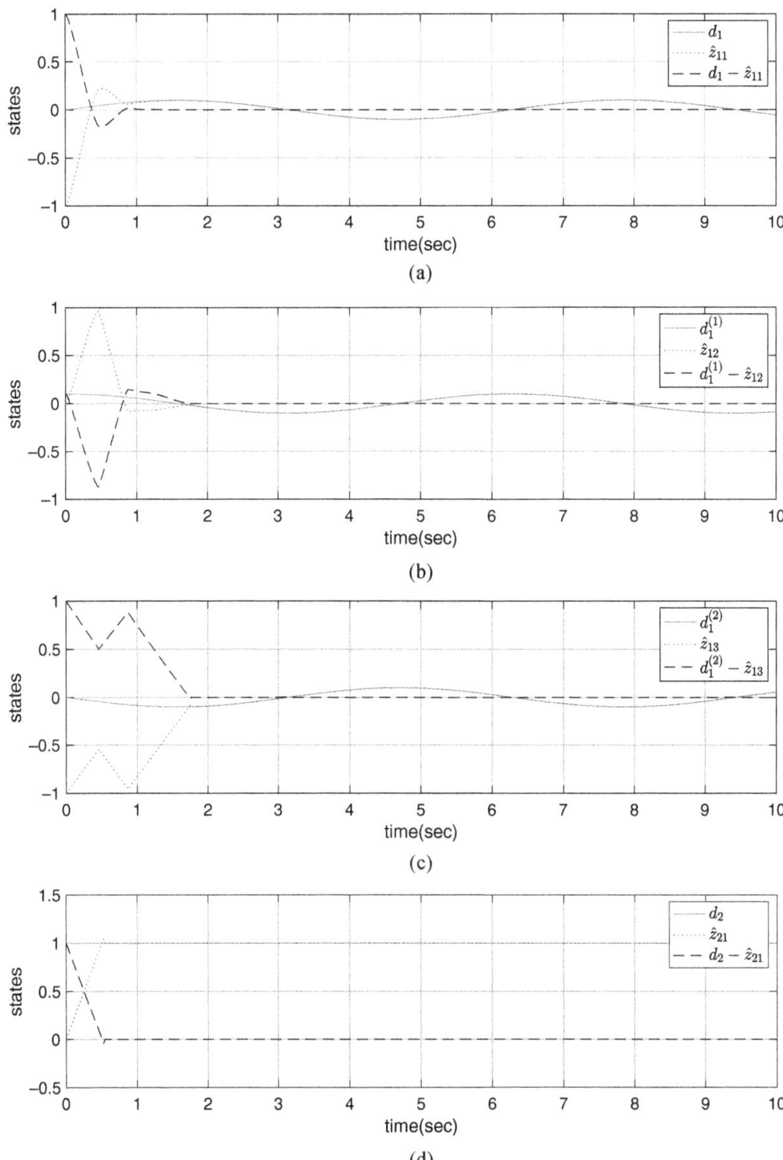

Figure 5.8 Disturbance observation performance under the HOSM disturbance observer. (a) d_1; (b) $d_1^{(1)}$; (c) $d_1^{(2)}$; (d) d_2.

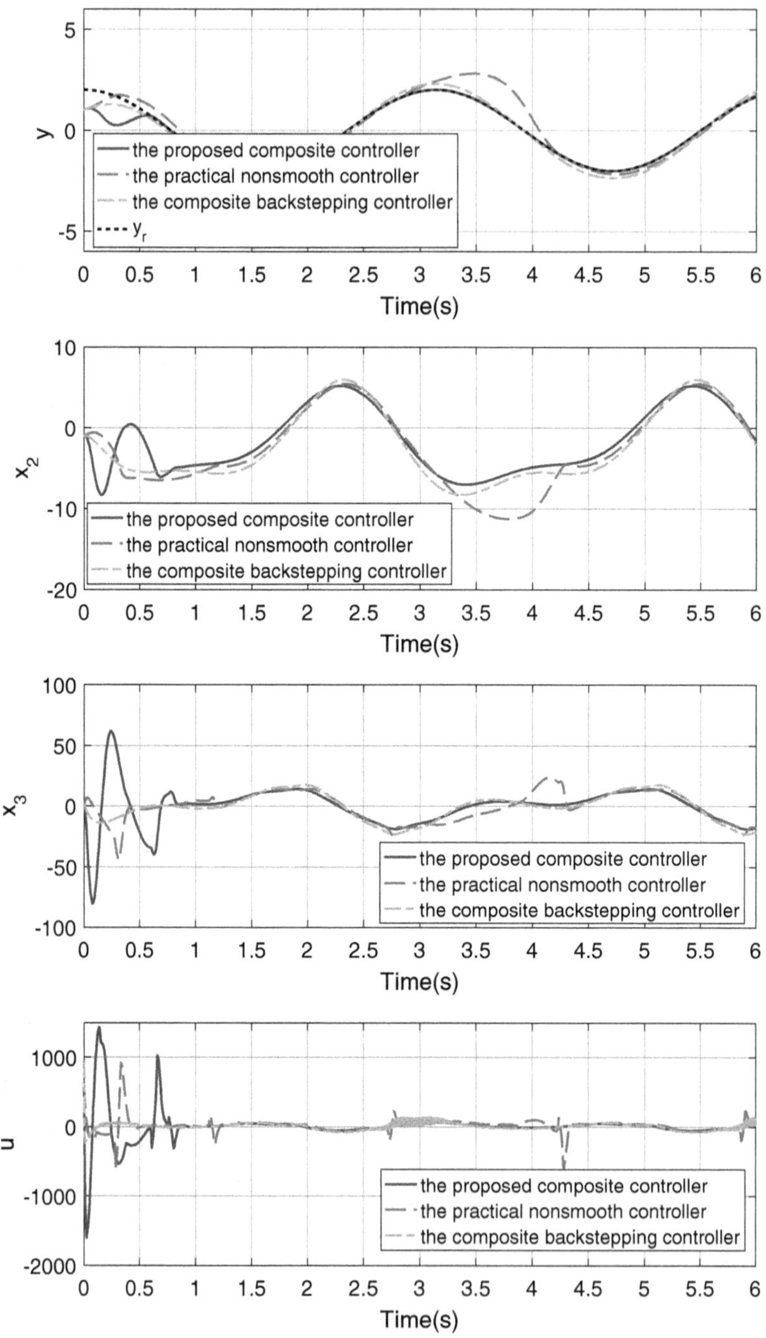

Figure 5.9 Control performance comparisons.

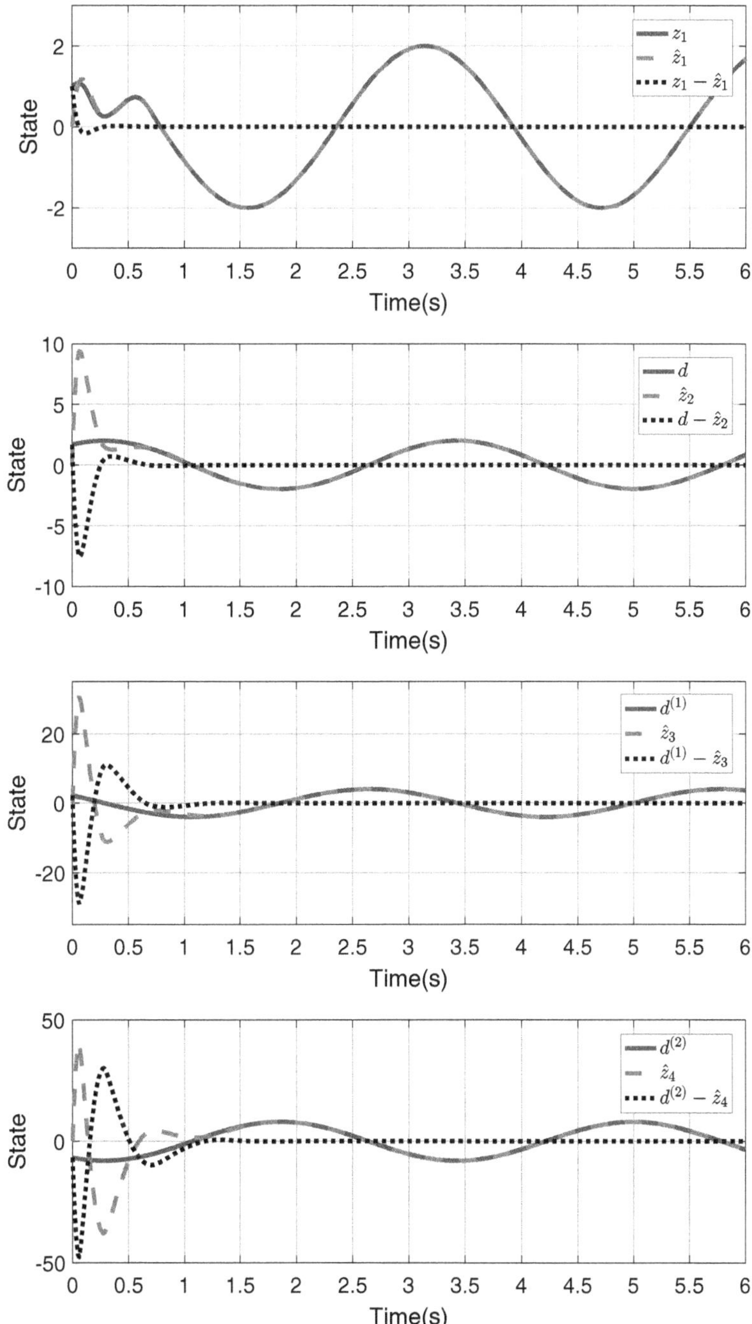

Figure 5.10 Disturbance estimation performance.

Chapter 6

Nonrecursive Adaptive Output Regulation for Nonlinear Systems with Mismatched Disturbances

In Chapter 5, for nonlinear systems with mismatched disturbances, a nonrecursive robust output regulation law design method has been proposed. Notably, it embodies a natural robust handling strategy in which the controllers are normally determined by the worst scenario aiming to maintain the system stability under various disturbance variation conditions. Consequently, the overconservative gain-configuration mechanism may lead to robustness redundancy of the closed-loop system, yielding a fact that the transient-time control performance may degrade with the wide range of operating conditions switching.

In order to further clarify the motivation of this chapter, take a DC-DC converter control issue feeding constant power loads (CPLs) as an example. On the one hand, in the case that the CPL varies from a small-signal perturbation case to a large-signal case, setting a sufficiently large bandwidth could possibly meet the voltage regulation requirement and ensure system stability. On the other hand, if the working condition varies from a large-signal perturbation case into a small-signal case with a fixed bandwidth determined in a robust control manner, possible deterioration of the transient-time control performance would render the DC output voltage regulation result deviating from the demanded performance indexes. Whereas, for power electronics systems, large overshoot and fluctuation

DOI: 10.1201/9781003399230-6

are harmful to practical applications. Accordingly, a self-tuning bandwidth factor of the controller would largely benefit the practical implementation of such industrial regulators.

Motivated by the above discussions, this chapter presents an alternative adaptive tracking control design algorithm for a class of nonlinear systems with the presence of mismatched disturbances. As an alternative control design approach for practitioners, a simple nonsmooth tracking scheme with a self-tuning scaling gain is proposed under a nonrecursive synthesis framework. In reference to the existing related works, a main distinguishable feature is that the scaling gain of the proposed regulation law can be adaptively adjusted subject to different perturbation levels, therefore, the control system is able to achieve an improved transient-time control performance while an accurate tracking objective is guaranteed. In addition, the proposed nonrecursive design philosophy is able to obtain a simplest state-feedback expression of the controller through essentially detaching the stability analysis from the controller design procedure, i.e., without going through the recursive determination steps of classical virtual controllers and tuning functions.

6.1 Linear Nonrecursive Adaptive Output Regulation

In this section, we consider a linear nonrecursive adaptive output regulation problem for the disturbed nonlinear system of form (5.1) under Assumptions 3.1.1 and 5.2.2.

6.1.1 System pre-treatment

Following a similar coordinate transformation to (5.19), the following change of coordinates is presented:

$$\eta_i = \frac{x_i - x_i^*}{L^{\rho+i-1}},\ i \in \mathbb{N}_{1:n},\ \upsilon = \frac{u - u^*}{L^{\rho+n}}, \tag{6.1}$$

where $L \geq 1$ is a dynamic scaling gain, υ is an interaction controller, and $\rho > 0$ is a design parameter determined later on.

To proceed, define (A,B) as the nth-order controllable canonical matrix pair, and a positive definite, symmetrical matrix $P \in \mathbb{R}^{n\times n}$ satisfying

$$(A - BK)^\top P + P(A - BK) \leq -I_n,$$

where $K = (k_1, k_2, \cdots, k_n)$ is the coefficient of a Hurwitz polynomial $H(s) = s^n + k_n s^{n-1} + \cdots + k_1$, and I_n is the identity matrix.

In what follows, the symbol Θ is further denoted as $\Theta \triangleq \text{diag}\{0,1,\cdots,n-1\}$. Then, the restriction of design parameter ρ can be derived:

$$\rho > \max\left\{0, -\frac{\lambda_{\min}(\Theta P + P\Theta)}{2\lambda_{\min}(P)}\right\}. \tag{6.2}$$

According to (6.1), one can get the following equivalent system:

$$\dot{\eta} = L(A\eta + B\upsilon) - (\rho I_n + \Theta)\frac{\dot{L}}{L} + F(\cdot) + \varepsilon, \tag{6.3}$$

where $\eta = (\eta_1, \eta_2, \cdots, \eta_n)^\top$, $\varepsilon = (\varepsilon_1/L^\rho, \varepsilon_2/L^{\rho+1}, \cdots, \varepsilon_n/L^{\rho+n-1})^\top, \varepsilon_i = f_i(x_1^*, x_2^*, \cdots, x_i^*) + x_{i+1}^* - \dot{x}_i^* - \left(f_i(\bar{\varsigma}_i) - \dot{\varsigma}_i + \varsigma_{i+1}\right), i \in \mathbb{N}_{1\to n}$, and

$$F(\cdot) = \begin{pmatrix} (f_1(x_1) - f_1(x_1^*))/L^\rho \\ (f_2(x_1,x_2) - f_2(x_1^*,x_2^*))/L^{\rho+1} \\ \vdots \\ (f_n(x) - f_n(x^*))/L^{\rho+n-1} \end{pmatrix}.$$

6.1.2 Controller construction

At this stage, without going through any complex calculations, one can obtain the following implementable adaptive controller:

$$\begin{cases} \dot{L} = c_1 L \max\{0, \hat{\mu} - c_2 L + c_3\},\ L_0 = 1 \\ \dot{\hat{\mu}} = c_4 \|\eta\|^2, \end{cases} \tag{6.4}$$

where the design parameters are subject to the following guidelines:

$$\begin{cases} c_1 \geq 2n\lambda_{\max}(P)/\left(2\rho\lambda_{\min}(P) + \lambda_{\min}(\Theta P + P\Theta)\right), \\ c_2 \in \left(0, \dfrac{1}{2n\lambda_{\max}(P)}\right), \\ c_3 = \dfrac{\lambda_{\max}(P)}{2}, \\ c_4 > 0. \end{cases}$$

Remark 6.1 For parameter selection, the following principles or steps can be followed: 1) Obtain the control gain K through the pole configuration, and configure the pole of the closed-loop system to the desired pole; 2) Calculate the linear matrix inequality $(A-BK)^\top P + P(A-BK) \leq -I_n$, so as to obtain a positive definite symmetric matrix P; 3) According to the selection criteria of the parameter ρ, determine its range; 4) One can calculate the range of c_1, c_2, c_3, c_4 according to their corresponding selection criteria, and they can be simply selected by "trial and error" procedures.

6.1.3 Stability analysis

The main result in this section can be summarized by the following theorem.

Theorem 6.1

Under Assumptions 3.1.1 and 5.2.2, assuming the system initial values satisfying $[x(0),\hat{\mu}(0)]^\top \in \mho \triangleq [-\rho,\rho]^{n+1}$ *for any constant* $\rho > 0$*, consider the closed-loop system including the original system* (5.1)*, the finite-time disturbance observer* (5.15) *and the dual-layer dynamic controller* (6.4)*. The following statements hold:*

- *All the signals in the closed-loop system are uniformly bounded.*
- $\lim_{t\to\infty} y = y_r$.

Proof: To begin with, build a candidate Lyapunov function as

$$V(\eta,\tilde{\mu}) \triangleq \eta^\top P\eta + \tilde{\mu}^2/(2\varpi_1),$$

where $\tilde{\mu} = \mu - \hat{\mu}$, ϖ_1 is an auxiliary design parameter, μ is an unknown parameter, and $\hat{\mu}$ is the estimation of μ. Then, taking the time derivative of V along (6.3) yields

$$\begin{aligned}\dot{V}(\eta) =& \frac{\partial V(\eta)}{\partial \eta^\top} L(A-BK)\eta - \frac{\partial V(\eta)}{\partial \eta^\top}(\rho I_n + \Theta)\frac{\dot{L}}{L}\eta \\ &+ \frac{\partial V(\eta)}{\partial \eta^\top} F(\cdot) + \frac{\partial V(\eta)}{\partial \eta^\top}\varepsilon - \frac{\tilde{\mu}}{\varpi_1}\dot{\hat{\mu}}.\end{aligned} \tag{6.5}$$

Define $\eta^* = x - x^*$ and $\eta^* \in \Omega_M \triangleq \left\{\eta^* \in \mathbb{R}^n | V(\eta,\tilde{\mu}) \le M + NE^2\right\}$ where

$$M \triangleq \sup_{\eta^* \in \Omega_{\eta^*} \triangleq [-(\bar{\rho}+\rho),\bar{\rho}+\rho],\mu\in\Omega_{\tilde{\mu}}} \{V(\eta,\tilde{\mu})\},$$

with $\bar{\rho} > 0$ satisfying $\max_{i\in\mathbb{N}_{1:n}}\{\sup_{t\ge 0}\{x_i^*(t)\}\} \le \bar{\rho}$, and $N = \frac{4}{(1-\varpi_1)\lambda_{\max}^{-1}(P)}$, μ is bounded and satisfies $\mu \in [\underline{\mu},\bar{\mu}]$, and $\tilde{\mu}(0) \in \Omega_{\tilde{\mu}} \triangleq [\underline{\mu}-\rho,\bar{\mu}+\rho]$.

After defining the symbols above, in what follows, we will estimate the right term of (6.5). First, there must exist a suitable $\sigma_0 \in (0, \frac{M+NE^2}{2})$ such that

$$E^2 < \frac{L(1-\varpi_3)\lambda_{\max}^{-1}(P)}{2}\sigma_0. \tag{6.6}$$

where $\varpi_3 \in (0,1)$ is a constant.

Subsequently, the series of estimations of (6.5) are analogous to (5.24), (5.26) and (5.27), i.e.,

$$\begin{aligned}
&\frac{\partial V(\eta)}{\partial \eta^\top} L(A-BK)\eta \leq -L\|\eta\|^2, \\
&\frac{\partial V(\eta)}{\partial \eta^\top} F(L,\rho,x,x^*) \leq 2n\mu\lambda_{\max}(P)\|\eta\|^2, \eta^* \in \Omega_M, \\
&\frac{\partial V(\eta)}{\partial \eta^\top} \varepsilon \leq n\lambda_{\max}^2(P)\|\eta\|^2 + E^2.
\end{aligned} \tag{6.7}$$

Keeping the selection guideline (6.2) in mind, we arrive at

$$\begin{aligned}
\frac{\partial V(\xi)}{\partial \eta^\top}(\rho I_n + \Theta)\frac{\dot{L}}{L}\eta &\geq \left(2\rho\lambda_{\min}(P) + \lambda_{\min}(\Theta P + P\Theta)\right)\frac{\dot{L}}{L}\|\eta\|^2 \\
&\geq \varpi_2 \frac{\dot{L}}{L}\|\eta\|^2,
\end{aligned} \tag{6.8}$$

where $\varpi_2 \in (0, 2\rho\lambda_{\min}(P) + \lambda_{\min}(\Theta P + P\Theta))$.

Up to now, for $t < T_1$, and $\eta^* \in \Omega_M$, (6.5) can be rewritten as

$$\begin{aligned}
\dot{V}(\eta,\bar{\mu})|_{\eta^*\in\Omega_M} \leq& -L\|\eta\|^2 + 2n\mu\lambda_{\max}(P)\|\eta\|^2 + E^2 + n\lambda_{\max}^2(P)\|\eta\|^2 \\
& - \varpi_2\frac{\dot{L}}{L}\|\eta\|^2 - \frac{\tilde{\mu}}{\varpi_1}\dot{\hat{\mu}} \\
\leq& -L(1-\varpi_3)\|\eta\|^2 - (\dot{\hat{\mu}} - 2n\lambda_{\max}(P)\varpi_1\|\eta\|^2)\frac{\tilde{\mu}}{\varpi_1} + E^2 \\
& - \frac{\varpi_2}{L}(\dot{L} - \frac{L}{\varpi_2}2n\lambda_{\max}(P)(\hat{\mu} - \frac{\varpi_3}{2n\lambda_{\max}(P)}L + \frac{\lambda_{\max}(P)}{2}))\|\eta\|^2 \\
:=& -L(1-\varpi_3)\|\eta\|^2 - \frac{\tilde{\mu}}{\varpi_1}(\dot{\hat{\mu}} - c_4\|\eta\|^2) + E^2 \\
& - \frac{\varpi_2}{L}(\dot{L} - c_1 L(\hat{\mu} - c_2 L + c_3))\|\eta\|^2.
\end{aligned} \tag{6.9}$$

where $c_1 = \frac{2n\lambda_{\max}(P)}{\varpi_2} \geq 2n\lambda_{\max}(P)/(2\rho\lambda_{\min}(P) + \lambda_{\min}(\Theta P + P\Theta))$, $c_2 = \frac{\varpi_3}{2n\lambda_{\max}(P)} \in (0, \frac{1}{2n\lambda_{\max}(P)})$, $c_3 = \lambda_{\max}(P)/2$, $c_4 > 0$.

Combined with (6.7), one can get that $\forall(\eta^{*\top}, \tilde{\mu})^\top \in \Omega_M, \exists\varpi_3 \in (0,1)$, $t \in [0, T_1]$, such that

$$\dot{V}(\eta,\tilde{\mu}) \leq -L(1-\varpi_3)\|\eta\|^2 + E^2. \tag{6.10}$$

a) Uniform boundness of the signals in the closed-loop system:

For $L(0) = 1$, we have $(\eta(0)^\top, \tilde{\mu}(0))^\top \in \Omega_M$, and

$$V(0) \leq M + NE^2, \dot{V}(0) \leq -L(1-\varpi_3)\|\eta\|^2 + E^2.$$

First of all, the following relationship holds:

$$V(\eta,\tilde{\mu} \leq \lambda_{\max}(P)\|\eta\|^2+\gamma(t)$$

where $\gamma(t)=\tilde{\mu}^2/(2\varpi_1)$.

1) Denote a set as $\Omega_1=\{\eta\in\mathbb{R}^n|V(\eta,\tilde{\mu})\leq\sigma_0\}$. Then, there must exist $\Omega_1\subset\Omega_M$. Besides, from (6.6), one can get that for any state trajectory $(\eta,\tilde{\mu})$ satisfying $\tilde{\mu}(0)\in\Omega_\mu,\eta(0)\in\Omega_{\eta^*}\cap(\Omega_M\backslash\Omega_1)$ must stay in Ω_M, and will be captured by Ω. Otherwise, there exists a finite time instant $t_2>0$, such that

$$\begin{aligned}&i)\ V(0)\leq V(t_2)=M+NE^2\\&ii)\ \dot{V}(t_2)\geq 0.\end{aligned}$$

Then for $t\in[0,t_2]$, the inequality (6.10) holds, and further

$$\begin{aligned}\dot{V}(\eta,\tilde{\mu})|_{\eta^*\in\Omega_M}&\leq -L(1-\varpi_3)\|\eta\|^2+E^2\\&\leq -L(1-\varpi_3)\lambda_{\max}^{-1}(P)(V_2-\gamma-\sigma_0/2)\\&<0,\end{aligned}$$

which clearly contradicts the hypothesis, i.e., $0\leq\dot{V}(t_2)<0$. Then it is clear that:

$$\forall\tilde{\mu}(0)\in\Omega_\mu,\eta(0)\in\Omega_{\eta^*}\cap(\Omega_M\backslash\Omega_1)\Rightarrow(\eta,\tilde{\mu})\in\Omega_M,\forall t\geq 0,$$

that is, when Ω_M is a fixed set, $\eta(t)$ will be captured by Ω_1.

2) When $t\geq T_1$, (6.10) can be rewritten as

$$\dot{V}\leq -L(1-\varpi_3)\|\eta\|^2\leq 0,$$

which also concludes at the boundness of $(\eta,\tilde{\mu})$.

Now we are able to conclude that η and $\tilde{\mu}$ are uniformly bounded. Subsequently, the boundness of the scaling gain L will be claimed.

Notably, $\dot{L}\geq 0$ is constantly established. For ease of description, suppose that L is unbounded, that is, there must exist a finite time instant t_3, such that $c_2L(t_3)>\bar{\bar{\mu}}+c_3$, where $\bar{\bar{\mu}}$ is the maximum value of $\hat{\mu}$, hence $L(t_3)>(\bar{\bar{\mu}}+c_3)/c_2$.

1) In the case that $\bar{\bar{\mu}}+c_3\leq c_2$, $\bar{\bar{\mu}}+c_3-c_2L(t_3)\leq\bar{\bar{\mu}}+c_3-c_2\leq 0$ holds. From (6.4), $\dot{L}=0$, i.e., $\forall t>0,L=L(0)\equiv 1$.

2) In the case that $\bar{\bar{\mu}}+c_3>c_2$, $\bar{\bar{\mu}}+c_3-c_2L(0)>0$, $\bar{\bar{\mu}}+c_3-c_2L(t_3)<0$ exists. Then there exists a time instant $t_0\in(0,t_3)$, such that $\bar{\bar{\mu}}+c_3-c_2L(t_0)=0$, i.e.,

$$\forall t\in(t_0,t_3),\ \bar{\bar{\mu}}+c_3-c_2L(t)<0,\ \dot{L}=0,\ L\equiv L(t_0)=\frac{\bar{\bar{\mu}}+c_3}{c_2},$$

which is clearly inconsistent with the assumption. In this case, one gets

$$\forall t\in[t_0,t_3),\ L\equiv L(t_0)=(\bar{\bar{\mu}}+c_3)/c_2.$$

Summarily, L is uniformly bounded, i.e., L has an upper value $\bar{L}$ satisfying $\bar{L} = \max\{1, (\bar{\hat{\mu}} + c_3)/c_2\}$.

*b) **States converge asymptotically to the reference value:***

Defining $Q \triangleq \int_0^t \|\eta\|^2 dt$, then

$$\begin{aligned}\lim_{t\to\infty} Q &\le -\int_{T_1}^{\infty} \frac{V(\eta,\tilde{\mu})}{1-\varpi_3} dt - \int_0^{T_1} \frac{\dot{V}(\eta,\tilde{\mu}) - E^2}{1-\varpi_3} dt \\ &\le \frac{V(0) + E^2 T_1}{1-\varpi_3} < \infty. \end{aligned} \tag{6.11}$$

Hence, combining with (6.3), (6.7) and $\eta \in \Omega_M$, the conclusion that $\dot{\eta}$, $\ddot{Q} = 2(\eta_1\dot{\eta}_1 + \eta_2\dot{\eta}_2 + \cdots + \eta_n\dot{\eta}_n)$ are uniformly bounded. Then we have $\lim_{t\to\infty} \dot{Q} = 0 \Leftrightarrow \lim_{t\to\infty} \eta = 0$, i.e., $\lim_{t\to\infty} y = y_r$ by Lemma 1.9.

This completes the proof of Theorem 6.1. ■

6.1.4 Numerical simulations

In this subsection, we consider the system in Example 5.3.1. Based on the proposed adaptive controller design procedure, the nonrecursive adaptive controller can be designed as follows:

$$\begin{aligned}&\dot{L} = c_1 L \max\{0, \hat{\mu} - c_2 L + c_3\},\ L_0 = 1, \\ &\dot{\hat{\mu}} = c_4(\eta_1^2 + \eta_2^2 + \eta_3^2). \end{aligned} \tag{6.12}$$

The parameters for the adaptive law are set as $[c_1, c_2, c_3, c_4, \rho; K^\top] = [10, 1, 0.02, 10, 3; 15, 18, 22]$ while all the other parameters are the same with Example 5.3.1.

According to Figure 6.1, one can observe that the controller proposed in this subsection can ensure that the system can be tracked to its expected reference signal x_1^*, x_2^*, x_3^*. For a time-varying disturbance, the dynamic scaling gain L starting from 1 converges to a value of 2.2. Figure 6.1(e) shows that the nonrecursive adaptive control can achieve the control goal of the system under the condition of consuming less control energy.

6.2 Nonrecursive $\mathbb{C}^0$ Adaptive Output Regulation

6.2.1 Problem formulation

For system (5.1), this section aims to present a nonsmooth composite control strategy with a self-tuning scaling gain based on the homogeneous domination theory, which is employed to realize three promising goals, i.e., mismatched disturbance attenuation, practical reference trajectory tracking, and achieving

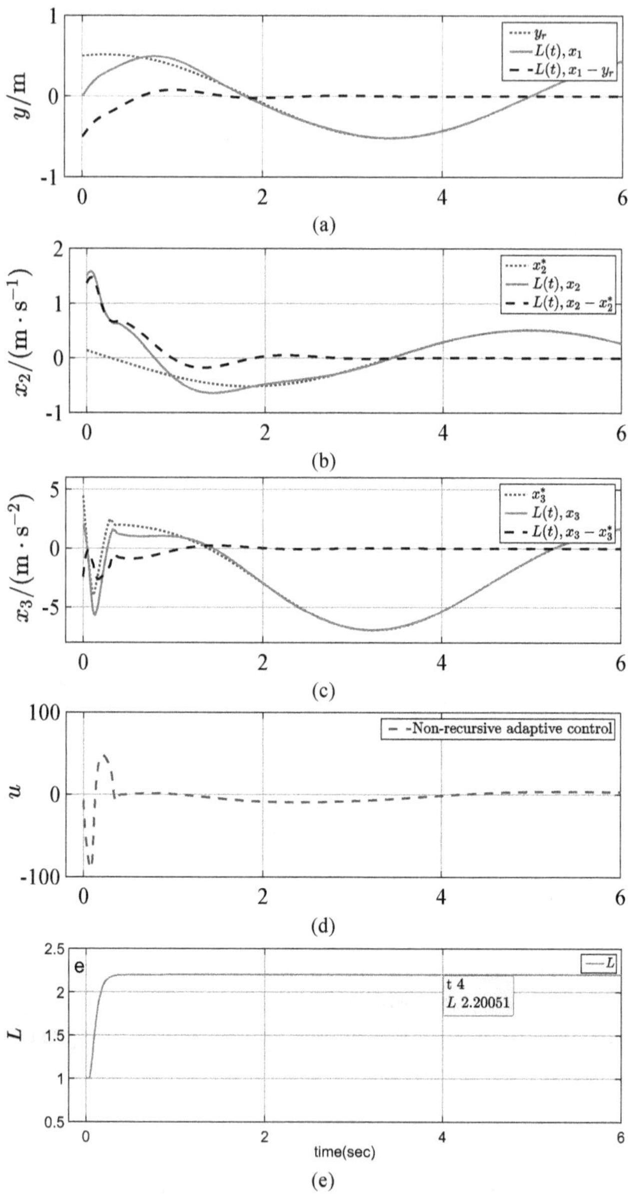

Figure 6.1 The control performance of a one-link manipulator with time-varying disturbances. (a) Output response; (b) State x_2; (c) State x_3; (d) The control effort; (e) Dynamic scaling gain.

satisfactory transient-time performance when the working conditions fluctuate widely.

To these aims, the composite dynamic controller proposed in this section can be depicted as the following compact form:

$$\dot{\hat{z}} = W(x,\hat{z}) \in \mathbb{C}^0,\ u = u(L,x,\hat{z},y_r) \in \mathbb{C}^0$$
$$\dot{L} = \phi(L,x,\hat{z},y_r),\ L(0) = L_0 = 1. \tag{6.13}$$

By this means, for any $x(0) \in \mho_x \triangleq [-\rho,\rho]^n$ with $\rho \in \mathbb{R}_+$ being a given constant and $\delta \in \mathbb{R}_+$ being an arbitrary small tolerance, such that under Assumption 3.1.1, the resulting closed-loop system (5.1)–(6.13) can be rendered semi-globally practically stable.

In what follows, we will present an explicit nonrecursive controller construction procedure, which is composed of disturbance observation, the reconstruction of output regulation function, and the design of a one-step control law.

6.2.2 Disturbance reconstruction

In order to achieve a fast performance recovery ability with the presence of unknown disturbances, a practical high-gain disturbance observer will be constructed. By setting $x_i = z_{i,1}$, $d_i^{(j)} = z_{i,j+2}$, $j \in \mathbb{N}_{0:n-i+1}$, $e_{i,1} = z_{i,1} - \hat{z}_{i,1}$, similar to (5.12), the observer can be designed in the following form :

$$\begin{cases} \dot{\hat{z}}_{i,1} = x_{i+1} + f_i(\bar{x}_i) + \hat{z}_{i,2} + l_{i,1}\beta_i \lfloor e_{i,1} \rceil^{1+\tau} \\ \dot{\hat{z}}_{i,j} = \hat{z}_{i,j+1} + l_{i,j}\beta_i^j \lfloor e_{i,1} \rceil^{1+j\tau},\ j \in \mathbb{N}_{2:n-i+1} \\ \dot{\hat{z}}_{i,n-i+2} = l_{i,n-i+2}\beta_i^{n-i+2} \lfloor e_{i,1} \rceil^{1+(n-i+2)\tau}, \end{cases} \tag{6.14}$$

where $\hat{z}_i = (\hat{z}_{i,1}, \hat{z}_{i,2}, \cdots, \hat{z}_{i,n-i+2})^\top$, and $\hat{z}_{i,j}, j \in \mathbb{N}_{1:n-i+2}$ denotes the estimation of corresponding $z_{i,j}$, β_i is a design parameter, $\tau \in (-\frac{1}{n+1}, 0)$ is the homogeneous degree, $L_i = (l_{i,1}, l_{i,2}, \cdots, l_{i,n-i+2})^\top$ is the observer gain vector while $l_{i,j}$ is the corresponding coefficient of a Hurwitz polynomial $R(s) = s^{n-i+2} + l_{i,1}s^{n-i+1} + \cdots + l_{i,n-i+1}s + l_{i,n-i+2}$.

6.2.3 Coordinates transformation

By following the similar handling process to (5.17)–(5.18), the following series of implementable steady-state reference functions can also be achieved with the disturbance observer (6.14):

$$\begin{cases} x_1^* = \chi_1 = y_r \\ x_i^* = \chi_i(\hat{z}, \bar{y}_r),\ i \in \mathbb{N}_{2:n+1}, \end{cases} \tag{6.15}$$

where $x_{n+1}^* = u^*$.

In what follows, let $L(t) \in \mathbb{C}^1$ being a time-varying scaling gain function satisfying $L_0 = 1, \dot{L}(t) \geq 0$. Then, employ $L(t)$ into system (5.1) with the subsequent change of coordinates:

$$\eta_i = \frac{x_i - x_i^*}{L^{\rho+i-1}},\ i \in \mathbb{N}_{1:n},\ \upsilon = \frac{u - u^*}{L^{\rho+n}}, \tag{6.16}$$

where $\rho \in \mathbb{R}_+$ is a design parameter and will be determined later on.

It is clear that $1 - \frac{\kappa}{2} > 0$ holds with $\kappa \in (-\frac{1}{n}, 0)$ being the homogeneous degree. Then, denote two auxiliary matrices as:

$$\Lambda_1 \triangleq \text{diag}\left\{\frac{1-\kappa/2}{r_1}, \cdots, \frac{1-\kappa/2}{r_n}\right\},$$
$$\Lambda_2 \triangleq \text{diag}\left\{0, \frac{1-\kappa/2}{r_2}, \cdots, \frac{(n-1)(1-\kappa/2)}{r_n}\right\}.$$

According to Lemma 1.8, there exist a matrix $P \in \mathbb{R}^{n\times n}$ satisfying $P = P^\top > 0$, and a gain vector $K = [k_1, k_2, \cdots, k_n]$ being the coefficient of a Hurwitz polynomial $H(s) = s^n + k_n s^{n-1} + \cdots + k_2 s + k_1$, such that

$$(A - BK)^\top P + P(A - BK) \leq -I_n,\ \Lambda_1 P + P\Lambda_1 > 0,$$

where A, B are the n-th order controllable canonical matrix pair. Then one can select the design parameter ρ as:

$$\rho > \max\left\{0, -\frac{\lambda_{\min}(\Lambda_2 P + P\Lambda_2)}{\lambda_{\min}(\Lambda_1 P + P\Lambda_1)}\right\}. \tag{6.17}$$

With the transformation (6.16), the original system (5.1) is equivalent to the following stabilizable system:

$$\dot{\eta} = L(A\eta + B\upsilon) - (\rho I_n + \Theta)\dot{L}/L\eta + F(\cdot) + \varepsilon, \tag{6.18}$$

where

$$\begin{aligned}
&\eta = (\eta_1, \eta_2, \cdots, \eta_n)^\top,\ \Theta = \text{diag}\{0, 1, \cdots, n-1\},\\
&F(\cdot) = \left((f_1(x_1) - f_1(x_1^*))/L^\rho, \cdots, (f_n(x) - f_n(x^*))/L^{\rho+n-1}\right)^\top,\\
&\varepsilon = (\varepsilon_1, \varepsilon_2, \cdots, \varepsilon_n)^\top,\ \varepsilon_i = \left(f_i(\bar{x}_i^*) - f_i(\bar{\chi}_i) + x_{i+1}^* - \chi_{i+1} + \dot{\chi}_i - \dot{x}_i^*\right)/L^{\rho-i+1},\\
&\bar{x}_i^* = (x_1^*, x_2^*, \cdots, x_i^*)^\top,\ i \in \mathbb{N}_{1:n},\ x^* = \bar{x}_n^*.
\end{aligned}$$

6.2.4 Nonsmooth dynamic tracking control law design

Up to now, without any recursive determination of virtual controllers and tuning functions, a nonsmooth dynamic tracking controller is able to be given as:

$$\begin{cases} \dot{L}(t) = \lambda \|\eta\|_{\Delta,2}^{2} \left(1 + \text{sign}(|e_s| - \delta)\right) \\ \upsilon = -K \lfloor \eta \rceil_{\Delta}^{r_n+\kappa}, \ u = L^{\rho+n}\upsilon + u^*, \end{cases} \tag{6.19}$$

where $\lambda \in \mathbb{R}_+$ is a design parameter aiming to restrain the growth rate of $L(t)$. The block diagram of the proposed method is depicted in Figure 6.2.

6.2.5 Stability analysis

At this stage, we are ready to present the following theorem.

Theorem 6.2

Considering the closed-loop system (5.1)–(6.14)–(6.19) satisfying Assumptions 3.1.1 and 5.2.2 and $x(0) \in \mho_x$, the following statements hold:

- *All the signals in the closed-loop system are semi-globally uniformly bounded.*
- *There exists a finite time $T > 0$ such that $|e_s| \leq \delta$, $\forall t \geq T$.*

Proof: In what follows, we will present the main stability analysis of this section step by step.

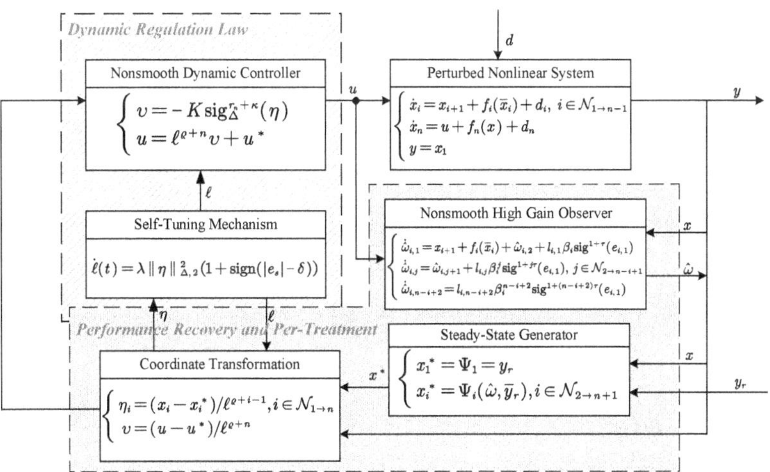

Figure 6.2 Block diagram of the proposed nonsmooth dynamic output regulation strategy

To begin with, the Lyapunov function candidate $V(\eta)$ is constructed as

$$V(\eta) = (\lfloor \eta \rceil_{\Delta}^{1-\frac{\kappa}{2}})^{\top} P \lfloor \eta \rceil_{\Delta}^{1-\frac{\kappa}{2}}.$$

It is clear that $V(\eta) \in \mathbb{C}^1$ holds. Moreover, one can verify $V(\eta) \in \mathbb{H}_{\Delta^r}^{2-\kappa}$, and $\|\eta\|_{\Delta,2}^2 \geq \alpha V^{\frac{2}{2-\kappa}}(\eta)$ with $\alpha \in \mathbb{R}_+$ according to Lemma 1.1. Additionally, for any well-defined $x(t)$, there must exist a certain positive constant $\mathcal{D}$ satisfying $\sup_{i \in \mathbb{N}_{1:n}} \{|\varepsilon_i|\} \leq \mathcal{D}$.

Taking the time derivative of $V(\eta)$ along (6.18), we arrive at

$$\begin{aligned} \dot{V}(\eta) = & \frac{\partial V(\eta)}{\partial \eta^{\top}} L(A\eta + Bv) + \frac{\partial V(\eta)}{\partial \eta^{\top}} F(\cdot) + \frac{\partial V(\eta)}{\partial \eta^{\top}} \varepsilon \\ & - \frac{\partial V(\eta)}{\partial \eta^{\top}} (\rho I_n + \Theta) \frac{\dot{L}}{L} \eta. \end{aligned} \tag{6.20}$$

With the selection guideline of ρ, the relation $(\Lambda_1 P + P\Lambda_1)\rho + \Lambda_2 P + P\Lambda_2 > 0$ exists. Besides, combining with the fact $\frac{\dot{L}}{L} \geq 0$, the following relationship holds:

$$\begin{aligned} \frac{\partial V(\eta)}{\partial \eta^{\top}} \frac{\dot{L}}{L} (\rho I_n + \Theta)\eta = & \frac{\dot{L}}{L} (\lfloor \eta \rceil_{\Delta}^{1-\frac{\kappa}{2}})^{\top} ((\Lambda_1 P + P\Lambda_1)\rho + \Lambda_2 P + P\Lambda_2) \lfloor \eta \rceil_{\Delta}^{1-\frac{\kappa}{2}} \\ \geq & a_1 \frac{\dot{L}}{L} \|\eta\|_{\Delta,2}^{2-\kappa} \geq 0, \end{aligned} \tag{6.21}$$

where a_1 is a positive constant.

Moreover, one can obtain the following statements whose derivations are similar to the corresponding part in Section 5.4 and are omitted here:

$$\begin{aligned} & \frac{\partial V(\eta)}{\partial \eta^{\top}} L(A\eta + Bv) \leq -La_2 \|\eta\|_{\Delta^r}^2 \\ & \frac{\partial V(\eta)}{\partial \eta^{\top}} F(\cdot)|_{\eta \in \Gamma_\eta} \leq a_3 \|\eta\|_{\Delta^r}^2 \\ & \sum_{i=1}^{n} \frac{\partial V}{\partial \eta_i} \frac{\varepsilon_i}{L^{\rho+i-1}} \leq a_4 \|\eta\|_{\Delta^r}^2 + \mathcal{D}^*, \end{aligned} \tag{6.22}$$

where Γ_η can be any given compact set, a_3 is a positive constant which is dependent on Γ_η, a_2, a_4 are two positive constants and $\mathcal{D}^*$ is a positive constant related to $\mathcal{D}$ but independent of L.

For the purpose of facilitating the organization, the detailed analysis is listed in the following steps.

A: The Boundness of L

In this regard, we shall figure out the boundness of $L(t)$. As depicted in dynamic updating mechanism (6.19), $L(t)$ is a monotone non-decreasing gain function, and specifically, the discussion about the growth rate of L should be split into the following two parts.

Part I: In the case when $|e_s| \geq \delta$, assume that $L(t)$ will escape at a finite time t_f, i.e., $\lim_{t:t_f} L(t) = \infty$. For arbitrary large compact set Γ_η and $\eta \in \Gamma_\eta$, there should exist a time instant $t_* \in [0, t_f)$, such that $L(t_*) = \frac{a_3+a_4+1}{a_2}$. Then for $t \in [t_*, t_f)$, substituting (6.21) and (6.22) into (6.20), it yields

$$\begin{aligned}\dot{V}(\eta)|_{\eta\in\Gamma_\eta} &\leq -(a_2L - a_3 - a_4)\|\eta\|_{\Delta,2}^2 - a_1\frac{\dot{L}}{L}\|\eta\|_{\Delta,2}^{2-\kappa} + \mathcal{D}^* \\ &\leq -(a_2L - a_3 - a_4)\|\eta\|_{\Delta,2}^2 + \mathcal{D}^* \\ &\leq -\alpha V^{\frac{2}{2-\kappa}}(\eta) + \mathcal{D}^*,\ \forall t \in [t_*, t_f).\end{aligned} \tag{6.23}$$

Then, taking the assumption listed above into consideration, the following expression gives:

$$\begin{aligned}\infty = &L(t_f) - L(t_*) = \int_{t_*}^{t_f} \dot{L}dt = 2\lambda \int_{t_*}^{t_f} \|\eta\|_{\Delta,2}^2 dt \\ \leq &2\lambda \int_{t_*}^{t_f} (-\dot{V} + \mathcal{D}^*)dt \\ &= 2\lambda\mathcal{D}^*(t_f - t_*) + 2\lambda\left(V(\eta(t_*)) - V(\eta(t_f))\right) = \text{constant},\end{aligned} \tag{6.24}$$

which results in a clear contradiction. Therefore, a preliminary conclusion can be drawn that for $|e_s| \geq \delta$, the scaling gain $L(t)$ is bounded.

Part II: For $|e_s| < \delta$, the direct relationship can be depicted as $\dot{L} = 0$. Accordingly, it is obvious that $L(t)$ can converge immediately to a certain constant.

Conclusively, at this point, we are able to declare that $L(t)$ is well defined and bounded in the time interval $t \in [0, \infty)$.

B: Semi-global Uniform Boundness

In what follows, we will prove that for $t \in [t_*, \infty)$, all signals of the closed-loop system are semi-globally uniformly bounded. To see why, noting from the uniform boundness feature of system (5.13), for any well-defined $x(t) \in [-\rho^*, \rho^*]^n$, there exists a positive constant $\bar{\rho}^*$ such that $\max_{i\in\mathbb{N}_{1:n}}\{\sup_{t\geq t_*}|x_i^*(t)|\} \leq \bar{\rho}^*$. Subsequently, we are capable of defining a level set as

$$\Omega \triangleq \left\{\xi \in \mathbb{R}^n | V(\xi) \leq V_{\max} \triangleq \sup_{\xi\in\Gamma_\xi \triangleq [-(\rho^*+\bar{\rho}^*),\ (\rho^*+\bar{\rho}^*)]^n} (\lfloor\xi\rceil_\Delta^{1-\frac{\kappa}{2}})^\top P \lfloor\xi\rceil_\Delta^{1-\frac{\kappa}{2}}\right\},$$

where $\xi_i = x_i - x_i^*$, $\xi = [\xi_1,\ \xi_2,\ \cdots, \xi_n]^\top$.

Simultaneously, with $L \geq 1$ and the transformation that $\eta_i = \xi_i / L^{\rho+i-1}$, we get that $\forall\xi(t) \in \Omega \Rightarrow \eta(t) \in \Omega$.

In what follows, by employing the contradiction argument, suppose that $\eta(t)$ will escape Ω within a finite time instant. According to the finite-time escaping

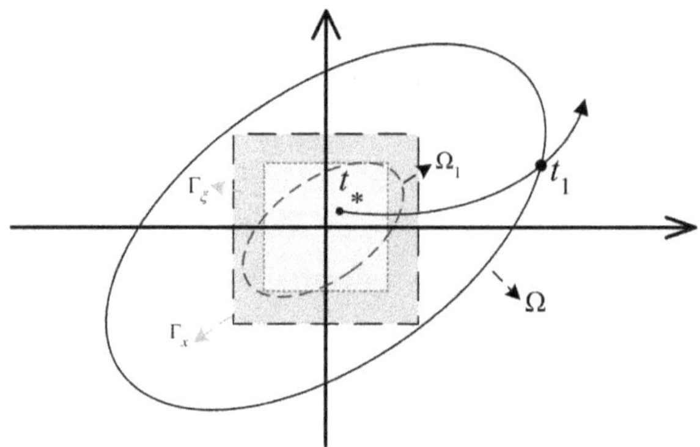

Figure 6.3 Sketch figure of the finite-time escaping phenomenon.

phenomenon, as sketchily depicted in Figure 6.3, there should exist a certain time constant $t_1 \geq t_*$, such that

$$(\lfloor \eta(t_1) \rceil_\Delta^{1-\frac{\kappa}{2}})^\top P \lfloor \eta(t_1) \rceil_\Delta^{1-\frac{\kappa}{2}} = V_{\max},\ \dot{V}(\eta(t_1)) > 0. \tag{6.25}$$

Thereafter, for any tolerance $m \in [(\frac{2\mathcal{D}^*}{\alpha})^{\frac{2-\kappa}{2}}, \frac{V_{\max}}{2}]$, combining with (6.23), we have

$$\dot{V}(\eta(t_1)) \leq -\alpha V_{\max}^{\frac{2}{2-\kappa}} + \mathcal{D}^* < -\alpha m^{\frac{2}{2-\kappa}} + \mathcal{D}^* \leq -2\mathcal{D}^* + \mathcal{D}^* < 0, \tag{6.26}$$

which leads to an obvious contradiction with (6.25).

In summary of the statements above, the conclusion can be depicted that

$$\forall x(0) \in \Gamma_x \Rightarrow \xi \in \Omega \Rightarrow \eta \in \Omega, \forall t \geq 0.$$

C: Local Practical Stability

Define a set as $\Omega_1 = \{\xi \in \mathbb{R}^n | V(\xi) \leq m\}$. Clearly, $\Omega_1 \subset \Omega$ holds. Subsequently, we shall show that for any state satisfying $\xi(t_*) \in \Gamma_\xi \cap (\Omega \backslash \Omega_1)$, the corresponding trajectory of $\xi(t)$ will stay in Ω forever and be captured by Ω_1.

From Step 2, we know that at the certain time constant t_1, the contradiction between (6.25) and (6.26) similarly occurs once applying the contradiction argument for the statement above. Hence, based on the semi-global boundness, we further know that the trajectory $\xi(t)$ will be captured by the level set Ω_1. Then, it is clear that the local practical convergence can be guaranteed, that is, there exists a finite-time $T > 0$, such that $|e_s| \leq \delta, \forall t \geq T$.

This completes the proof of Theorem 6.2. ■

6.2.6 An illustrative example and numerical simulations

Aiming to demonstrate the simplicity and effectiveness of the proposed controller construction, a typical third-order nonlinear system subject to time-varying disturbances d_1, d_2 is presented as an illustrative example:

$$\begin{cases} \dot{x}_1 = x_2 + x_1^2 + d_1 \\ \dot{x}_2 = x_3 + \sin(x_1 x_2) \\ \dot{x}_3 = u + \ln(|x_2| + 1) + d_2 \\ y = x_1. \end{cases} \tag{6.27}$$

According to the proposed nonrecursive nonsmooth dynamic control design framework, the candidate controller can be depicted as the following steps:

Step 1. Disturbance Observation: The nonsmooth disturbance observer is built to estimate d_1, d_2:

$$\begin{aligned} d_1 &: \begin{cases} \dot{\hat{z}}_{1,1} = x_2 + x_1^2 + \hat{z}_{1,2} + l_{1,1}\beta_1 \lfloor z_{1,1} - \hat{z}_{1,1} \rceil^{1+\tau} \\ \dot{\hat{z}}_{1,2} = \hat{z}_{1,3} + l_{1,2}\beta_1^2 \lfloor z_{1,1} - \hat{z}_{1,1} \rceil^{1+2\tau} \\ \dot{\hat{z}}_{1,3} = \hat{z}_{1,4} + l_{1,3}\beta_1^3 \lfloor z_{1,1} - \hat{z}_{1,1} \rceil^{1+3\tau} \\ \dot{\hat{z}}_{1,4} = l_{1,4}\beta_1^4 \lfloor z_{1,1} - \hat{z}_{1,1} \rceil^{1+4\tau}; \end{cases} \\ d_2 &: \begin{cases} \dot{\hat{z}}_{2,1} = u + \ln(|x_2| + 1) + l_{2,1}\beta_2 \lfloor z_{2,1} - \hat{z}_{2,1} \rceil^{1+\tau} + \hat{z}_{2,2} \\ \dot{\hat{z}}_{2,2} = l_{2,2}\beta_2^2 \lfloor z_{2,1} - \hat{z}_{2,1} \rceil^{1+2\tau}. \end{cases} \end{aligned} \tag{6.28}$$

Step 2. Coordinates Transformation: From (6.15), the following steady-state reference functions can be calculated:

$$\begin{aligned} & x_1^* = y_r,\ x_2^* = \dot{y}_r - y_r^2 - \hat{z}_{1,2}, \\ & x_3^* = y_r^{(2)} - 2y_r\dot{y}_r - \hat{z}_{1,3} - \sin(x_1^* x_2^*), \\ & u^* = y_r^{(3)} - 2\dot{y}_r^2 - 2y_r y_r^{(2)} - \hat{z}_{1,4} - \cos(x_1^* x_2^*)(\dot{y}_r x_2^* + y_r(y_r^2 - 2y_r\dot{y}_r - \hat{z}_{1,3})) \\ & \quad - \ln(|x_2^*| + 1) - \hat{z}_{2,2}, \end{aligned}$$

where $\hat{z}_{1,2}, \hat{z}_{1,3}, \hat{z}_{1,4}$, and $\hat{z}_{2,2}$ represent the estimation of $d_1, \dot{d}_1, d_1^{(2)}$, and d_2, respectively. Then, the coordinates transformation gives:

$$\begin{aligned} & \eta_1 = (x_1 - x_1^*)/L^\rho,\ \eta_2 = (x_2 - x_2^*)/L^{\rho+1}, \\ & \eta_3 = (x_3 - x_3^*)/L^{\rho+2},\ \upsilon = (u - u^*)/L^{\rho+3}. \end{aligned} \tag{6.29}$$

Step 3. Controller Construction: Up to now, one can straightforwardly present the nonsmooth dynamic tracking controller as the following simple form:

$$\begin{cases} \upsilon = -k_1 \lfloor \eta_1 \rceil^{1+3\kappa} - k_2 \lfloor \eta_2 \rceil^{\frac{1+3\kappa}{1+\kappa}} - k_3 \lfloor \eta_3 \rceil^{\frac{1+3\kappa}{1+2\kappa}} \\ u = L^{\rho+3}\upsilon + u^* \\ \dot{L} = \lambda(|\eta_1|^2 + |\eta_2|^{\frac{2}{1+\kappa}} + |\eta_3|^{\frac{2}{1+2\kappa}})(1 + \mathrm{sign}(|e_s| - \delta)). \end{cases} \tag{6.30}$$

To clarify the advantages of the proposed controller, the simulation performance comparison tasks are divided into two cases.

Case 1. Comparison with the backstepping-based control method: In this case, a nonrecursive nonsmooth robust composite controller (5.38) and a recursive composite controller are employed for performance comparisons. The former controller is implemented by setting $\dot{L}$ in (6.30) as zero. Notably, owing to the worst-case design strategy, the fixed L should be a sufficiently large positive value. Based on the result developed in [69], a composite backstepping controller employing a nonlinear disturbance observer (NDO) can be constructed as:

$$\begin{aligned}
\hat{d}_1 =& c_1(x_1 - p_1),\ \dot{p}_1 = x_2 + x_1^2 + \hat{d}_1, \\
\hat{d}_2 =& c_2(x_3 - p_2),\ \dot{p}_2 = u + \ln(|x_2|+1) + \hat{d}_2, \\
x_2^* =& -\sigma_1(x_1 - y_r) - x_1^2 - \hat{d}_1 + \dot{y}_r, \\
x_3^* =& -\sigma_2(x_2 - x_2^*) - \sin(x_1x_2) - (\sigma_1 + 2x_1)(x_2 + x_1^2 + \hat{d}_1) + \sigma_1\dot{y}_r + y_r^{(2)}, \\
u =& -\sigma_3(x_3 - x_3^*) - \ln(|x_2|+1) - \hat{d}_2 + \sigma_1\sigma_2\dot{y}_r + y_r^{(3)} \\
& - (x_2 + x_1^2 + \hat{d}_1)\big[\sigma_1\sigma_2 + 2\sigma_2 x_1 + \cos(x_1x_2)x_2 + 2x_2 \\
& + 2x_1^2 + 2\hat{d}_1 + 2(\sigma_1 + 2x_1)x_1\big] - (x_3 + \sin(x_1x_2))\cdot(\sigma_1 \\
& + \sigma_2 + 2x_1 + x_1\cos(x_1x_2)) + (\sigma_1 + \sigma_2)y_r^{(2)},
\end{aligned} \tag{6.31}$$

where $c_1, c_2, \sigma_1, \sigma_2, \sigma_3$ are positive design parameters.

In the simulation, the disturbances are set as $d_1 = \frac{1}{2} + \frac{1}{2}\cos(2t), d_2 = \frac{1}{2}\sin(t), 0 < t \leq 1; d_1 = 30\cos(2t), d_2 = 20, 1 < t \leq 2.2$. Meanwhile $y_r = 2\sin(6t) + \frac{1}{2}\cos(6t) + 2$.

The parameters of this simulation are chosen as: $\tau = \kappa = -0.1$, the disturbance observer: $L_1 = [4,6,4,1]^\top, \beta_1 = 20; L_2 = [2,1]^\top, \beta_2 = 20$, the controller: $[k_1,k_2,k_3] = [27,27,9]$; $[\rho,\lambda,\delta] = [0.01, 0.008, 0.005]$. Further, the design parameters of the nonsmooth robust composite controller are similar to the proposed controller and the fixed L is set as 8. The control gains of the composite backstepping controller are chosen as $[c_1,c_2] = [80,42]$; $[\sigma_1,\sigma_2,\sigma_3] = [36,36,12]$. To carry out the simulation, the initial values are given as $x(0) = [3.5,0,0]^\top$; $\hat{\omega}(0) = [0,0,0,0,0,0]^\top$.

The output tracking control results are depicted in Figure 6.4. The response curves of the states x_2, x_3 are shown in Figure 6.5. Figure 6.6 shows the control input signals and the convergence trajectory of the dynamic gain L. Clearly, the proposed tracking control objective can be realized via all candidate approaches. Compared with the nonsmooth robust controller, the proposed algorithm could yield a smoother transient-time control performance at the small disturbance working condition. With the presence of large signal disturbances, the nonsmooth robust controller and the proposed strategy have shown a very similar control performance owing to the fact the dynamic scaling gain L has converged to a similar value with the case in the robust controller. Compared with the composite backstepping controller, the proposed controller expression is much simpler, as can be observed from (6.30) and (6.31). Moreover, the track-

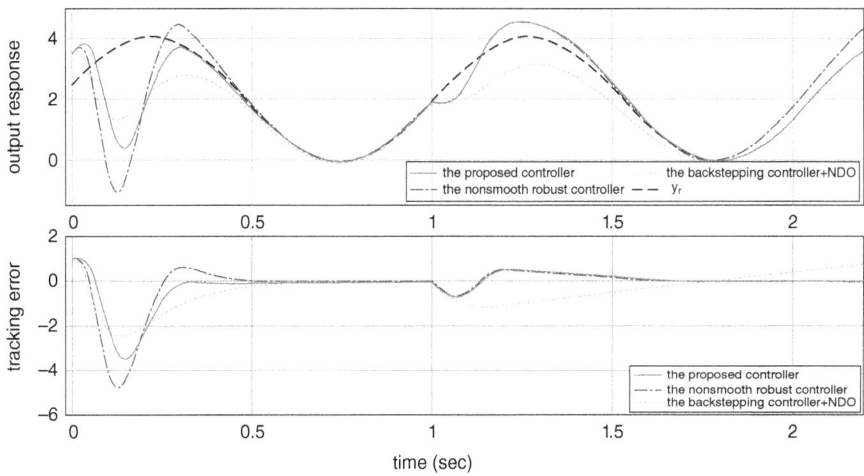

Figure 6.4 Output tracking performance.

ing accuracy and convergent rate of the proposed method also outperform the composite backstepping controller.

Case 2. The homogeneous degree selection: In this case, we conduct a series of simulation studies to illustrate the control performance affected by the selection of different homogeneous degrees in the proposed nonsmooth controller. On the one hand, as depicted in Figure 6.7, for the large level of disturbance signals, a smaller κ will clearly lead to a faster convergence speed and a lower steady error. However, for smaller disturbances, a smaller κ may result in a more aggressive transient-time response. On the other hand, the nonsmooth dynamic controller tends to approach the smooth linear controller in the case when the

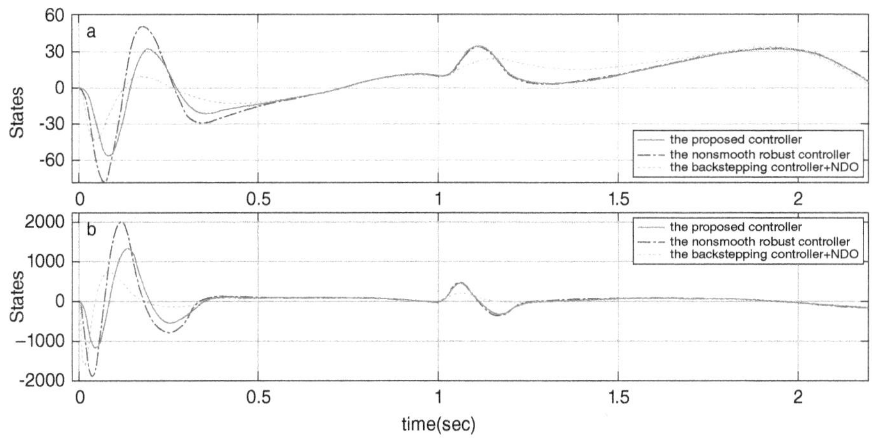

Figure 6.5 State response curves. (a) x_2; (b) x_3.

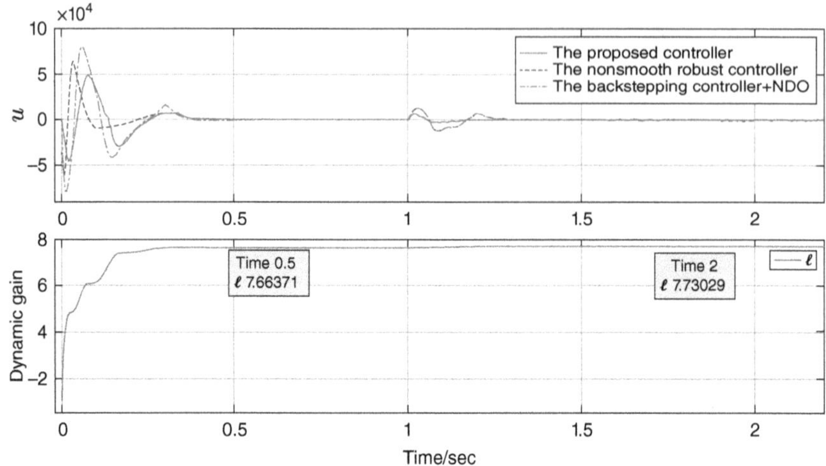

Figure 6.6 The control efforts and dynamic gain.

homogeneous degree κ is set close to zero, which will lead to a larger settling time but smoother transient-time control performance.

Remark 6.2 As illustrated in the numerical simulation process, the parameter selection guidelines of the proposed dynamic controller can be briefly discussed as the following procedures: a) The homogeneous degrees are chosen as a value approximate to zero, first; b) The traditional linear pole placement method can be employed to determine the control gain K; c) By solving a joint linear matrix inequality (LMI) related to P as stated above, matrix P can be subsequently calculated; d) Tune the homogeneous degrees to be slightly smaller according to design requirement; e) The range of ρ can be calculated by (6.17), and generally, it can be chosen as a small positive number. Additionally, the scaling gains of $\dot{L}(t)$, i.e., λ, δ which are critical

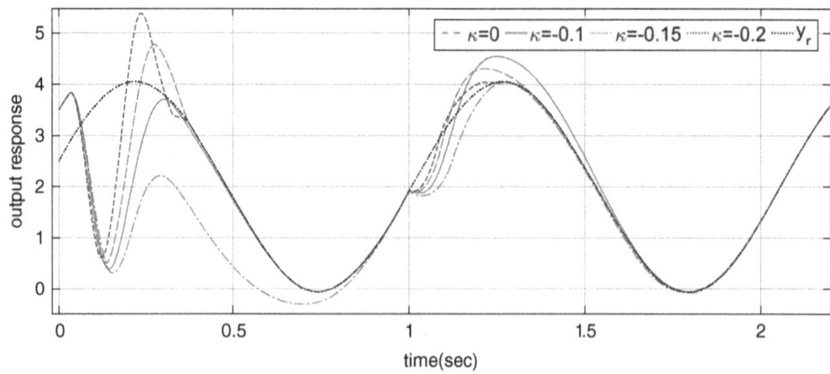

Figure 6.7 Output tracking performance with different homogeneous degrees.

the convergence rate of L, can be selected with the "trial and error" method. Specifically, in industrial application scenarios, the selection of δ should be larger than the steady-state error caused by measurement noise to suppress the growth rate of L.

Chapter 7

Application to Permanent Magnet Synchronous Motor Systems

7.1 Introduction

Permanent magnet synchronous motor (PMSM) has been widely applied in electric vehicles, robots, servo systems, and aerospace applications due to its simple design, compact structure, and high energy density [75], [65]. Specifically, as the electric vehicle market has grown rapidly in recent years, PMSM has gradually replaced AC induction motors as the mainstream electric drive system. However, due to the fact that PMSM has multivariable, strong coupling, and nonlinear characteristics, it is a challenging job to achieve high accuracy and performance speed closed loop control of the PMSM systems. Therefore, in recent years, in order to improve the dynamic performance and disturbance rejection capability of PMSM drive systems, abundant advanced control methods have been proposed in the literature, which can be roughly divided into cascade and non-cascade control strategies according to the control structure. The cascade control consists of an outer speed loop and an inner current loop, the speed control strategies based on this structure are currently the main choice. It is worth noting that in the non-cascade control structure, the current loop is omitted, which avoids the bandwidth limitation problem in cascade control and thus improves the system's dynamic performance. However, under the non-cascade structure, disturbances in the current loop are mismatched, which enter the system via a different channel from the control input. Mainly caused by external load torques, unmodeled

DOI: 10.1201/9781003399230-7

dynamics and parameter perturbations, the mismatched disturbances have negative impact on system performance and stability [10]. To improve the system's ability to suppress disturbances and uncertainties, several control strategies combined with disturbance observer have been proposed [70, 84]. However, how to achieve the balance between system robustness while maintaining the dynamic control performance is still a significant issue.

In this chapter, we will apply the proposed theoretical nonrecursive control results to provide a simple controller design method considering mismatched disturbances for the speed control problem of PMSM drive systems under a non-cascaded control structure, which can significantly improve the dynamic performance of the system while guaranteeing the system robustness against external disturbances and uncertainties. Meanwhile, the parameter tuning is intuitive and computational burden is light, which naturally is very easy to be implemented in low-cost microprocessors.

In addition, to provide engineers with a clearer and more intuitive understanding of the structure and parameter design of the controller, each control parameter's effect on system performance is simulated in this chapter via the MATLAB/Simulink platform, which is a helpful instruction for the further application of the proposed control method to practical engineering.

7.2 Modeling of PMSM

To begin with, the three phase voltage and flux chain equations are given as follows:

$$\begin{cases} u_A = R_s i_A + \dot{\Psi}_A \\ u_B = R_s i_B + \dot{\Psi}_B \\ u_C = R_s i_C + \dot{\Psi}_C, \end{cases} \tag{7.1}$$

$$\begin{cases} \Psi_A = L_s i_A + M(i_B + i_C) + \psi_f \cos\theta \\ \Psi_B = L_s i_B + M(i_A + i_C) + \psi_f \cos(\theta - \frac{2\pi}{3}) \\ \Psi_C = L_s i_C + M(i_A + i_B) + \psi_f \cos(\theta - \frac{4\pi}{3}), \end{cases} \tag{7.2}$$

where i_A, i_B, and i_C are stator winding current of the A, B, and C phases respectively; R_s is the stator resistance; L_s is the stator inductance; M is the stator mutual inductance; ψ_f is the rotor flux linkage; and θ is the electric angle.

Aiming at the three-phase stator winding adopting the Y-connection, the stator current satisfies the following relationship:

$$i_A + i_B + i_C = 0. \tag{7.3}$$

Combining with (7.1), (7.2), (7.3), one can get that:

$$\begin{cases} \Psi_A = (L_s - M)i_A + \psi_f \cos\theta \\ \Psi_B = (L_s - M)i_B + \psi_f \cos(\theta - \frac{2\pi}{3}) \\ \Psi_C = (L_s - M)i_C + \psi_f \cos(\theta - \frac{4\pi}{3}). \end{cases} \tag{7.4}$$

Substituting (7.4) in (7.1) yields:

$$\begin{cases} u_A = R_s i_A + (L_s - M)\dot{i}_A - \psi_f \omega_e \sin\theta \\ u_B = R_s i_B + (L_s - M)\dot{i}_B - \psi_f \omega_e \sin(\theta - \frac{2\pi}{3}) \\ u_C = R_s i_C + (L_s - M)\dot{i}_C - \psi_f \omega_e \sin(\theta - \frac{4\pi}{3}), \end{cases} \tag{7.5}$$

where ω_e is the electrical angular velocity.

The transformation from a three-phase stationary coordinate system to α-β two-phase one can be performed by adopting the Clarke coordinate transformation with the following relations:

$$\begin{cases} \alpha = \frac{2}{3}(A - \frac{1}{2}B - \frac{1}{2}C) \\ \beta = \frac{2}{3}(\frac{\sqrt{3}}{2}B - \frac{\sqrt{3}}{2}C). \end{cases} \tag{7.6}$$

Subsequently, the voltage equation in α-β axis can be attained by substituting (7.5) in (7.6)

$$\begin{cases} u_\alpha = R_s i_\alpha + L_s \dot{i}_\alpha - w_e \psi_f \sin\theta \\ u_\beta = R_s i_\beta + L_s \dot{i}_\beta + w_e \psi_f \cos\theta. \end{cases} \tag{7.7}$$

Furthermore, transform α-β axis to the rotational one d-q utilizing the following Park transformation:

$$\begin{cases} d = \alpha\cos\theta + \beta\sin\theta \\ q = -\alpha\sin\theta + \beta\cos\theta. \end{cases} \tag{7.8}$$

Combining (7.7) with (7.8), one can obtain the voltage equation of PMSM in d-q axis in the following form:

$$\begin{cases} u_d = R_s i_d + L_s \dot{i}_d - \omega_e L_s i_q \\ u_q = R_s i_q + L_s \dot{i}_q + \omega_e (L_s i_d + \psi_f). \end{cases} \tag{7.9}$$

Consequently, the equations of electromagnetic torque and motor motion in d-q axis system are expressed in the following form:

$$\begin{cases} T_e = J\dot{\omega} + T_L + B\omega \\ T_e = \dfrac{3}{2} n_p \psi_f i_q, \ n_p\omega = \omega_e, \end{cases} \tag{7.10}$$

where J is the rotor inertia, n_p is the number of pole pairs, B is the viscous frictional coefficient, T_L is the load torque, and ω is the mechanical angle speed.

Combining (7.9) with (7.10), the mathematical model of a surface-mounted motor in the d-q coordinates frame is further stated as:

$$\begin{cases} \dot{\omega} = \dfrac{1}{J}(\dfrac{3}{2} n_p \psi_f i_q - B\omega - T_L) \\ \dot{i}_d = \dfrac{1}{L}(-R_s i_d + n_p \omega L i_q + u_d) \\ \dot{i}_q = \dfrac{1}{L}(-R_s i_q - n_p \omega L i_d - n_p \psi_f \omega + u_q). \end{cases} \tag{7.11}$$

7.3 Nonsmooth Robust Control for Speed Regulation of PMSM Systems

7.3.1 Controller design

Under a semi-global stability criterion, one can utilize the proposed exact tracking controller proposed in Chapter 5 while several auxiliary variables are calculated as:

$$\begin{cases} i_d^* = 0, \\ u_d^* = L_s i_d^{*(1)} + R_s i_d^* - n_p L_s \omega^* i_q^* - z_{3,1}, \end{cases} \tag{7.12}$$

$$\begin{cases} \xi_d = i_d - i_d^*, \\ v_d = (u_d - u_d^*)/L_d; \end{cases} \tag{7.13}$$

$$\begin{cases} \omega^* = \omega_r, \\ i_q^* = \dfrac{2}{3n_p\psi_f}\left(J\omega_r^{(1)} + B\omega_r - Jz_{1,1}\right), \\ u_q^* = \dfrac{2L_s}{3n_p\psi_f}\left(J\omega_r^{(2)} + B\omega_r^{(1)} - Jz_{1,2}\right) + R_s i_q^* + n_p L_s \omega_r i_d^* + n_p \psi_f \omega_r - z_{2,1}; \\ \xi_\omega = \omega - \omega^*, \ \xi_q = (i_q - i_q^*)/L_q, \ v_q = (u_q - u_q^*)/L_q^2. \end{cases} \tag{7.14}$$

Hence, a robust composite control scheme is then explicitly presented as:

$$T_L:\begin{cases}\dot{z}_{1,0}=\dfrac{1}{J}\left(\dfrac{3}{2}n_p\psi_f i_q-B\omega\right)+\hbar_{1,0}\\ \dot{z}_{1,1}=\hbar_{1,1},\ \dot{z}_{1,2}=\hbar_{1,2}\\ \hbar_{1,0}=-l_{1,0}\lambda_1^{1/3}\lfloor z_{1,0}-\omega\rceil^{2/3}+z_{1,1}\\ \hbar_{1,1}=-l_{1,1}\lambda_1^{1/2}\lfloor z_{1,1}-\hbar_{1,0}\rceil^{1/2}+z_{1,2}\\ \hbar_{1,2}=-l_{1,2}\lambda_1\lfloor z_{1,2}-\hbar_{1,1}\rceil^{0};\end{cases} \tag{7.15}$$

$$d_1:\begin{cases}\dot{z}_{2,0}=\dfrac{1}{L_s}\left(-R_s i_q-n_p\omega L_s i_d-n_p\psi_f\omega+u_q\right)+\hbar_{2,0}\\ \dot{z}_{2,1}=\hbar_{2,1}\\ \hbar_{2,0}=-l_{2,0}\lambda_2^{1/2}\lfloor z_{2,0}-i_q\rceil^{1/2}+z_{2,1}\\ \hbar_{2,1}=-l_{2,1}\lambda_2\lfloor z_{2,1}-\hbar_{2,0}\rceil^{0};\end{cases} \tag{7.16}$$

$$u_q:\begin{cases}v_q=-K_q\left[\lfloor\xi_\omega\rceil^{1+2\tau_q},\lfloor\xi_q\rceil^{\frac{1+2\tau_q}{1+\tau_q}}\right]^{\top},\\ u_q=L_q^2 v_q+u_q^*;\end{cases} \tag{7.17}$$

$$d_2:\begin{cases}\dot{z}_{3,0}=\dfrac{1}{L_s}\left(-R_s i_d+n_p\omega L_s i_q+u_d\right)\\ \dot{z}_{3,1}=\hbar_{3,1}\\ \hbar_{3,0}=-l_{3,0}\lambda_3^{1/2}\lfloor z_{3,0}-i_d\rceil^{1/2}+z_{3,1}\\ \hbar_{3,1}=-l_{3,1}\lambda_3\lfloor z_{3,1}-\hbar_{3,0}\rceil^{0};\end{cases} \tag{7.18}$$

$$u_d:\begin{cases}v_d=-K_d\lfloor\xi_d\rceil^{1+\tau_d},\\ u_d=L_d v_d+u_d^*,\end{cases} \tag{7.19}$$

where $K_q=[k_{q1},k_{q2}]$ with $k_{q1}>0$, $k_{q2}>0$, and $K_d>0$.

7.3.2 Simulation studies

Parameters selection: In this simulation, the observer gains are configured as: $l_{1,0}=10$, $l_{1,1}=7$, $l_{1,2}=4$, $\lambda_1=1.9\times10^7$; $l_{2,0}=11$, $l_{2,1}=3$, $\lambda_2=1.55\times10^5$; $l_{3,0}=4$, $l_{3,1}=3$, $\lambda_3=1.35\times10^4$. The controller gains are configured as: $K_q=[0.02,0.01]$, $L_q=1.1$, $\tau_q=-0.05$, $K_d=0.1$, $L_d=1.1$, $\tau_d=-0.05$.

The test scenario is that the system reference speed is switched from 500 rpm to 1000 rpm at 1s, followed by a sudden load disturbance of 30% of the nominal torque (0.08 Nm) at 1.5 s, where the parameters of PMSM are listed in Table 7.1 and the control frequency is set as 20 kHz.

As can be observed from Figure 7.1, the response speed and disturbance rejection ability of the control system are markedly improved as k_{q1} increases while

Table 7.1 Parameters of the PMSM platform

Parameter description (symbol)	Value
number of pole pairs (n_p)	4
stator resistance (R_s)	$0.36\,\Omega$
stator inductance (L_s)	2×10^{-4} mH
rotor flux linkage (ψ_f)	0.0064 Wb
rotor inertia (J)	7.066×10^{-6} kg·m^2
viscous frictional coefficient (B)	2.637×10^{-6} N·m·s/rad

other parameters remain the same, but an excessive k_{q1} may cause the system chattering phenomenon. From Figure 7.2, one can naturally find that the parameter k_{q2} can regulate the system damping, and the larger the parameter, the stronger the damping, which can effectively alleviate the issue of large overshoot of the system caused by the k_{q1} being oversized. Nevertheless, it will deteriorate the dynamic performance of the system when the parameters are excessively large. Consequently, the satisfactory dynamic performance can be achieved by choosing a appropriate group of k_{q1} and k_{q2}. The parameter L_q can tune the system bandwidth, and its simulation results are shown in Figure 7.3, which demonstrates that the increase of L_q can improve the response speed and disturbance rejection ability. τ_q provides a new dimension of the regulatory control performance, and the simulation results are shown in Figure 7.4. It is straightforward to see that the presence of the parameter τ_q can significantly affect the system dynamic response. While due to the mathematical properties of the fractional order itself, compared to the traditional linear feedback control laws, a significant improvement can be achieved in the system robustness against uncertainties and disturbances.

Performance verification: In this simulation, we choose a relatively suitable set of composite controller parameters as possible, where the observer

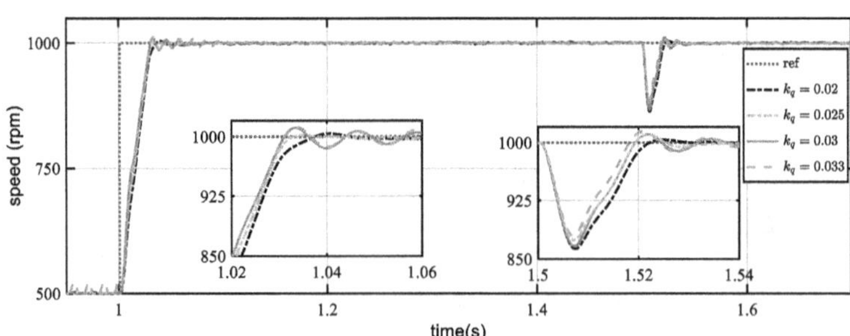

Figure 7.1 Simulation speed regulation results under different parameter k_q.

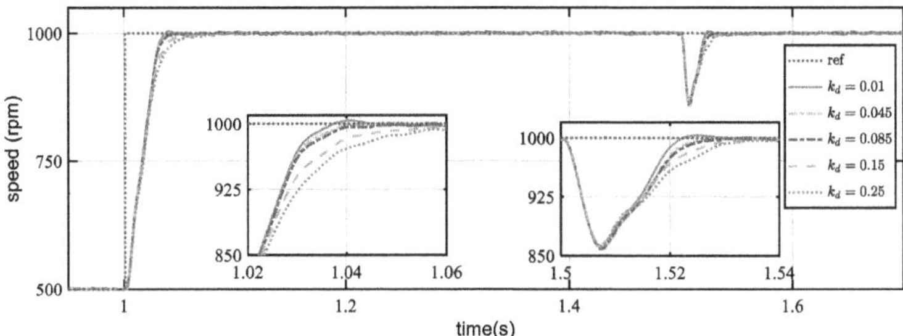

Figure 7.2 Simulation speed regulation results under different parameter k_d.

parameters are set as: $l_{1,0} = 10$, $l_{1,1} = 7$, $l_{1,2} = 4$, $\lambda_1 = 1.9 \times 10^7$; $l_{2,0} = 11$, $l_{2,1} = 3$, $\lambda_2 = 1.55 \times 10^5$; $l_{3,0} = 4$, $l_{3,1} = 3$, $\lambda_3 = 1.35 \times 10^4$. The control parameters are set as: $K_q = [0.02, 0.03]$, $L_q = 1.25$, $\tau_q = -0.1$, $K_d = 0.05$, $L_d = 1.2$, $\tau_d = -0.1$, and the PMSM parameters are given in Table 7.1 with the sampling frequency of 20 kHz. The test scenario is divided into the following stages: at first, the speed reference is switched from 500 rpm to 1000 rpm at 1 s, followed by a sudden load disturbance of 0.08 Nm on the motor at 2 s, then a sinusoidal load disturbance of $0.08\sin(2\pi t)$ Nm on the motor at 3 s, and finally the sinusoidal disturbance is removed at 5 s. The initial values of the closed-loop system are chosen as 0.

As is clearly depicted by Figure 7.5 (a), in the presence of unknown load torque variation, the speed regulation objective can still be well achieved under the proposed robust controller. The response curves of i_q, i_d and the time histories of two control inputs u_q, u_d are presented in Figures 7.5 (b) and (c), which clearly demonstrates the effectiveness of the designed HOSM disturbance observer depicted in Eqs. (7.15) and (7.16).

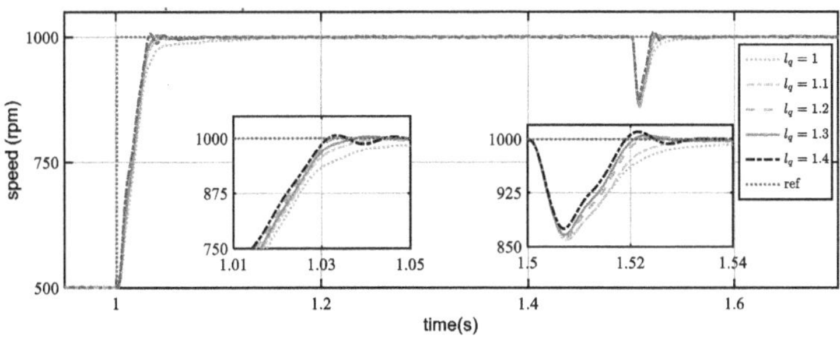

Figure 7.3 Simulation speed regulation results under different parameter l_q.

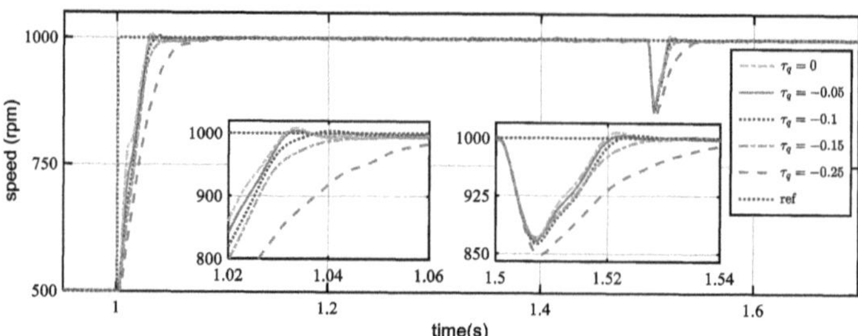

Figure 7.4 Simulation speed regulation results under different parameter τ_q.

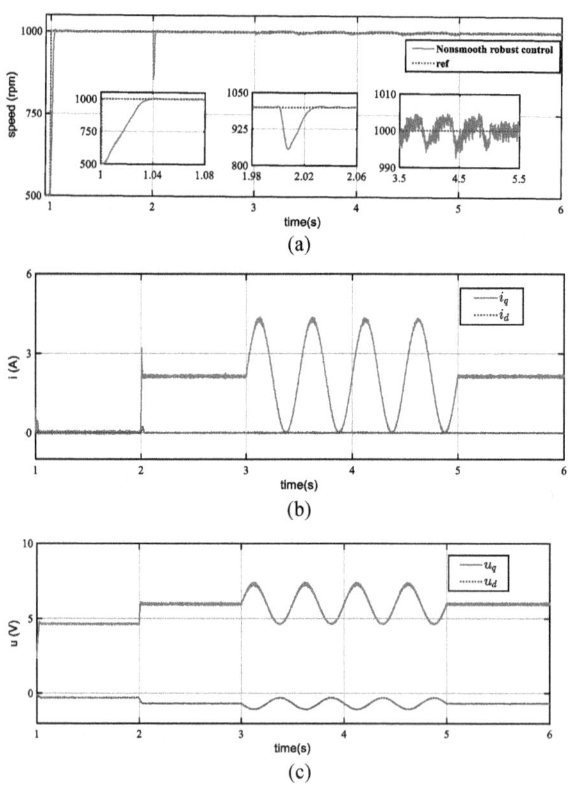

Figure 7.5 (a) Speed tracking performance; (b) Current response curves; (c) Voltage response curves.

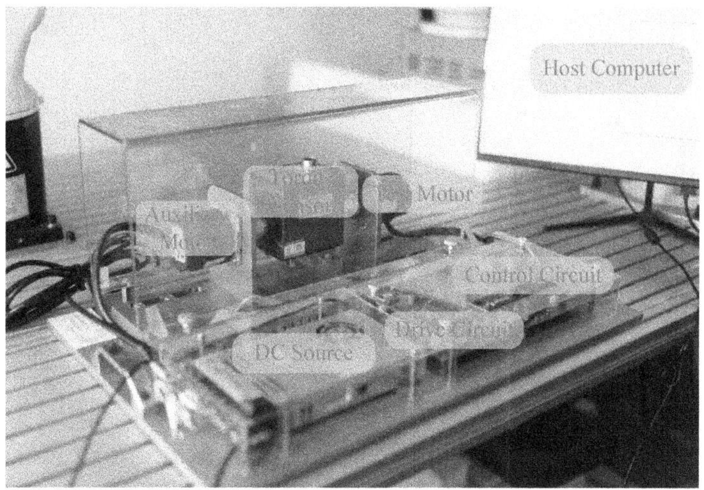

Figure 7.6 Experiment setup of a PMSM system.

7.3.3 Experimental tests

To further validate the effectiveness of the designed controller, the experimental setup has been constructed and its setting is shown in Figure 7.6. The control and driver circuits adopted are the F28379D Launched Pad and DRV-8305 developed by Texas Instrument. The PMSM parameters are listed in Table 7.1, and the sampling frequency is 20 kHz, where a conventional cascaded PI speed controller is used as the comparison group. Specifically, the control parameters are given as: $l_{1,0} = 13$, $l_{1,1} = 7$, $l_{1,2} = 3$, $\lambda_1 = 4.85 \times 10^5$; $l_{2,0} = 11$, $l_{2,1} = 6$, $\lambda_2 = 6 \times 10^4$; $l_{3,0} = 4$, $l_{3,1} = 3$, $\lambda_3 = 3500$. $K_q = [0.04, 0.025]$, $L_q = 1.2$, $\tau_q = -0.15$, $K_d = 0.1$, $L_d = 1.2$, $\tau_d = -0.15$ for the proposed robust controller; $k_{dp} = 2.31$, $k_{di} = 0.28$, $k_{pp} = 2$, $k_{pi} = 6.66 \times 10^{-4}$ for the cascaded PI controller.

For more detailed demonstration of dynamic performance and disturbance rejection of the proposed controller, the experimental validation is considered as two scenarios: the first test scenario is that the reference speed switches from 500 rpm to 1000 rpm at 5 s followed by a sudden load disturbance of 0.1 Nm at 7 s, similarly, the second test scenario is that the reference speed switches from 500 rpm to 1000 rpm at 5 s, while a sinusoidal load disturbance of form $0.1\sin(4\pi t)$ Nm is applied at 7 s.

As can be visually observed in Figure 7.7, the dynamic performance of the proposed controller is significantly improved compared to the cascaded PI controller. Further, after the disturbance is added at $t = 7$ s, the PI controller suffers a large speed drop, while the proposed nonsmooth composite controller still works well with only a slight drop. Subsequently, in Figure 7.8, the time-varying disturbance is applied at 7 s. It can be clearly seen that the conventional PI controller

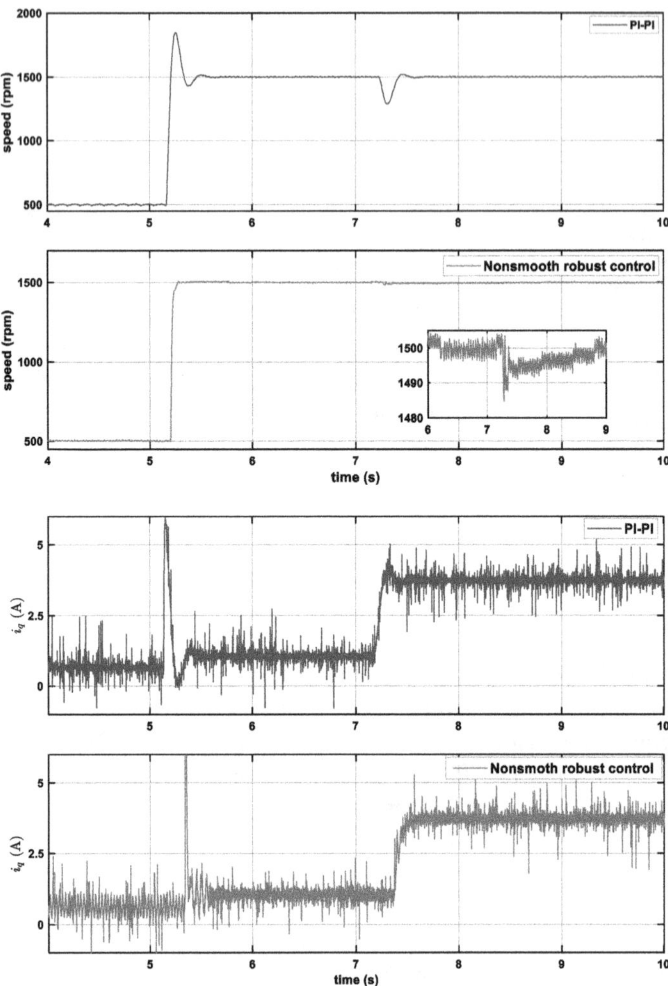

Figure 7.7 Experiment results of the reference speed switches from 500 rpm to 1500 rpm with $T_L = 0$ Nm at 5 s, and the T_L switches to 0.1 Nm at 7 s.

shows large fluctuation in that the frequency is the same as the applied load disturbance. In contrast, the proposed controller shows desirable robustness to the time-varying disturbance, with only tiny fluctuation. Consequently, one can conclude that the proposed control method is able to improve the dynamic performance, the ability to suppress disturbances and system uncertainties compared to the traditional cascaded PI control method.

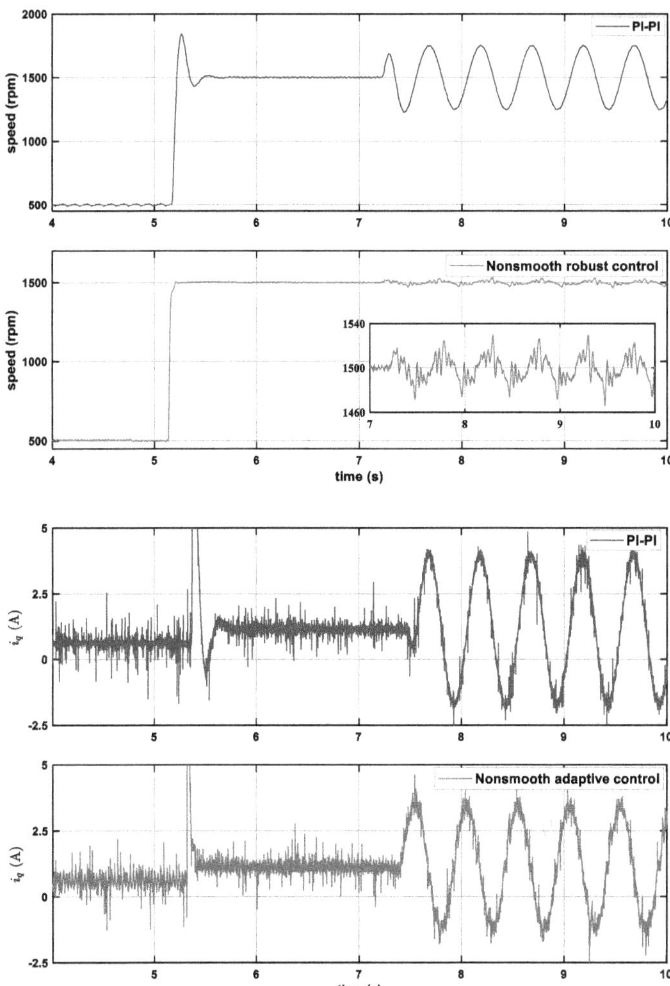

Figure 7.8 Experiment results of the reference speed switches from 500 rpm to 1500 rpm with $T_L = 0$ Nm at 5 s, and the T_L switches to $0.1\sin(4\pi t)$ Nm at 7 s.

7.4 Adaptive Control for Speed Regulation of PMSM Systems

7.4.1 Linear adaptive composite control

7.4.1.1 Controller design

For the PMSM system (7.11), consider a more ideal case, i.e., $i_d \doteq i_d^* = 0$. Furthermore, we introduce the following transformation $x_1 = \omega_{\text{ref}} - \omega$, $x_2 = \frac{1}{J}(-n_p\psi_f i_q + B\omega_{\text{ref}})$. Then, the control-oriented model is written as:

$$\begin{cases} \dot{x}_1 = x_2 - Bx_1/J + d_1, \\ \dot{x}_2 = u - \dfrac{3}{2}\left(n_p^2\psi_f^2 x_1 - (n_p^2\psi_f^2 + \dfrac{2}{3}R_s B)\omega_{\text{ref}}\right)/(JL_s) - R_s x_2/L_s + d_2 \\ y = x_1, \end{cases}$$

where $u = -\frac{n_p\psi_f}{JL}u_q$ is the control input, y is the system output, $d_1 = \frac{1}{J}T_L$ is regarded as the mismatched disturbance, and d_2 is the matched disturbance which contains unmodeled dynamic or measurement errors, etc.

At this point, we have transferred the rotating speed regulation problem into a stabilizable one. By the proposed means, one can calculate the steady-state signal as:

$$\begin{aligned} x_1^* &= y_r = 0, \\ x_2^* &= -z_{1,1}, \\ x_3^* &= -z_{1,2} - \frac{1}{JL_s}\left(R_s J z_{1,1} + (n_p^2\psi_f^2 + R_s B)\omega_{\text{ref}}\right) - z_{2,1}, \end{aligned}$$

where $z_{1,1}, z_{1,2}, z_{2,1}$ are the disturbance estimations derived from (5.3). Hence, the coordinate transformations are

$$\eta_1 = x_1 - x_1^*,\ \eta_2 = x_2 - x_2^*,\ \upsilon = u - x_3^*. \tag{7.20}$$

Then, one is able to formulate the controller in the following form:

$$\begin{cases} u = -(k_1\eta_1 + k_2\eta_2) + x_3^*, \\ \dot{L} = c\left(\dfrac{\eta_1^2}{L^{2\rho}} + \dfrac{\eta_2^2}{L^{2\rho+2}}\right)(1 + \text{sign}(|e_s| - \delta)). \end{cases}$$

7.4.1.2 Experimental tests

In the following experimental tests, the parameters of the PMSM are listed in Table 7.1, and the control switching frequency is 15 kHz. The experiments are implemented from two perspectives. On the one hand, in addition to the proposed linear dynamic nonrecursive controller, the other two existing control

Table 7.2 Control design parameters of the linear adaptive controller

Parameter description (symbol)	Value
observer #1 gains ($\lambda_1, l_{1,1}, l_{1,2}, l_{1,3}$)	1200, 40, 48, 54
observer #2 gains ($\lambda_2, l_{2,1}, l_{2,2}$)	1×10^7, 72, 84
controller gains (k_1, k_2)	10/3, 2.5
design parameters (ρ, c, δ)	1, 0.5, 15
PI gains of i_d-axis (k_{dp}, k_{di})	2.3197, 0.2817
PI gains of i_q-axis (k_{pp}, k_{pi})	2, 6.6675×10^{-4}

methodologies are chosen as the benchmark control group, that are, the traditional double-loop PI control and the Generalized Predictive Control (GPC) method [87], aiming to conduct the performance comparison. On the other hand, the speed regulation working conditions of the PMSM are essential to be rigorously designed. Concretely, the test procedures are classified as the following cases: the reference speed switches from 500 rpm to 1500 rpm, from 500 rpm to 3000 rpm, respectively, with the presence of $T_L = 0.018$ Nm. Additionally, for the purpose of validating the anti-disturbance ability of the candidate controllers, the T_L switches from 0.018 Nm to $0.018\sin(t)$ Nm in the last cases under a fixed speed of 1500 rpm. The control parameters for the linear adaptive controller are presented in Table. 7.2.

Scenario 1. the reference speed switches from 500 rpm *to 1500* rpm *with* $T_L = 0.018$ Nm*:* The experimental results of this case can be referred to Figure 7.9. In this case, the worst speed response performance is clearly the PI controller, whether settling time (0.592 s) or maximum overshoot (129.71 rpm).

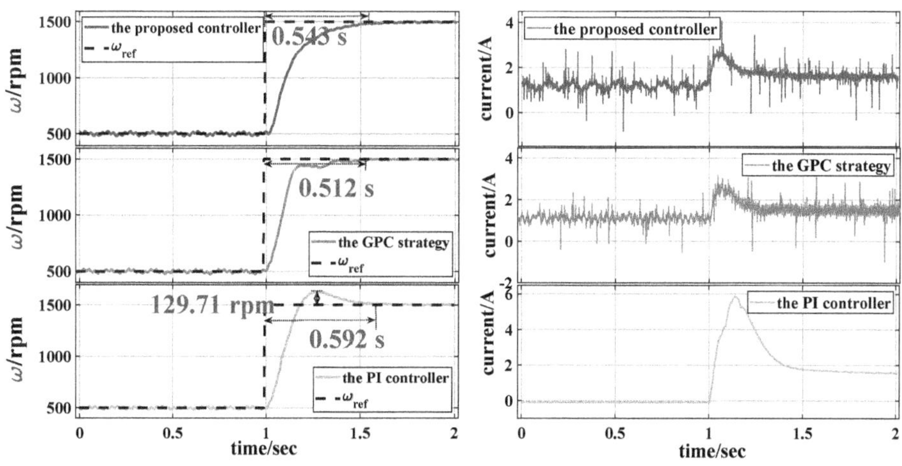

Figure 7.9 Experiment results of the reference speed switches from 500 rpm to 1500 rpm with $T_L = 0.02$ Nm.

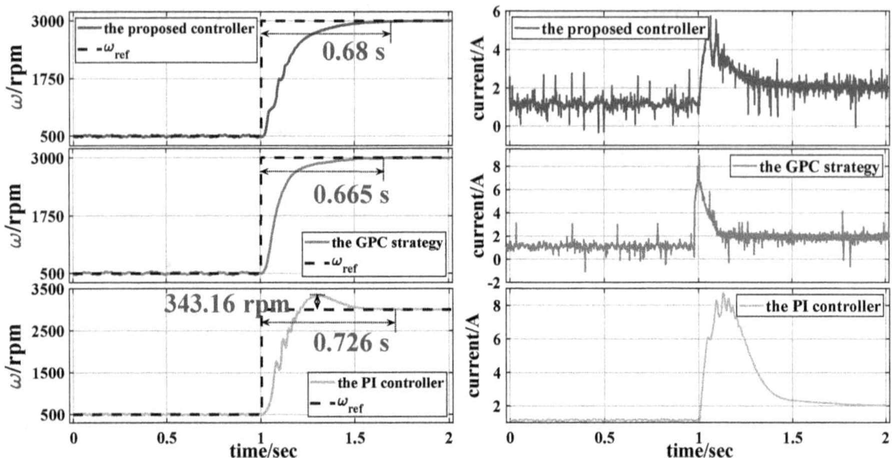

Figure 7.10 Experiment results of the reference speed switches from 500 rpm to 3000 rpm with $T_L = 0.02$ Nm.

Besides, compared with the GPC strategy, the proposed controller has a smoother response curve, but a slower regulation time, which causes by over strong robustness of the GPC controller form, i.e., unsurprisingly, the conservative horizon determination leads to a faster response time but a more intense process. The dynamic scaling gain converges from 1 to 140.45 to 161.03.

Scenario 2. the reference speed switches from 500 rpm *to 3000* rpm *with* $T_L = 0.018$ Nm: Clearly, from Figure 7.10, the phenomenon occurring to the PI controller is consistent with Scenario 1, i.e., longest settling time (0.726 s) and largest overshoot (343.16 rpm). From the perspective of nonrecursive control, one can find that the output speed behaviors of the GPC controller and linear dynamic composite controller are quite similar in this case. The philosophy of this manifestation is that the fixed horizon (bandwidth factor) is artificially chosen as the most suitable value to confront this case (the worst case in these experiments), hence, the dynamic performance can be guaranteed. For the proposed dynamic gain design, in this case, the dynamic scaling gain has to tune to the most conservative value (commonly close to the fixed gain) according to the self-configuration mechanism, and only in this way can the sudden change of this severe working condition be suppressed, detailedly, the dynamic gain switches from 1 to 16.86 to 36.13.

Scenario 3: T_L *switches from* 0.02 Nm *to* $0.02\sin(t)$ Nm *with the fixed speed 1500* rpm. The experimental results are shown in Figure 7.11. For this scenario, the output response performance of all candidate controllers has a huge difference. First of all, the PI controller is unsatisfactory as always with the T_L presence of frequent switching, and the steady-state speed fluctuation seems to be hard to be accepted. The steady-state speed of the dynamic gain design is relatively more

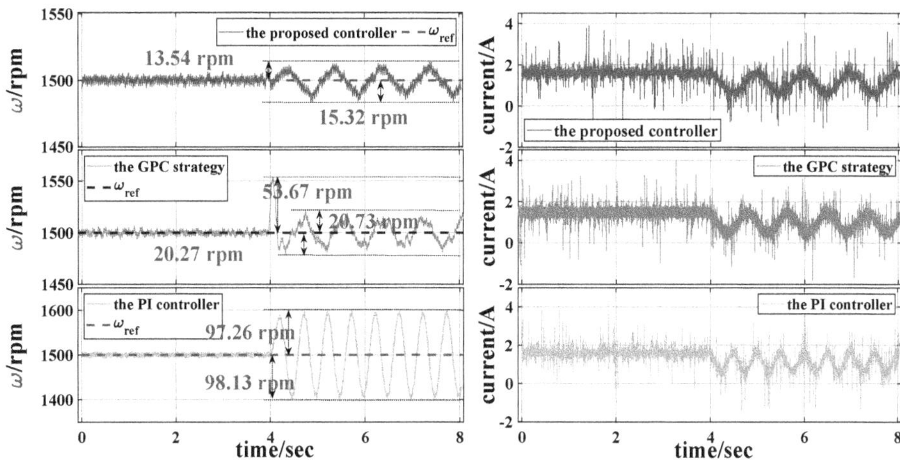

Figure 7.11 Experiment results of T_L switches from 0.02 Nm to 0.02 sin(t)Nm under the fixed speed of 1500 rpm.

desirable than the GPC approach, due to the self-tuning gain generated seems to be more suitable in this condition, which switches from 1 to 78.37 to 142.05, as shown in Figure 7.12. At the T_L operation condition switching moment, the maximum overshoot of the GPC controller is larger than the proposed controller, and the obvious speed fluctuation occurs in the PMSM drive system.

The detailed behavior data of the candidate controllers are listed in Table 7.3 for reference. In summary, the controlled design expectancy is basically achieved in the experimental verification.

7.4.2 Nonsmooth adaptive composite control

7.4.2.1 Controller design

Assuming d-axis current loop works well and satisfies $i_d = i_d^* = 0$. Based on the defined auxiliary variables ω^*, i_q^* and u_q^* in (7.14) and the built disturbance observer in (7.15) and (7.16), a nonsmooth nonrecursive adaptive controller can be constructed in the following form:

$$\begin{cases} u = -L^{2+\rho}\left(k_1\lfloor\xi_\omega\rceil^{1+2\kappa} + k_2\lfloor\xi_q\rceil^{\frac{1+2\kappa}{1+\kappa}}\right) + u^* \\ \dot{L} = \lambda\left(|\xi_\omega|^2 + |\xi_q|^{\frac{2}{1+\kappa}}\right)\left(1+\mathrm{sign}(|\omega-\omega^*|-\delta)\right), \end{cases}$$

where $\xi_\omega = (\omega-\omega^*)/L^\rho$, $\xi_q = (i_q - i_q^*)/L^{\rho+1}$ and L is the adaptive gain.

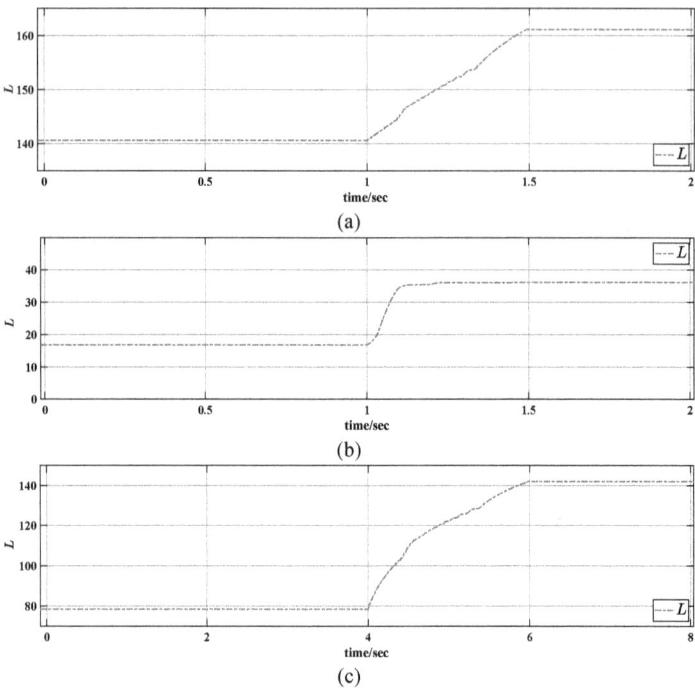

Figure 7.12 Dynamic scaling gain curves. (a) Scenario 1; (b) Scenario 2; (c) Scenario 3.

7.4.2.2 Simulation studies

In this simulation, the controller parameters are set as: $l_{1,0} = 9$, $l_{1,1} = 6$, $l_{1,2} = 3$, $\lambda_1 = 1.5 \times 10^7$, $l_{2,0} = 8$, $l_{2,1} = 2.5$, $\lambda_2 = 1 \times 10^5$, $l_{3,0} = 4$, $l_{3,1} = 5$, $\lambda_3 = 9.55 \times 10^3$, $k_1 = 0.03$, $k_2 = 0,05$, $\kappa = -0.15$, $\lambda = 5$, $\rho = 1.1$, $\delta = 0.5$, and the control frequency is 20 kHz. The PMSM parameters are given in Table 7.1. The test scenario is that the reference speed is switched from 500 rpm to 1000 rpm at 1 s, then the 30% nominal torque, i.e., 0.08 Nm is applied to motor at 1.5 s.

Nonsmooth adaptive control makes parameter tuning easier compared to the design robust controller in (7.17) due to the introduction of a self-tuning mechanism to adjust the system bandwidth. λ is used to tune the adaptive growth rate, and the simulation results are presented in Figure 7.13, it can be observed that by choosing a larger λ the system has a faster response rate. The response curves for L at different λ are shown in Figure 7.14.

Note that the parameter κ is included in both the adaptive mechanism and the designed control law. As can be observed in Figure 7.16, the smaller the choice of κ, the larger the steady value of L. Furthermore, from Figure 7.14, one can learn that the increase of L can enhance system response, while from Figures 7.15 and 7.16, one can observe that the increase of L does not have much

Table 7.3 Performance indexes of three candidate controllers

Controller	Working condition	Maximum deviation/rpm	Settling time/s
the proposed controller	case 1	0	0.543
	case 2	0	0.68
	case 3	15.32	/
the GPC strategy	case 1	0	0.512
	case 2	0	0.665
	case 3	53.67	/
the PI controller	case 1	129.71	0.592
	case 2	343.16	0.726
	case 3	98.13	/

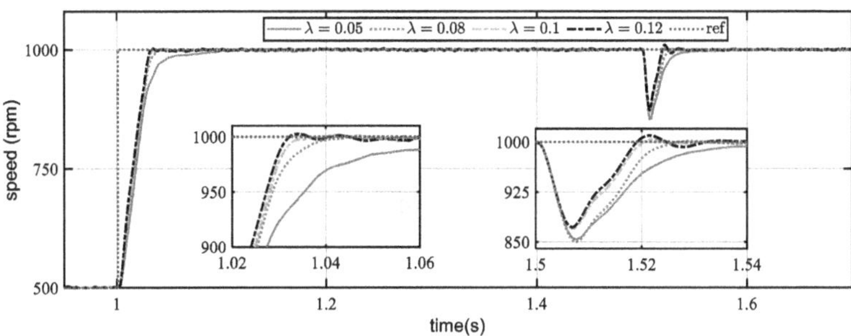

Figure 7.13 Speed regulation results under different parameter λ.

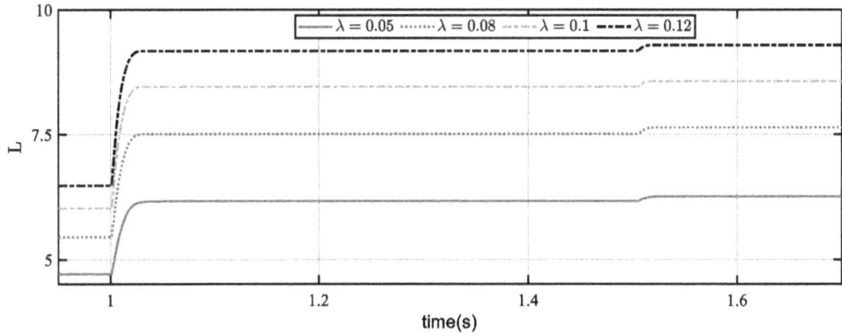

Figure 7.14 Response curves of L under different parameter λ.

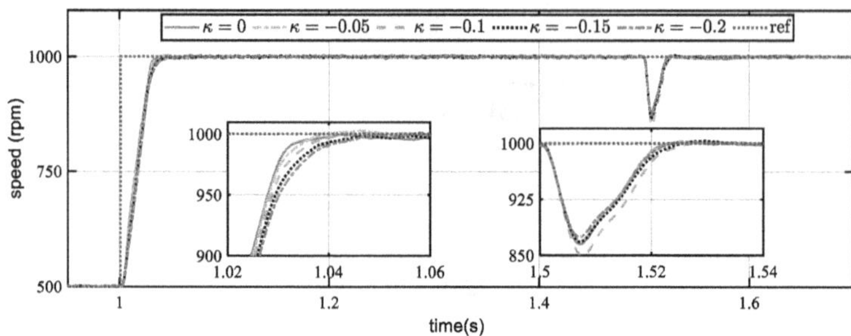

Figure 7.15 Speed regulation results under different parameter κ.

impact on dynamic performance of system. As a result, by adjusting κ, it can not only alleviate the system overshoot problem caused by the excessive L, but also suppress the defect that introducing fractional order reduces the system response speed. Naturally, with the appropriate adjustment of the parameter L, one can achieve the introduction of fractional orders to improve the robustness of system to external disturbances without deteriorating the dynamic performance of the system.

ρ is the parameter to regulate the system bandwidth from a new dimension, and the simulation results are displayed in Figures 7.17 and 7.18, from which it can be observed that the ρ can suppress the growth rate of the adaptive scaling gain. With λ, κ chosen appropriately, a slight change in ρ can enable the system to achieve a better control performance.

7.4.2.3 Experimental tests

In this experiment, the adopted experimental configuration is the same as in the robust control experiment above as shown in Figure 7.6, and control frequency is 20 kHz. The cascaded PI speed controller is used as the benchmark controller for comparison.

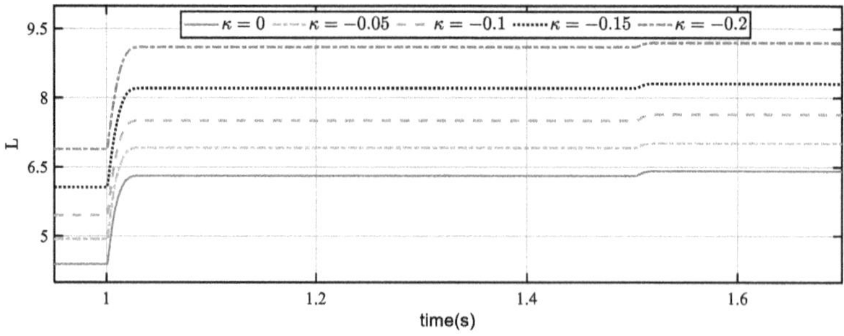

Figure 7.16 Response curves of L under different parameter κ.

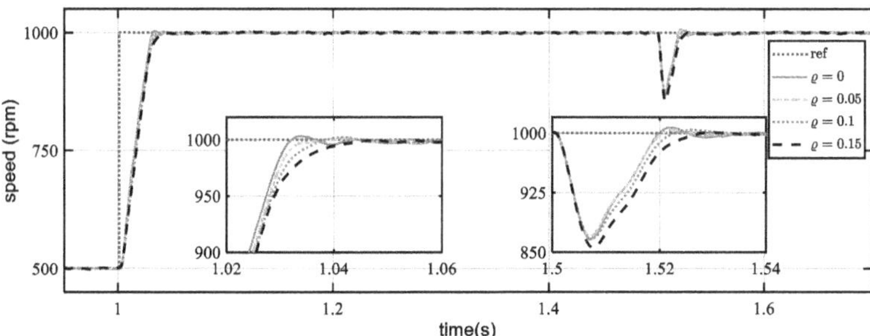

Figure 7.17 Speed regulation results under different parameter ρ.

Specifically, the control parameters are given as: $l_{1,0} = 10$, $l_{1,1} = 6$, $l_{1,2} = 3.5$, $\lambda_1 = 6 \times 10^5$, $l_{2,0} = 11$, $l_{2,1} = 6.5$, $\lambda_2 = 7 \times 10^4$, $l_{3,0} = 4$, $l_{3,1} = 3$, $\lambda_3 = 3500$, $k_1 = 0.1$, $k_2 = 0.07$, $\delta = 1$, $\lambda = 0.25$, $\rho = 1.1$, $\kappa = -0.05$ for the proposed robust controller; $k_{pp} = 2$, $k_{pi} = 6.66 \times 10^{-4}$, $k_{dp} = 2.31$, $k_{di} = 0.28$ for the cascaded PI controller. Similarly, the experimental tests are divided into the following two scenarios: the first scenario is that at 5s, the reference speed is switched from 500 rpm to 2000 rpm, following the load disturbance of 0.12 Nm is applied at 7 s. In the second one, the reference speed is still switched from 500 rpm to 2000 rpm at 5 s, while a sinusoidal disturbance of $0.12\sin(4\pi t)$ Nm is introduced to the motor at 7 s.

As can be observed from Figure 7.19, the proposed nonrecursive adaptive controller can quickly track to the new speed reference with minimal overshoot after the reference switching, and the dynamic performance is satisfactory to the traditional cascaded PI controller. With the disturbance enforced at 5 s, it can be observed from Figure 7.19 that the cascaded PI controller shows a large speed drop, while the proposed nonsmooth adaptive nonrecursive control shows only a slight speed drop. Similarly, in Figure 7.20, the cascaded PI controller exhibits

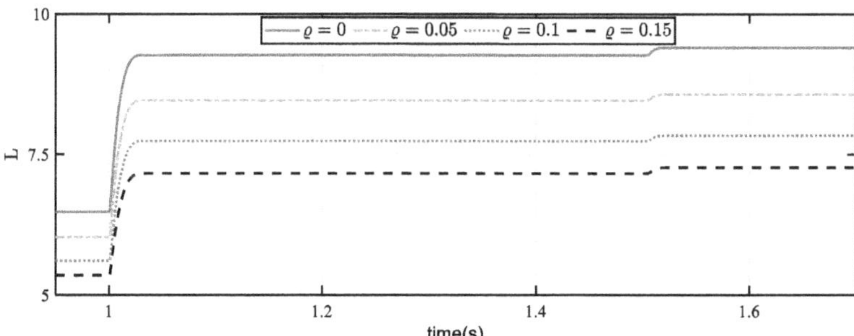

Figure 7.18 Response curves of L under different parameter ρ.

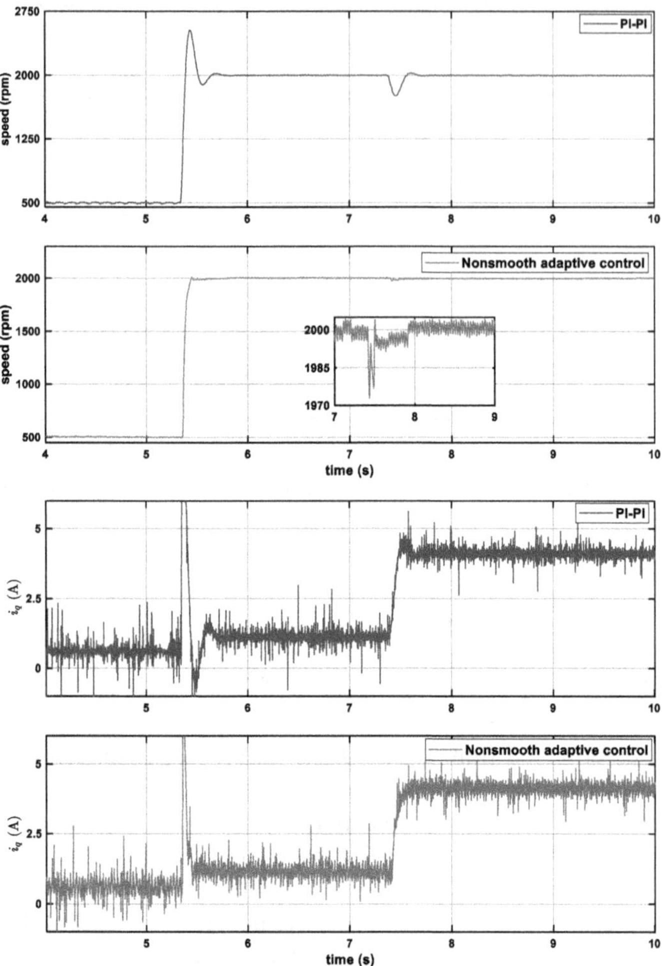

Figure 7.19 Experiment results of the reference speed switches from 500 rpm to 1500 rpm with $T_L = 0$ Nm at 5 s, and the T_L switches to 0.12 Nm at 7 s.

large speed oscillation with the presence of a sinusoidal load disturbance, while the proposed control can regulate speed fluctuations to within 40 rpm.

Consequently, we can obtain from the experiment results above that the proposed composite control method can notably improve the dynamic performance of the system under the non-cascaded structure, achieving accurate compensation of mismatched torque disturbance. Compared with the conventional cascaded PI controller, a significant performance enhancement in the dynamic performance and disturbance rejection capability of the closed-loop control system are achieved.

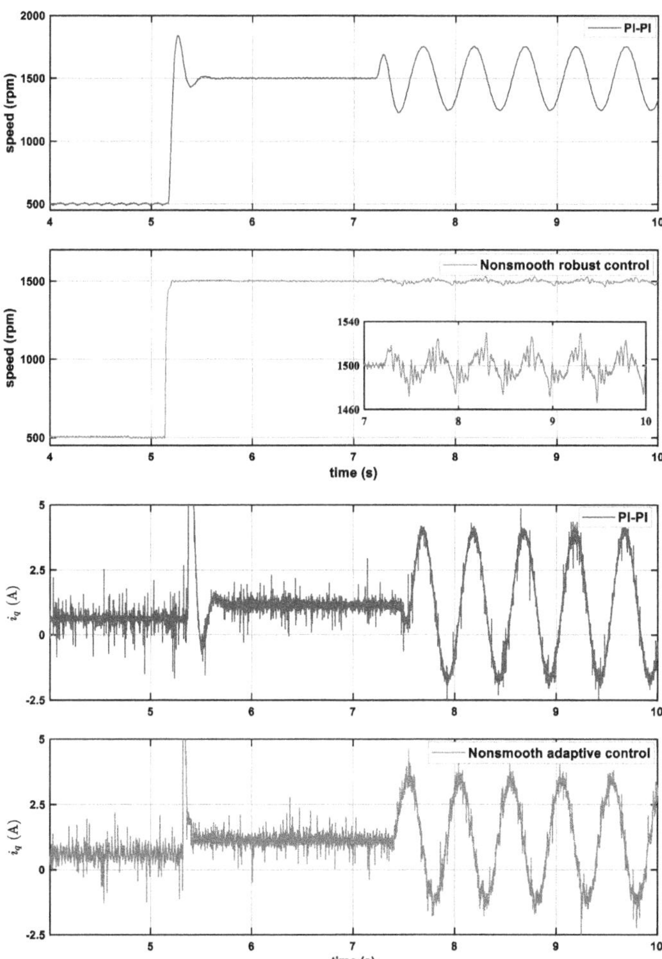

Figure 7.20 Experiment results of the reference speed switches from 500 rpm to 1500 rpm with $T_L = 0$ Nm at 5 s, and the T_L switches to $0.12\sin(4\pi t)$ Nm at 7 s.

Chapter 8

Application to Series Elastic Actuator Systems

8.1 Introduction

Series Elastic Actuators (SEAs) are widely applied in advanced robot applications due to their advantages over conventional stiff and non-back derivable actuators in force control, e.g., high fidelity, low cost, low stiction, etc. [41, 57, 88]. As a sketch review of the related literature, linear controllers constitute a main choice. Furthermore, the performances can also be improved by using nonlinear control strategies and adding feedforward control loops, see for examples [51,53,64], only mention a few. However, it is worthy of pointing out that almost all those existing control applications to SEAs are based on an asymptotical control result. Due to the apparent advantage of fast convergence rate and stronger robustness, finite-time output regulation is of significance in the robust motion control problem for SEAs. One recent work [78] in the literature addresses the finite-time control problem for SEA by using a terminal sliding-mode control scheme, where the bothersome chattering issue cannot be avoided. In this section, we will show that the proposed theoretical result will provide a much easier control implementation, and meanwhile significant control performance improvements can be achieved compared with the conventional PD and linear state feedback controllers, while there are not much added complexities of the gain tuning mechanism.

In this chapter, we study the advanced control problem for a class of SEA system with a novel design (as depicted by Figure 8.1) that gives the actuator

DOI: 10.1201/9781003399230-8

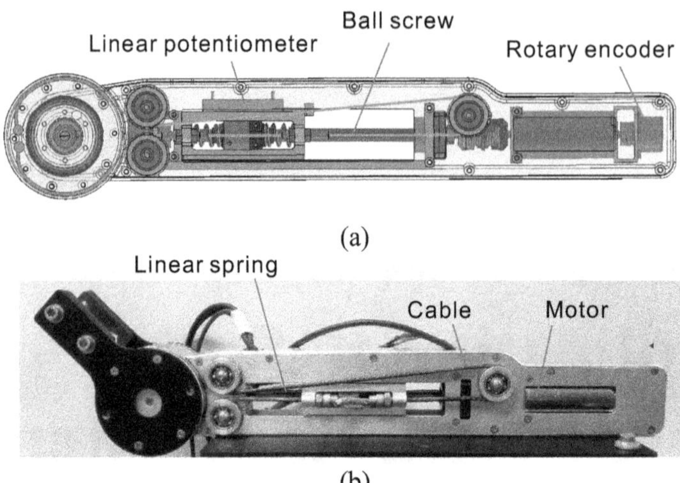

Figure 8.1 Series Elastic Actuator (SEA).

different impedances at different force ranges. The actuator has two series of elastic elements: a linear spring with a low stiffness and a torsional spring with a high stiffness. Figure 8.1 (a) is a cross section showing the structure of the studied actuator. The motor (Maxon EC-4-pole brushless DC motor operating at 200 W) shown is coupled to a ball screw through a torsional spring. Two incremental encoders (Renishaw RM22IC) with resolutions of 2048 and 1024 pulses per revolution are used to measure the angular displacement of the motor shaft and lead screw respectively.

8.2 Modeling of SEAs

Using the analogy of two-mass-spring-damper system, by neglecting the inevitable unmodeled disturbances, one can obtain the nominal mathematical model of the following form [64]:

$$\begin{cases} m_m\ddot{q}_m + b_m\dot{q}_m = F_m - k(q_m - q_l) \\ m_l\ddot{q}_l + b_l\dot{q}_l = k(q_m - q_l), \end{cases} \tag{8.1}$$

where the descriptions of all involved parameters are listed in Table 8.1.

Table 8.1 Involved parameters of the SEA

Parameters	Description
m_m	inertia/mass of the motor
m_l	inertia/mass of the link
b_m	viscous friction coefficient of the motor
b_l	viscous friction coefficient of the link
q_m	angle/position of the motor
q_l	angle/position of the link
k	stiffness of the SEA
F_m	motor torque/force

Let $x = [x_1, x_2, x_3, x_4]^\top = [q_l, \dot{q}_l, q_m, \dot{q}_m]^\top$. System (8.1) can be expressed as the following general state-space form:

$$\begin{cases} \dot{x}_1 = x_2, \\ \dot{x}_2 = \dfrac{k}{m_l}x_3 - \dfrac{k}{m_l}x_1 + d_1, \\ \dot{x}_3 = x_4, \\ \dot{x}_4 = \dfrac{1}{m_m}F_m - \dfrac{k}{m_m}(x_3 - x_1) + d_2, \end{cases}$$

where d_1, d_2 are the lumped mismatched/matched disturbance torque/force which might consist of the unknown viscous friction effects, internal uncertainties and external disturbances, respectively.

8.2.1 Parameters in experimental setup

In the experimental setup, the control algorithms (8.3) and (8.6) are implemented in real-time at 1 KHz on a dSPACE DS1007 processor board with the DS3002 incremental encoder board for reading the encoders, while the motor is controlled using the Elmo Gold Whistle Servo Drive. The involved parameter values of the experimental SEA are identified as $m_m = 2.2 \times 10^{-6}(\mathrm{kg \cdot m^2})$, $m_l = 4 \times 10^{-6}(\mathrm{kg \cdot m^2})$, $k = 0.14(\mathrm{N \cdot m/rad})$.

8.3 Practically Oriented Finite-Time Control for SEA

8.3.1 Controller design

In this section, we first consider the practically oriented finite-time control problem for the SEA system. We consider a more practical control objective, namely, semi-global control, rather than the restrictive global control target. Hence, without any pre-verifications of certain nonlinearity growth conditions, it is straightforward to utilize the nonrecursive tracking control approach proposed in Chapter 3 to design a robust finite-time control law (the tracking reference is denoted by q_{lref}) to realize the accurate position control for the SEA system.

We start to address the controller by first neglecting the disturbance terms d_1, d_2. With a series of pre-calculations of the following form:

$$\begin{cases} x_1^* = q_{lref}, \\ x_2^* = q_{lref}^{(1)}, \\ x_3^* = \dfrac{m_l}{k}\left(q_{lref}^{(2)} + \dfrac{k}{m_l}q_{lref} + \dfrac{b_l}{m_l}q_{lref}^{(1)}\right), \\ x_4^* = x_3^{*(1)}, \\ F_m^* = m_m\left(x_4^{*(1)} + \dfrac{k}{m_m}(x_3^* - x_1^*) + \dfrac{b_m}{m_m}x_4^*\right), \end{cases}$$

and the following change of coordinates:

$$\begin{cases} z_1 = x_1 - x_1^*, \\ z_2 = (x_2 - x_2^*)/L, \\ z_3 = (x_3 - x_3^*)/L^2, \\ z_4 = (x_4 - x_4^*)/L^3, \end{cases} \tag{8.2}$$

then we are able to construct the following implementable finite-time control law

$$\begin{cases} v = -K\left[\lfloor z_1\rceil^{1+4\tau}, \lfloor z_2\rceil^{\frac{1+4\tau}{1+\tau}}, \lfloor z_3\rceil^{\frac{1+4\tau}{1+2\tau}}, \lfloor z_4\rceil^{\frac{1+4\tau}{1+3\tau}}\right]^\top, \\ F_m = L^4 v + F_m^*. \end{cases} \tag{8.3}$$

8.3.2 Experimental studies

In what follows, starting from PD control, we will show how to establish the proposed practically oriented finite-time controller and elaborate the control performance improvement by choosing appropriate parameters of the proposed finite-time controller.

Step 1: from PD control to state feedback control.

In the starting session, we first implement a conventional PD controller to the SEA. Figures 8.2–8.3 show the set-point and trajectory tracking performances under different proportional gains ($k_p = 0.3, 0.5, 0.7, 0.9$ while $k_d = 0.01$), where $q_{lref} = 0.5$(rad) in Figure 8.2 and $q_{lref} = 0.5\sin(5t + \varphi)$(rad) in Figure 8.3, respectively. Generally speaking, choosing a larger proportional gain will result in faster convergence rate, higher precision, but meanwhile, larger overshoot and

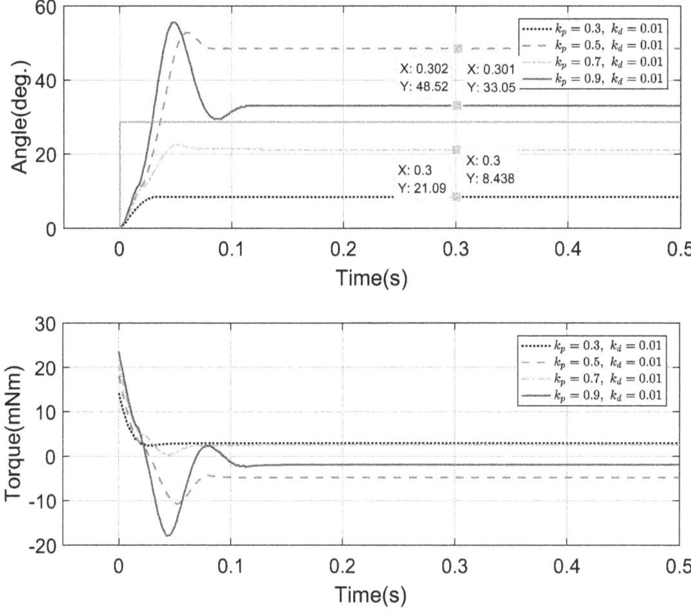

Figure 8.2 Set-point tracking performances under PD controller with different proportional gain k_p.

control energy consumption. Without loss of generality, we extend the PD coefficients of the dash lines in Figures 8.2 and 8.3 as the coefficients of linear state feedback controller, i.e., $K = [0.5, 0.01, 0.1, 0.01]$. Compared with the dot line in Figures 8.4 and 8.7, it is obvious to see the progressive tracking performances of state feedback controller.

Step 2: from state feedback control to finite-time control.

First notice that the proposed finite-time control law (8.3) reduces to a linear state feedback controller if we set the homogeneous degree $\tau = 0$. In this step, by simply modifying the homogeneous degree τ from 0 to several negative values gradually, we can implement the proposed finite-time controller to obtain a better control performance while the control gains can be set as fixed values. By understanding that in real-life systems, there are various disturbances/uncertainties, hence it is of significance that the proposed finite-time controller could reduce the settling time and improve the system robustness against the inevitable disturbances/uncertainties. As depicted by Figures 8.4 and 8.7, the control performance is significantly improved if τ is a negative value and moreover, a smaller τ will clearly lead to a faster convergence speed and lower steady error. Under the proposed finite-time controller (8.3), it can be observed from Figures 8.5 and 8.8 that the bandwidth factor L has also played a key role as a larger L will lead to an

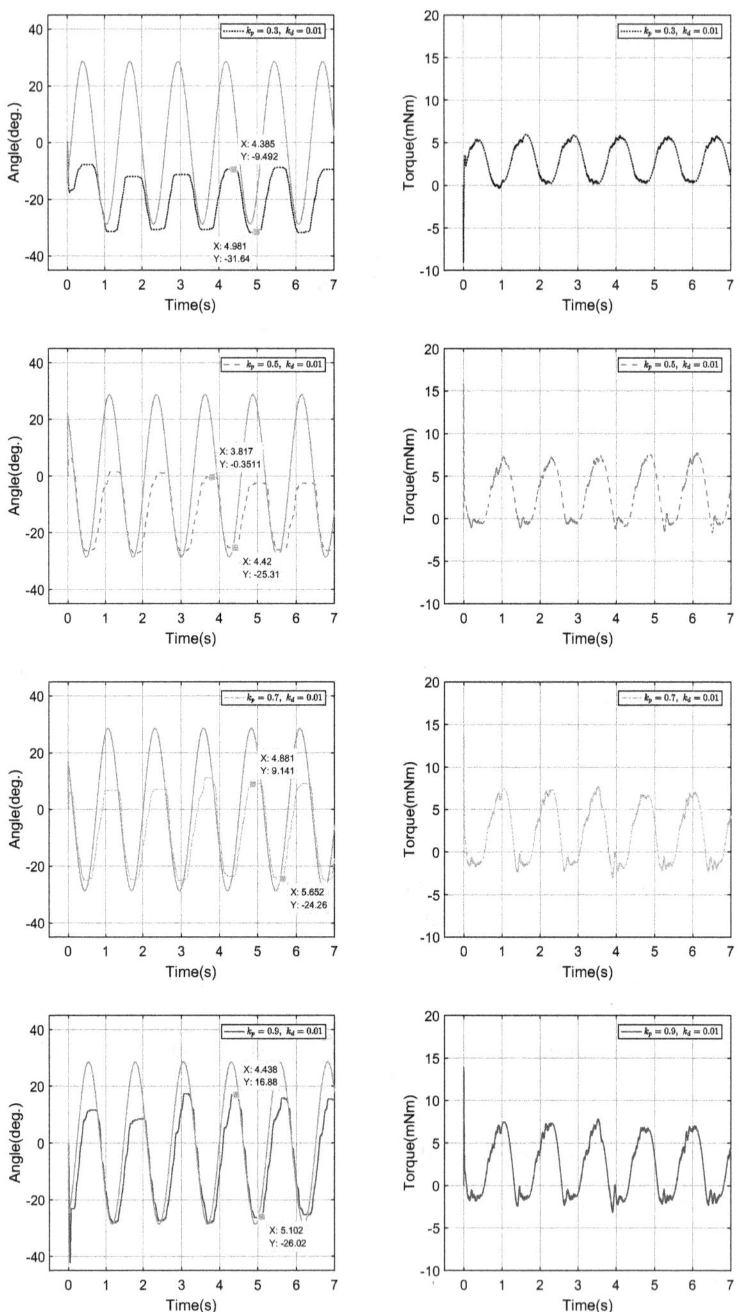

Figure 8.3 Trajectory tracking performances under PD controller with different proportional gain k_p.

obvious performance variation as well. However, it should be pointed out here that a larger L will cause a clear deterioration of the system robustness against the measurement noises, which is a common problem of the existing high gain control methods. In Figures 8.6 and 8.9, the control performance variation along with the control gain selection k_1 while the other control parameters are fixed is depicted. To make the comparisons clearer and more precise, the performance indexes (overshoot, offset) of set-point tracking and integral square error (ISE) index ($\int_{t_1}^{t_2} e^2(t)dt$ where $[t_1, t_2]$ is a period of the reference signal in the steady state and $e(t)$ is the tracking error) for trajectory tracking case are included in Table 8.2.

As a direct conclusion from the above illustrated figures, the proposed finite-time control strategy will clearly lead to a significant control performance improvement while the control gain selection guideline is as simple as conventional linear state feedback controllers. Moreover, the added negative homogeneous degree will endow the control engineers a much flexibility of tuning the control performances in practical implementations.

Performance comparison with optimal controller

In order to better demonstrate the control performance superiorities of the proposed nonsmooth controller with the existing asymptotical controllers, we present an experimental performance comparison with an optimal controller by predictive approach [8, 18, 87]. The controller is derived based on optimiz-

Table 8.2 Performance indexes of PD controller, linear state feedback controller, and finite-time controller

Methods	Parameters	Overshoot	Offset	ISE
PD	$k_p = 0.3$	0.00	−20.21	751.14
	$k_p = 0.5$	8.68	19.87	425.32
	$k_p = 0.7$	4.90	−7.56	176.71
	$k_p = 0.9$	68.08	4.40	114.31
FTC	$\tau = 0$	25.72	−4.04	133.68
	$\tau = -0.05$	26.39	−3.34	141.67
	$\tau = -0.1$	33.74	−2.63	46.29
	$\tau = -0.15$	44.97	−0.52	10.45
	$L = 1$	33.74	−2.63	46.29
	$L = 1.05$	73.30	−2.28	23.66
	$L = 1.1$	57.70	−1.23	10.77
	$L = 1.15$	88.33	−1.58	17.45
	$k_1 = 0.3$	42.05	2.29	56.88
	$k_1 = 0.4$	52.45	0.18	23.54
	$k_1 = 0.5$	57.70	−1.23	10.77
	$k_1 = 0.6$	66.28	0.53	9.26

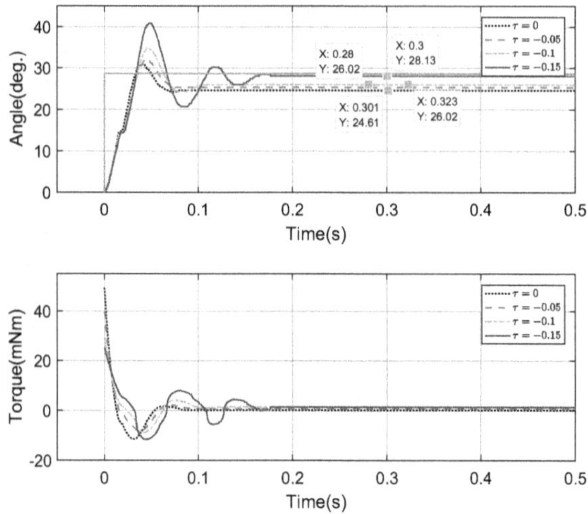

Figure 8.4 Set-point tracking performances under finite-time controller (8.3) with different homogeneous degree τ while $K = [0.5, 0.01, 0.1, 0.01]$ and $L = 1$.

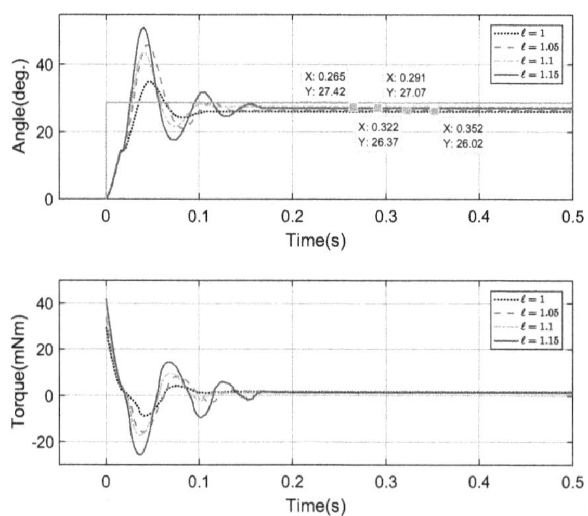

Figure 8.5 Set-point tracking performances under finite-time controller (8.3) with different bandwidth factor L while $\tau = -0.1$ and $K = [0.5, 0.01, 0.1, 0.01]$.

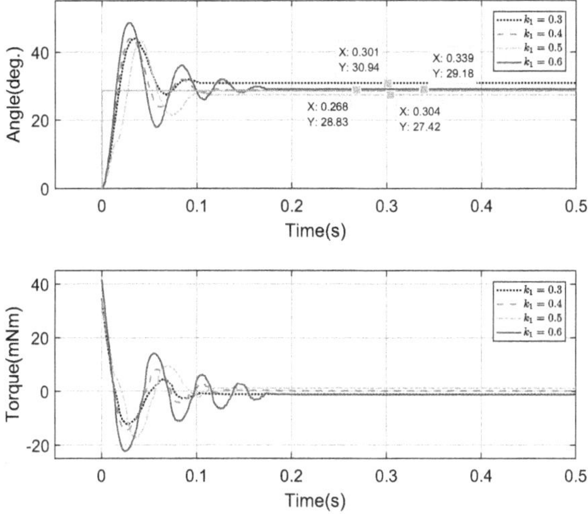

Figure 8.6 Set-point tracking performances under finite-time controller (8.3) with different control gain k_1 while $[k_2, k_3, k_4] = [0.01, 0.1, 0.01]$, $\tau = -0.1$, and $L = 1.1$.

ing a performance index $J(t) = \frac{1}{2}\int_0^T (x_1(t+\tau) - x_1^*(t+\tau))^2 d\tau$, where T is the predictive period. Utilizing a predictive approach associated with the Taylor expansion, the optimal controller is derived in the form of

$$F_m^{Opt} = -\sum_{i=1}^{4} k_i^{Opt}(x_i - x_i^*). \tag{8.4}$$

The detailed process of derivation can be found in [8]. It is worth noting that the optimal gains $k_i^{Opt}, i \in \mathbb{N}_{1:4}$ are only related to the predictive period T and the control order r. For simplicity, the control order r is set as 0 and the predictive period T is set as the only tunable parameter.

Under a set-point tracking control objective, the control performance comparisons of the proposed finite-time controller (5.22) and the optimal controller (8.4) are presented in Figure 8.10. By noting that the initial torque amplitude of the candidate controllers is placed in a similar level in order to make a fair comparison, the robustness exhibited by the proposed finite-time controllers is much stronger than the optimal controllers. A lower steady state error can be achieved with the import of a negative homogeneous degree. Similar conclusions can also be obtained from the case of trajectory tracking as shown in Figure 8.11. To make the comparison clearer, the detailed performance indexes of both set-point tracking and trajectory tracking cases are also included in Table 8.3.

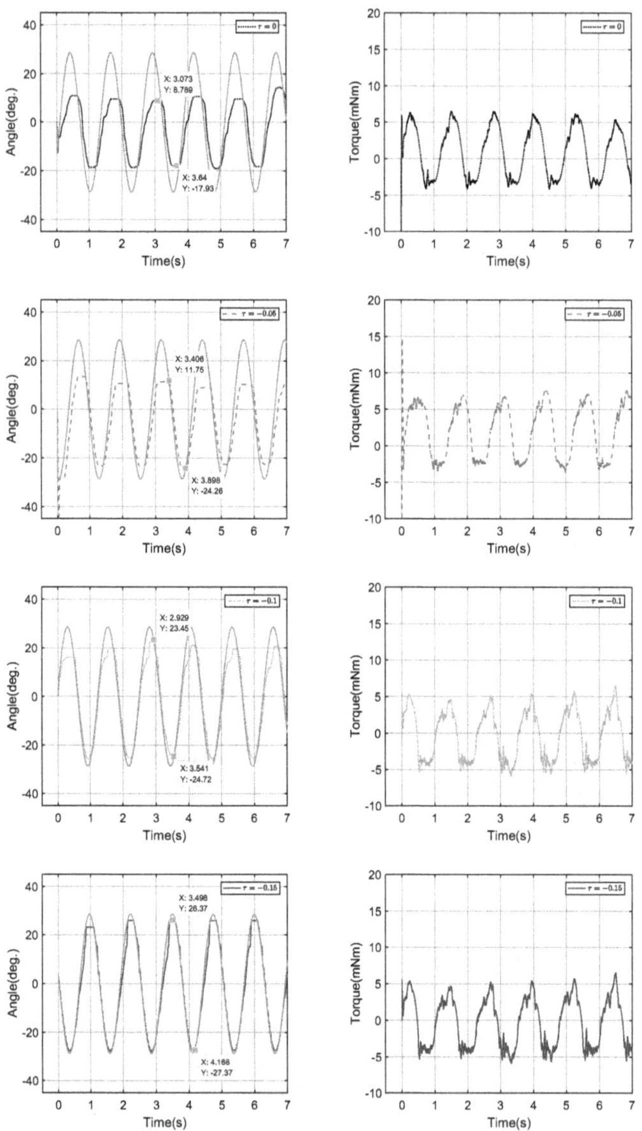

Figure 8.7 Trajectory tracking performances under finite-time controller (8.3) with different homogeneous degree τ while $K = [0.5, 0.01, 0.1, 0.01]$ and $L = 1$.

8.4 Nonsmooth Composite Control for SEA

In order to improve the output regulation precision for the position control of SEA with the presence of serious external disturbances, an active disturbance

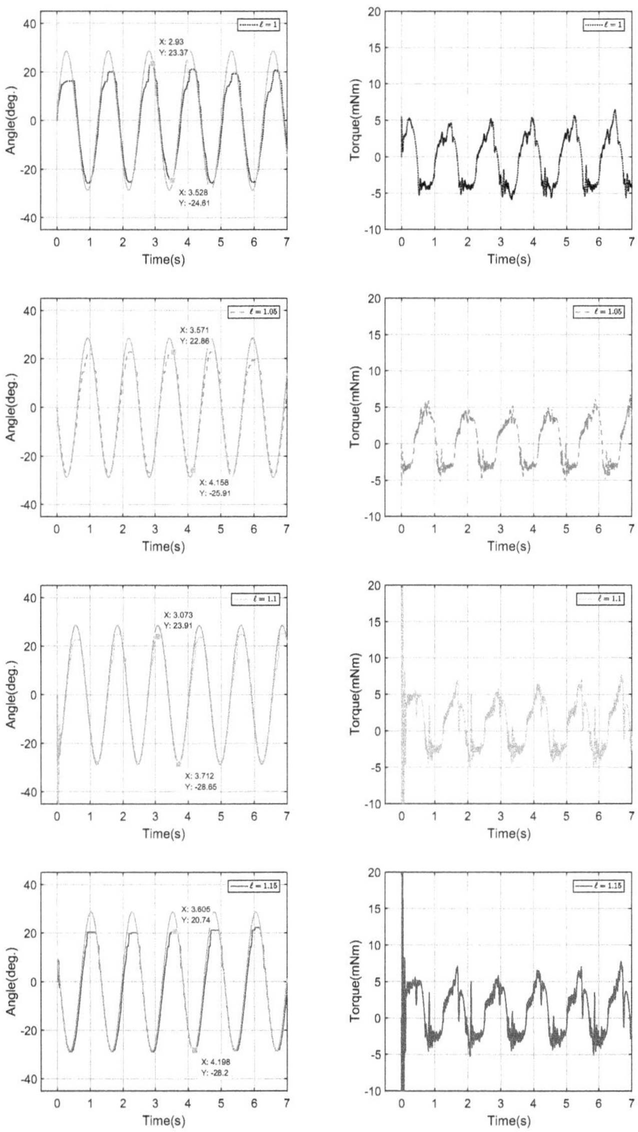

Figure 8.8 Trajectory tracking performances under finite-time controller (8.3) with different bandwidth factor L while $\tau = -0.1$ and $K = [0.5, 0.01, 0.1, 0.01]$.

estimation and feedforward attenuation strategy will be of significance. In this section, we will apply the nonrecursive composite output regulation method to further improve the control performance of the SEA system.

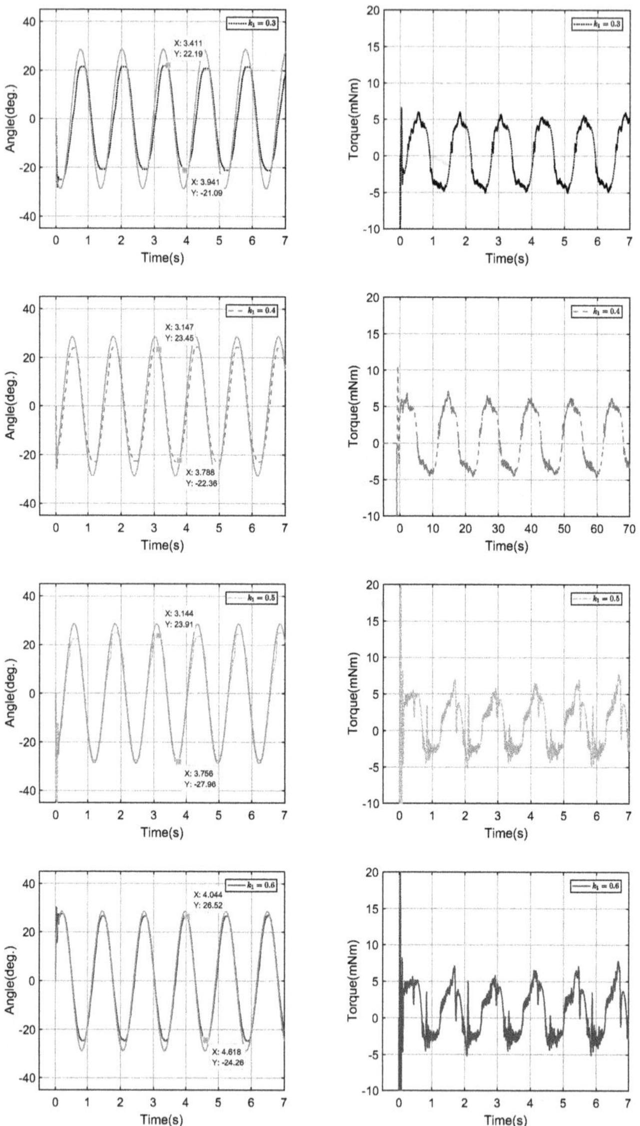

Figure 8.9 Trajectory tracking performances under finite-time controller (8.3) with different control gain k_1 while $[k_2, k_3, k_4] = [0.01, 0.1, 0.01]$, $\tau = -0.1$, and $L = 1.1$.

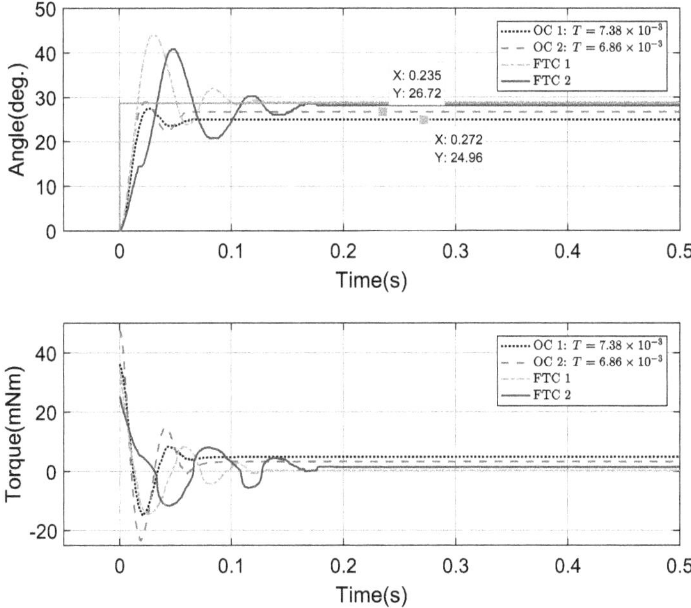

Figure 8.10 Set-point tracking performance comparisons under optimal controller and finite-time controllers (FTC 1: $K = [0.4, 0.01, 0.1, 0.01]$, $L = 1.1$, and $\tau = -0.1$; FTC 2: $K = [0.5, 0.01, 0.1, 0.01]$, $L = 1$, and $\tau = -0.15$).

8.4.1 Controller design

Following Theorem 5.5, a nonsmooth composite control law with the tracking reference, denoted by y_r, can be explicitly built. With a series of pre-calculations as

$$\begin{aligned}
x_1^* &= y_r, \\
x_2^* &= y_r^{(1)}, \\
x_3^* &= \frac{m_l}{k}\left(y_r^{(2)} + \frac{k}{m_l} y_r - \hat{z}_{2,2}\right), \\
x_4^* &= \frac{m_l}{k}\left(y_r^{(3)} + \frac{k}{m_l} y_r^{(1)} - \hat{z}_{2,3}\right), \\
F_m^* &= m_m\Big(\frac{m_l}{k}\big(y_r^{(4)} + \frac{k}{m_l} y_r^{(2)} - \hat{z}_{2,4}\big) + \frac{k}{m_m}(x_3^* - x_1^*) - \hat{z}_{4,2}\Big),
\end{aligned} \tag{8.5}$$

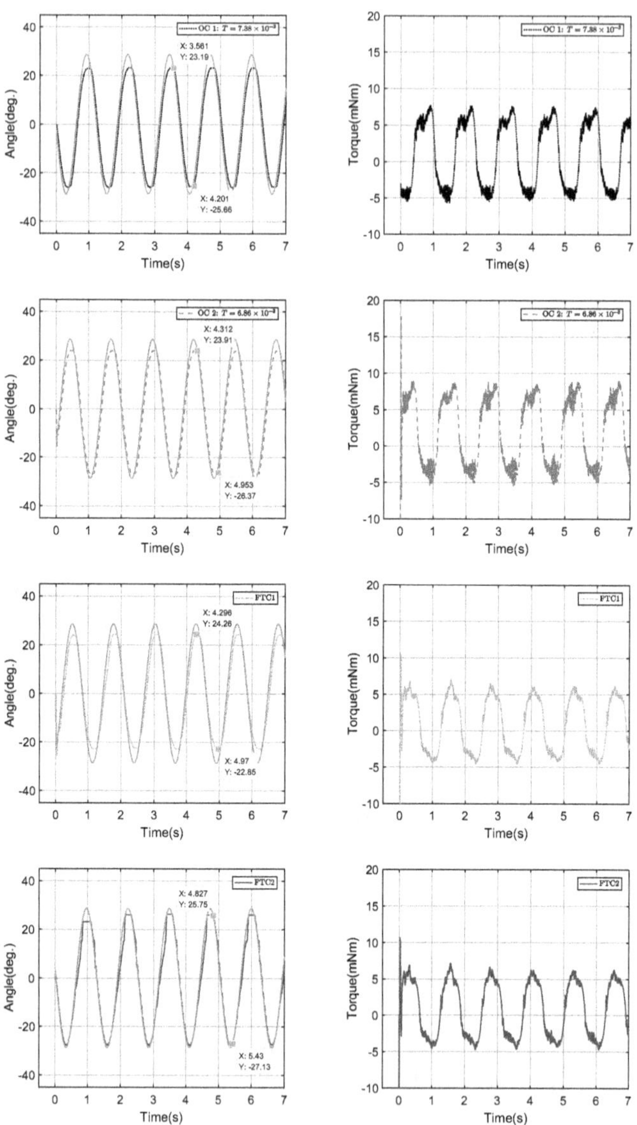

Figure 8.11 Trajectory tracking performance comparisons under optimal controllers and finite-time controllers (FTC 1: $K = [0.4, 0.01, 0.1, 0.01]$, $L = 1.1$, and $\tau = -0.1$; FTC 2: $K = [0.5, 0.01, 0.1, 0.01]$, $L = 1$, and $\tau = -0.15$).

Table 8.3 Performance indexes of the optimal controller and the finite-time controller

Methods	Parameters	Overshoot	Offset	ISE
OC	OC1	9.86	−3.69	27.18
	OC2	7.90	−1.93	18.53
FTC	FTC1	52.45	0.18	23.54
	FTC2	44.97	−0.52	10.45

we are able to obtain the following nonsmooth composite controller

$$
\begin{aligned}
&\begin{cases}
\dot{\hat{z}}_{2,1} = \frac{k}{m_l}x_3 - \frac{k}{m_l}x_1 + \hat{z}_{2,2} + \ell_{2,1}\sigma_2\lfloor z_{2,1} - \hat{z}_{2,1}\rceil^{1+\tau} \\
\dot{\hat{z}}_{2,2} = \hat{z}_{2,3} + \ell_{2,2}\sigma_2^2\lfloor z_{2,1} - \hat{z}_{2,1}\rceil^{1+2\tau} \\
\dot{\hat{z}}_{2,3} = \hat{z}_{2,4} + \ell_{2,3}\sigma_2^3\lfloor z_{2,1} - \hat{z}_{2,1}\rceil^{1+3\tau} \\
\dot{\hat{z}}_{2,4} = \ell_{2,4}\sigma_2^4\lfloor z_{2,1} - \hat{z}_{2,1}\rceil^{1+4\tau};
\end{cases} \\
&\begin{cases}
\dot{\hat{z}}_{4,1} = \frac{1}{m_m}F_m - \frac{k}{m_m}(x_3 - x_1) + \hat{z}_{4,2} + \ell_{4,1}\sigma_4\lfloor z_{4,1} - \hat{z}_{4,1}\rceil^{1+\tau} \\
\dot{\hat{z}}_{4,2} = \ell_{4,2}\sigma_4^2\lfloor z_{4,1} - \hat{z}_{4,1}\rceil^{1+2\tau};
\end{cases} \\
&F_m = -L^4K\left[\lfloor\zeta_1\rceil^{1+4\omega}, \lfloor\zeta_2\rceil^{\frac{1+4\omega}{1+\omega}}, \lfloor\zeta_3\rceil^{\frac{1+4\omega}{1+2\omega}}, \lfloor\zeta_4\rceil^{\frac{1+4\omega}{1+3\omega}}\right]^{\top} + F_m^*.
\end{aligned} \tag{8.6}
$$

8.4.2 Experimental studies

Case I: A linear composite control scenario

First, we consider a simple linear control case, that is, by intentionally setting the homogeneous degree as $\tau = \omega = 0$ in the control scheme (8.6), the proposed method reduces to a linear GPI observer based controller (GPIOBC) [48]. In the experimental test, the gain parameters for the GPIOBC are given as $K = [0.8, 0.01, 0.2, 0.01]$, $L = 1$, $[\ell_{2,1}, \ell_{2,2}, \ell_{2,3}, \ell_{2,4}] = [4, 6, 4, 1]$, $[\ell_{4,1}, \ell_{4,2}] = [2, 1]$, $\sigma_2 = 10$, $\sigma_4 = 50$. It is shown in Figures 8.12 and 8.13 that even in the linear composite control scenario, fine control performances of the SEA system under both set-point and trajectory tracking objective are well illustrated.

Case II: A nonsmooth composite control scenario

To better demonstrate the proposed nonsmooth control effectiveness and performance improvements of the proposed control algorithm, experimental performance comparisons are conducted among the proposed nonsmooth controller

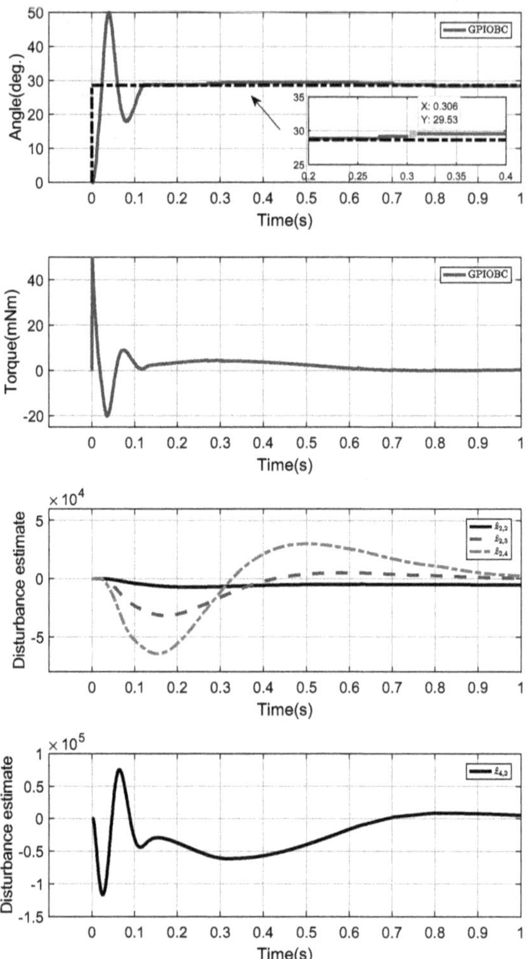

Figure 8.12 Set-point tracking control performances under the GPIOBC (i.e., controller (8.6) with $\tau = \omega = 0$).

(8.6), PID controller, and a finite-time controller (FTC) if the disturbances are not actively taken into consideration [90].

To conduct fair comparisons, lots of efforts have been made in the experimental tests where the control parameters for three candidate controllers have been tuned to achieve their best possible performances, respectively. Besides, we also present three different values of k_i in the PID controller for a better illustration of the steady state error regulation level. Note that the FTC can be regarded as an ideal control law for systems without external disturbances. Hence it can be treated as a special case of the proposed composite controller by ne-

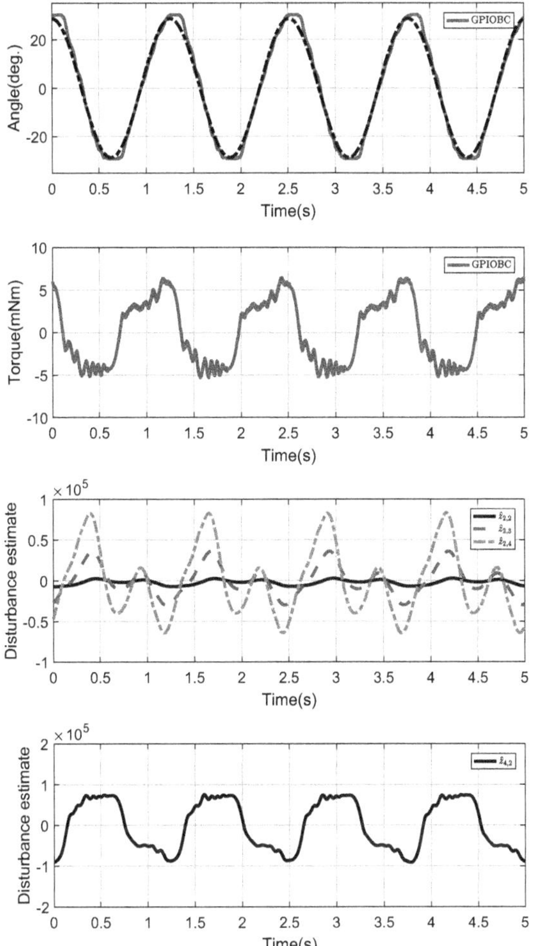

Figure 8.13 Trajectory tracking control performances under the GPIOBC (i.e., controller (8.6) with $\tau = \omega = 0$).

glecting the feedforward loops. As a consequence, the gain parameters of the PID controller are given as $k_p = 0.08$, $k_i = 0; 10; 30$, $k_d = 0.01$, respectively. The gain parameters of the FTC are given as $K = [0.660, 0.008, 0.018, 0.009], \omega = -0.1, L = 1.05$. The control parameters of the proposed controller are given as $[\ell_{2,1}, \ell_{2,2}, \ell_{2,3}, \ell_{2,4}] = [4, 6, 4, 1]$, $[\ell_{4,1}, \ell_{4,2}] = [2, 1], \sigma_2 = 50, \sigma_4 = 100, \tau = -0.1$ and K, ω, L are chosen the same with FTC.

In the set-point tracking case, as one can observe from Figures 8.14 and 8.15, the proposed nonsmooth composite controller (8.6) results in a shorter rising time than the PID controller. More obviously, it is shown that under the PID controller,

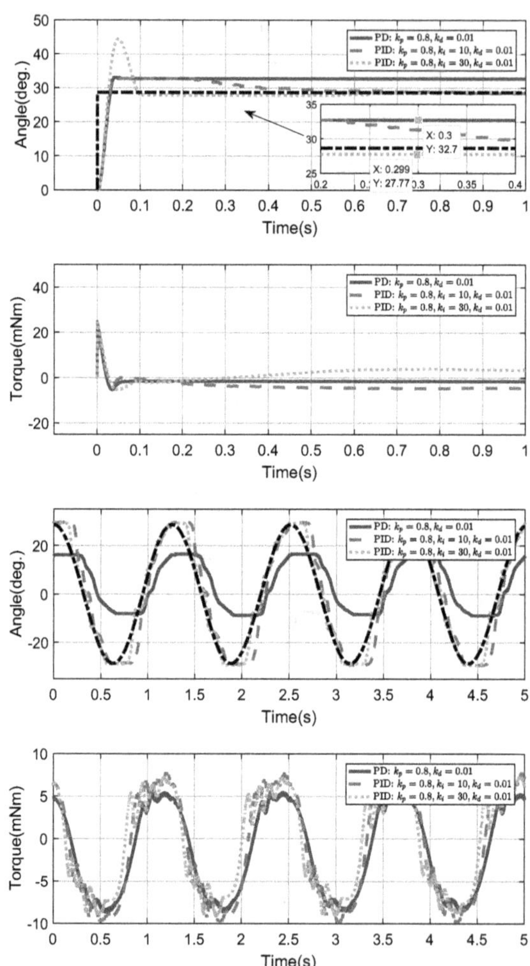

Figure 8.14 Set-point and trajectory tracking performances: PID controller.

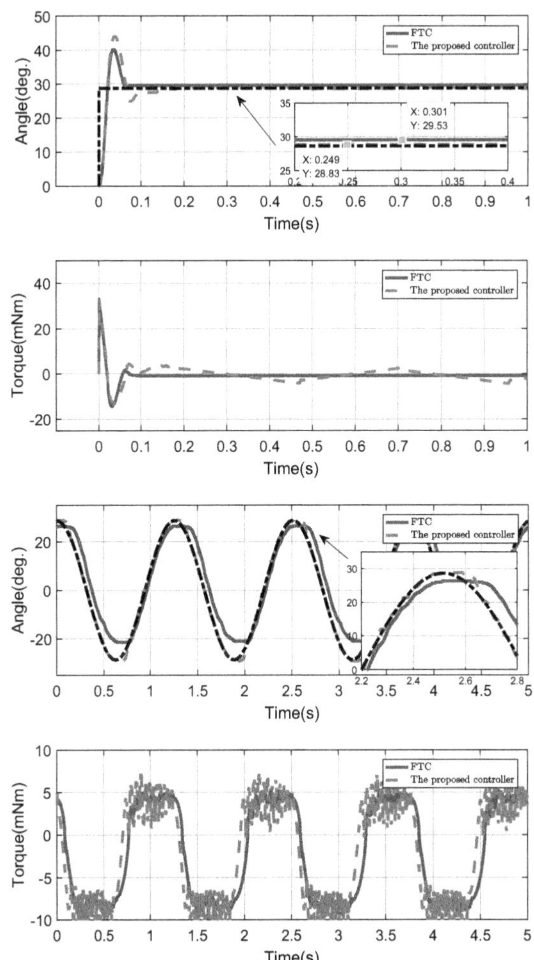

Figure 8.15 Set-point and trajectory tracking performances comparisons: FTC and the proposed controller.

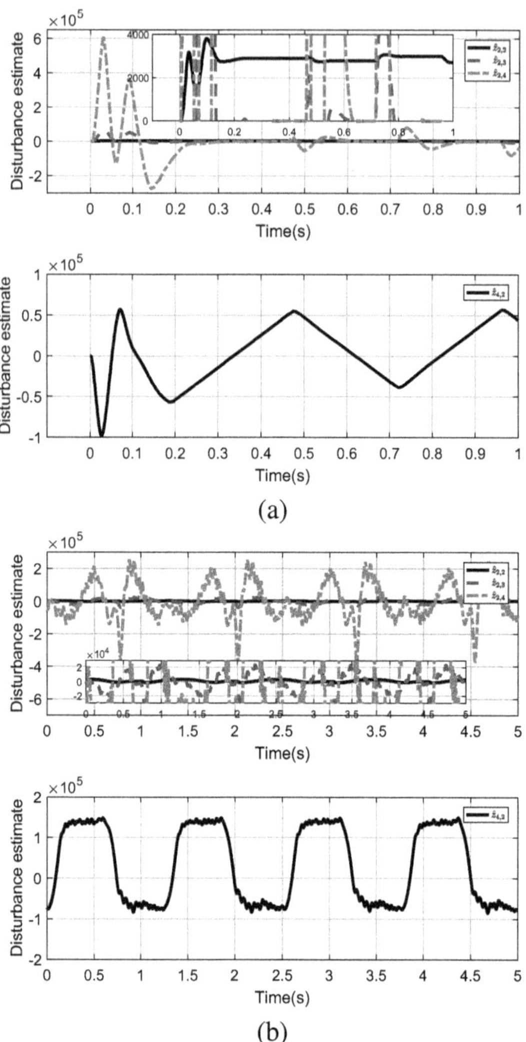

Figure 8.16 Disturbance observation performances under the proposed nonsmooth observer in (8.6). (a) Set-point tracking; (b) Trajectory tracking.

a bigger integral gain k_i could lead to a smaller steady state error. However, owing to the intentionally added feedfoward decoupling loops enabled by the designed nonsmooth observer, a smaller steady state error can be achieved than PID controller with all three difference gain parameter choices. It is shown that FTC can also provide a fast convergence speed owing to the same feedback control gain vector with the proposed controller. But on the other hand, the requirement of small steady state error cannot be guaranteed. In the trajectory tracking case,

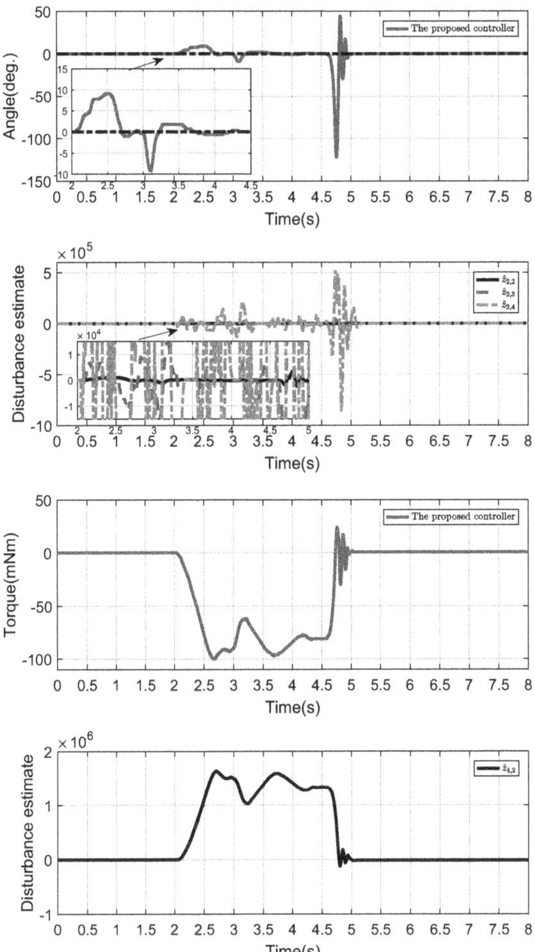

Figure 8.17 System performance recovery abilities under the proposed controller (8.6).

a consistent conclusion can be obtained that better steady state control performance under the proposed controller than both PID and FTC controllers, as is clearly observed in Figures 8.14 and 8.15. Figure 8.16 shows that the proposed disturbance observer could perform a satisfactory disturbance estimation result, which enables the feedforward compensation in each channel.

In order to test the performance recovery ability of the control system with the presence of large external disturbances, an artificial external torque is imposed and evacuated to the SEA system around $t = 2\,\text{s}$ and $t = 4.6\,\text{s}$, respectively. The position of the link will inevitably fluctuate, as illustrated by Figure

8.17. However the proposed composite controller could perform a rapid system performance recovery owing to the function of the embedded nonsmooth disturbance observer. This feature enhances the practical nature of the proposed composite control strategy.

Chapter 9

Applications to DC Microgrid Systems

9.1 Introduction of DC Microgrids

In recent years, both academia and industry take great interest in DC microgrids. As a kind of small-scale DC power distribution system, it provides a feasible and effective platform to integrate renewable energy sources (RESs), energy storage systems (ESSs) and different kinds of loads. Compared with AC microgrids, DC microgrids behave relatively simple topological structure, less energy conversion process, etc. Moreover, the regulations of reactive power/frequency are no longer demanded. Owing to the above advantages, DC microgrids are widely applied in electric vehicles, more electric aircraft, commercial buildings, rural areas, etc.

Maintaining stable operation of the bus voltage is the basic condition for DC microgrids. According to a particular DC microgrid, which is shown in Figure 9.1, the stability of its bus voltage is influenced by the power variation of input and output sides. On the input power side, RESs (e.g., photovoltaic and wind turbines) usually work in maximum power point tracking (MPPT) mode to largely harvest power without participating in the regulation of bus voltage. However, the output power of RESs always varies with solar irradiance, environment temperature, wind speed, weather conditions, etc. For the output side, the drastic changes in loads also affect the stability of the system. Besides conventional resistive loads, it is noted that power electronic converters and motor drives in advanced automotive systems, when tightly regulated, behave as constant power loads (CPLs). As reported in [17], they always show negative impedance characteristics at input terminals, which might affect power quality and even lead to

DOI: 10.1201/9781003399230-9

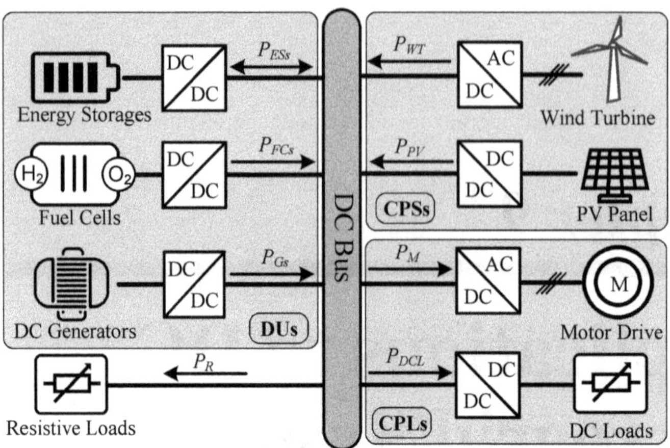

Figure 9.1 A generic layout of a DC microgrid.

unstable situations. Hence, various active control techniques have been investigated to mitigate the instability problems caused by CPLs. Furthermore, the high frequency characteristics of the switching power supply will increase the current/voltage rate of change and thus lead to electrical interactions. In this regard, multiple converters (especially DC-DC converters) usually produce adverse coupled interactions at the system level. These undesired interactions may result in global system instability, load imbalance for parallel converters, noise coupling and electromagnetic interference. Hence it is widely regarded as a critical issue to maintain the stability of DC microgrids when facing the inevitable system uncertainties and disturbances.

In order to handle the mismatched power issue with the presence of continuous variations of all kinds of loads and constant power sources (CPSs), dispatchable units (DUs), which usually refer to fuel cells, energy storage, micro-turbine, etc., are introduced into DC microgrids. By doing so, the mismatched power can be compensated/absorbed timely and the bus voltage fluctuation can be maintained within a proper range. For DC DUs, the realizations of constant voltage mode (CVM) and droop mode (DM) significantly rely on their properly controlled interfacing converters.

Regarding the stabilization issue for DC microgrids, a variety of studies can be found in the literature, which mainly falls into two categories: small-signal analysis (SSA) and large-signal analysis (LSA). On one hand, the principle of SSA is to calculate and locally linearize the system at the operating point. Then the stability can be studied via classical linear analysis tools. However, the small-signal model of the system will lose its accuracy if the operating condition is largely deviated, especially when plug-and-play (PnP) operation, RES fluctuations and load variations happen. On the other hand, LSA uses nonlinear tools, e.g., Lyapunov function based methods, to analyze the global system stability. It enables the system to resist large external disturbances caused by the reconnection/disconnection of DUs or the unpredictable variations of local/global loads.

To avoid unnecessary compromises to model completeness and make LSA more applicable to the DC MG with amounting complexities, in this chapter, we propose several new decentralized composite controller design methods to unify both CVM and DM. Each DU, as well as its interfacing converter, constitutes a subsystem. For easy explanation, RESs are emulated as constant power sources (CPSs). These CPSs together with resistive loads and CPLs are merged into the equivalent lumped load which would be fed by DUs. In this context, the electrical coupling of a particular DU subsystem with other subsystems/loads is first estimated by a certain disturbance observer, such as higher-order sliding mode observer (HOSMO), nonlinear disturbance observer (NDO), high gain disturbance observers, etc. Then, instead of directly dealing with the DC MG as a whole, the DCC is individually designed to counteract the observed coupling and simultaneously stabilize the internal states of each subsystem in the large-signal sense. By doing so, each DU subsystem can be decoupled from one another and operates in an isolated way from the perspectives of control and stability. When those subsystems are interconnected, the global large-signal stability of the entire MG can be achieved by simply ensuring the local stabilities of individual subsystems. Therefore, this chapter mainly discusses the decentralized based stabilization scheme without any requirement for communication links between DUs, which remarkably enhances the flexibility and scalability of the DC microgrid.

9.2 A Decentralized Composite Controller Design for DC Microgrids

9.2.1 Standardized modeling of DC microgrids

This section is meant to propose a universal control scheme for both a single DU subsystem and multiple DU subsystems. Without loss of generality, the averaged model of the ith boost converter in Figure 9.2 can be given as follows:

$$\begin{cases} L_i \dot{i}_{Li} = -(1-d_i)v_{Ci} + E_i, \\ C_i \dot{v}_{Ci} = (1-d_i)i_{Li} - \dfrac{P_{oi}}{v_{Ci}} \end{cases} \tag{9.1}$$

where L_i, C_i, i_{Li}, v_{Ci}, d_i, and E_i are the inductance value, capacitance value, instantaneous inductor current, instantaneous capacitor voltage, duty cycle, and DC source voltage of ith DU subsystem respectively, and P_{oi} represents the output power.

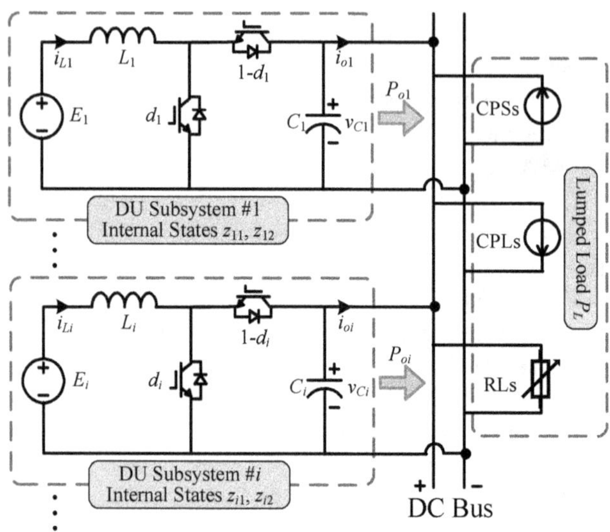

Figure 9.2 Detailed DU configurations feeding the lumped load.

Inspired by the exact feedback linearization reported in [27], the original bilinear system (9.1) can be transformed by the following coordinate transformation:

$$\begin{aligned} z_{i1} &= 0.5L_i i_{Li}^2 + 0.5C_i v_{Ci}^2, \\ z_{i2} &= E_i i_{Li}. \end{aligned} \tag{9.2}$$

Then, the first-order derivative of the above two equations can be obtained:

$$\begin{aligned} \dot{z}_{i1} &= L i_{Li} \dot{i}_{Li} + C_i v_{Ci} \dot{v}_{Ci} = E_i i_{Li} - P_{oi}, \\ \dot{z}_{i2} &= \frac{E_i^2}{L_i} - \frac{E_i v_{Ci}}{L_i}(1 - d_i). \end{aligned} \tag{9.3}$$

Combining (9.2) with (9.3), the following relationship can be obtained

$$\begin{cases} \dot{z}_{i1} = z_{i2} + \varsigma_i, \\ \dot{z}_{i2} = u_i, \end{cases} \tag{9.4}$$

where $u_i = \frac{E_i^2}{L_i} - \frac{E_i v_{Ci}}{L_i}(1 - d_i)$, $\varsigma_i = -P_{oi}$.

Through the above analysis, the reference value is obtained and the control objective of u_i is enable z_{i1} track its reference value z_{i1r}, which is expressed as

$$\begin{aligned} z_{i1r} &= 0.5L_i i_{Lir}^2 + 0.5C_i v_{Cir}^2 \\ &= 0.5L_i(-\varsigma_i/E_i)^2 + 0.5C_i v_{Cir}^2. \end{aligned} \tag{9.5}$$

In (9.5), v_{Cir} represents the output voltage reference. It could be a constant in CVM, or determined by a droop controller in DM. For the latter, v_{Cir} can be written as,

$$v_{Cir} = V^* - m_i P_{oi} = V^* + m_i \varsigma_i, \tag{9.6}$$

where V^* is the nominal DC bus voltage and m_i is the droop coefficient of the ith DU subsystem.

9.2.2 Decentralized composite controller design

Comparing (9.1) and (9.4), it is shown that the original nonlinear model of the boost converter has been converted into a linear one, which substantially facilitates the large-signal regulator design. However, the dynamics described in (9.4) are affected by ς_i that is exactly the additive inverse of the output power. The presence of ς_i may threaten the tracking performance and even impair the system stability. On these bases, it is imperative to know the precise value of ς_i and further process it appropriately in the control law. To this end, a disturbance observer can be constructed as the following form [37]:

$$\begin{cases} \dot{\hat{z}}_{i11} = z_{i2} + \hat{z}_{i12} + l_{i1}\sigma_i(z_{i1} - \hat{z}_{i11}), \\ \dot{\hat{z}}_{i12} = \hat{z}_{i13} + l_{i2}\sigma_i^2(z_{i1} - \hat{z}_{i11}), \\ \dot{\hat{z}}_{i13} = l_{i3}\sigma_i^3(z_{i1} - \hat{z}_{i11}), \end{cases} \tag{9.7}$$

where $\hat{z}_{i11}$, $\hat{z}_{i12}$, and $\hat{z}_{i13}$ are the estimates of z_{i1}, ς_i, and $\dot{\varsigma}_{i1}$, respectively $\sigma_i > 1$ is a scaling gain which will be identified later. $H_i = \text{col}\,(l_{i1}, l_{i2}, l_{i3})$ is an observer gain vector with its components being corresponding to the coefficients of a Hurwitz polynomial,

$$p_i(s) = s^3 + l_{i1}s^2 + l_{i2}s + l_{i3}. \tag{9.8}$$

From a practical point of view, P_{oi} has its physical meaning, and it would not go to infinity in the real engineering. Therefore, the derivatives of ς_i should be limited in bounded ranges, which can be mathematically delineated below,

$$\max_{i \le n, j=1,2} \left\{ \sup \left| \frac{\partial \varsigma_i^j}{\partial t^j} \right| \right\} \le D,\ D \in \mathbb{R}_+^+. \tag{9.9}$$

Combing (9.4) and (9.7) with the denotations $e_{i1} = z_{i1} - \hat{z}_{i11}$, $e_{i2} = (\varsigma_i - \hat{z}_{i12})/\sigma_i$, $e_{i3} = (\dot{\varsigma}_i - \hat{z}_{i13})/\sigma_i^2$, the error dynamics gives

$$\dot{\mathbf{e}}_i = \sigma_i \mathbf{A}_i \mathbf{e}_i + \mathbf{B}_i \ddot{\varsigma}_i, \tag{9.10}$$

where $\mathbf{e}_i = [e_{i1}, e_{i2}, e_{i3}]^\top$ is the error vector. $\mathbf{A}_i = [-l_{i1}, 1, 0; -l_{i2}, 0, 1; -l_{i3}, 0, 0;]$ and $\mathbf{B}_i = [0, 0, \sigma_i^{-2}]^\top$ represent the system matrices.

Hence, there must exist a symmetrical and positive definite matrix $\mathbf{Q}_i$ satisfying $\mathbf{Q}_i^\top \mathbf{A}_i + \mathbf{A}_i \mathbf{Q}_i + \mathbf{I} = 0$.

Next, it will be shown that error dynamics can be stabilized at the origin if σ_i is properly tuned, which suggests that $\hat{z}_{i11}$, $\hat{z}_{i12}$, and $\hat{z}_{i13}$ enable to practically track z_{i1}, ς_i, $\dot{\varsigma}_i$.

To this end, constructing a Lyapunov function $\mathbf{V}_{ei} = \mathbf{e}_i^\top \mathbf{Q}_i \mathbf{e}_i$ whose time derivative along the error dynamics is given by,

$$\begin{aligned}\dot{\mathbf{V}}_{ei} &= \sigma_i \frac{\partial \mathbf{V}_{ei}}{\partial \mathbf{e}_i^\top} \mathbf{A}_i \mathbf{e}_i + \frac{\partial \mathbf{V}_{ei}}{\partial \mathbf{e}_i^\top} \mathbf{B}_i \ddot{\varsigma}_i = -\sigma_i ||\mathbf{e}_i||^2 + 2\mathbf{e}_i^\top \mathbf{Q}_i \mathbf{B}_i \ddot{\varsigma}_i \\ &\leq -\sigma_i ||\mathbf{e}_i||^2 + 2\gamma_i ||\mathbf{e}_i||, \end{aligned} \tag{9.11}$$

where $\gamma_i = \sigma_i^{-2} \lambda_{\max}(\mathbf{Q}_i) D$, and $\lambda_{\max}(\mathbf{Q}_i)$ denotes the maximum eigenvalue of $\mathbf{Q}_i$.

With the completion of squares, the following inequality holds

$$2\gamma_i ||\mathbf{e}_i|| \leq ||\mathbf{e}_i||^2 + \gamma_i^2. \tag{9.12}$$

Substituting (9.12) into (9.11) yields,

$$\begin{aligned}\dot{\mathbf{V}}_{ei} &\leq -(\sigma_i - 1)||\mathbf{e}_i||^2 + \gamma_i^2 \\ &\leq -(\sigma_i - 1)\lambda_{\max}^{-1}(\mathbf{Q}_i)\mathbf{V}_{ei} + \gamma_i^2. \end{aligned} \tag{9.13}$$

Defining $\Omega_{\varepsilon i} = \{\mathbf{e}_i \in \mathbb{R}^3 | \mathbf{V}_{ei} \leq \varepsilon_i\}$ where $\varepsilon_i > 0$ is a constant which can be arbitrarily small, it is apparent that $\Omega_{\varepsilon i} \subset \mathbb{R}^3$. Noting that σ_i is independent of γ_i, a sufficiently large σ_i can be selected to satisfy

$$\gamma_i^2 \leq 0.5(\sigma_i - 1)\lambda_{\max}^{-1}(\mathbf{Q}_i)\varepsilon_i. \tag{9.14}$$

Then, for any $\mathbf{e}_i \in (\mathbb{R}^3 \setminus \Omega_{\varepsilon i})$ and thereby $\mathbf{V}_{ei} \geq \varepsilon_i$, (9.11) can be rearranged as

$$\dot{\mathbf{V}}_{ei} \leq -(\sigma_i - 1)\lambda_{\max}^{-1}(\mathbf{Q}_i)\varepsilon_i + \gamma_i^2. \tag{9.15}$$

With the substitution of (9.14) into (9.15), slightly manipulating (9.15) gives the following:

$$\dot{\mathbf{V}}_{ei} \leq -0.5(\sigma_i - 1)\lambda_{\max}^{-1}(\mathbf{Q}_i)\varepsilon_i < 0. \tag{9.16}$$

As implied by (9.16), the time derivative of Lyapunov function for errors can be rigorously negative in the case when $V_{ei} > \varepsilon_i$. Moreover, by scrutinizing (9.14), σ_i and ε_i are inversely correlated. When γ_i keeps unchanged, the increased σ_i leads to the decreased ε_i. Hence, it is possible to adequately magnify σ_i so that ε_i almost equals zero, which means that $\Omega_{\varepsilon i}$ can be infinitesimal and consequently $\mathbf{e}_i$ can be regulated as small as possible. In this sense, based on (9.7), it is rational to replace ς_i in (9.5) with the estimated value $\hat{z}_{i12}$, which can be expressed as

$$z_{i1r} = 0.5L_i(-\hat{z}_{i12}/E_i)^2 + 0.5C_i v_{Cir}^2. \tag{9.17}$$

Then, the estimated quantity can now be employed to accomplish the composite controller design. For that, intermediate states should be introduced first. These states are defined as

$$\begin{aligned} \xi_{i1} &= z_{i1} - z_{i1r}, \\ \xi_{i2} &= (z_{i2} - z_{i2r})/\beta_i, \\ v_i &= (u_i - u_{ir})/\beta_i^2, \end{aligned} \tag{9.18}$$

where $\beta_i > 1$ is a positive scaling factor which will be made precisely later on, v_i is an auxiliary control input, z_{i1r} has been given by (9.17), z_{i2r} and u_{ir} are the reference signals for z_{i2} and u_i, which are given below:

$$z_{i2r} = \dot{z}_{i1r} - \hat{z}_{i12}, \; u_{ir} = \ddot{z}_{i1r} - \hat{z}_{i13}. \tag{9.19}$$

It is evident that z_{i1} and z_{i2} are equal to their references z_{i1r} and z_{i2r} in the case that the intermediate states reduce to zero. Accordingly, the boost converter output voltage v_{Ci} can be forced to track v_{Cir}.

For a better understanding this scheme, the time derivatives of ξ_{i1} and ξ_{i2} can be computed as

$$\dot{\xi}_{i1} = \beta_i \xi_{i2} + \sigma_i e_{i2}, \; \dot{\xi}_{i2} = \beta_i v_i + l_{i2}\sigma_i^2 e_{i1}/\beta_i. \tag{9.20}$$

To stabilize the above dynamics, v_i could be designed as the linear combination of ξ_{i1} and ξ_{i2}, i.e., $-k_{i1}\xi_{i1} - k_{i2}\xi_{i2}$ where k_{i1} and k_{i2} are positive constants. Then the equivalent decentralized composite controller (DCC) u_i can be expressed as

$$u_i = -\beta_i^2 (k_{i1}\xi_{i1} + k_{i2}\xi_{i2}) + u_{ir}. \tag{9.21}$$

9.2.3 Experimental tests

To verify the effectiveness and feasibility of the proposed DCC, an in-house experimental platform is built up, as displayed in Figure 9.3. The platform consists of a dSPACE controller, two boost converters, a resistive load and an electronic load. In this section, the electronic load acting as a CPL is integrated to the DC bus, and for operating safety, the resistor is also incorporated into the system. It should be noted that the damping effect devoted by the resistor is minor because the resistance is purposely tuned to be excessively large, i.e., 1698 Ω. Hence, the CPL still overwhelmingly dominates the lumped load. Six cases are arranged to compare the voltage regulation performances of the converters controlled under the DCC and the PI. Experimental results will show that the former method helps to stabilize the DC MG and contributes larger operation range than the latter one in both CVM and DM.

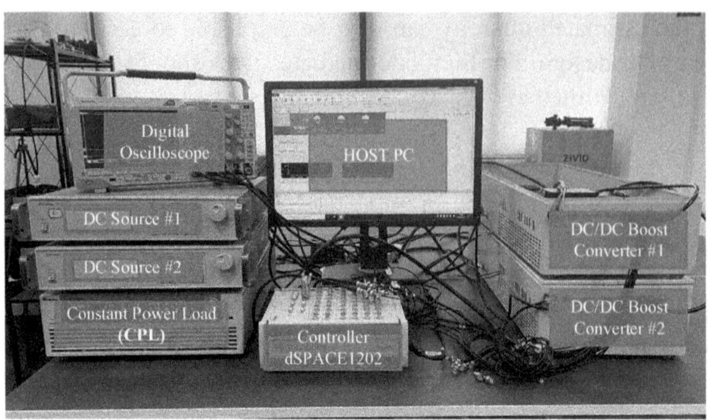

Figure 9.3 Experimental setup.

Constant voltage mode test:

This case investigates normal system operations in CVM. The key parameters in the DCC have been provided in Table 9.1. PI parameters (voltage loop: k_p=0.5, k_i=15.75; current loop: k_p=0.0775, k_i=24.35;) are carefully tuned so that the voltage deviations resulting from load changes are identical to that with the DCC. Hence, it is fair to conduct necessary comparisons of DCC and PI controllers.

As shown in Figure 9.4, DC bus is rigorously regulated at 170 V at the beginning. When a sudden CPL step-up from 50 W to 350 W is activated, identical voltage drops (5 V) are observed for both PI controller and DCC. It is conspicuous that the DCC is competent to restore the bus voltage to its nominal value (170 V) within about 9 ms, whereas the voltage recovery time of PI is estimated as 70 ms. A slight overshoot on the inductor current i_{L1} is due to the fact that the DCC makes efforts to transfer more power from the source to the load for

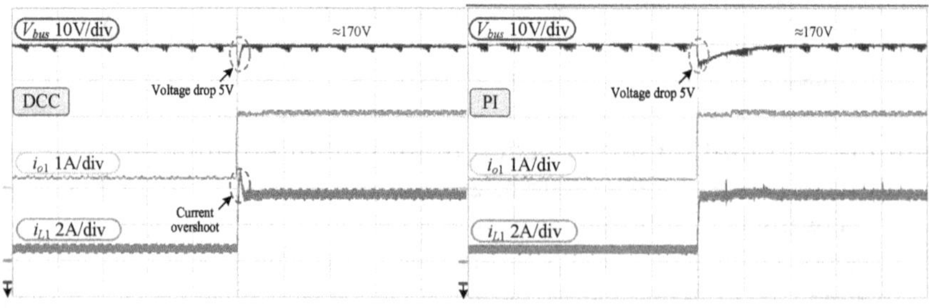

Figure 9.4 Experimental results of case 1: CVM with CPL from 50 W to 350 W.

Table 9.1 System and control parameters configuration

Parameters	Description	Value
V^*	nominal bus voltage	170 V
E_1, E_2	converter input voltage	100 V
L_1, L_2	nominal inductance value	2 mH
C_1, C_2	nominal capacitance value	470 μF
$l_{i,1},\ l_{i,2},\ l_{i,3}$	observer gains ($i = 1,2$)	3, 3, 1
$\sigma_1,\ \sigma_2$	observer scaling gains	3000
f_{sw}	switching frequency	20 kHz
$k_{11},\ k_{12}$	DCC parameters for #1 converter	1, 2
$k_{21},\ k_{22}$	DCC parameters for #2 converter	1, 2
$\beta_1,\ \beta_2$	DCC scaling gains	650

shortening the transient duration. Hence, the load-changing effects on the bus voltage can be minimized.

It is well-known that a DC system integrated with an invariant CPL and a resistor may suffer from more serious stability problems when the bus voltage is lowered. Motivated by this assertion, different from case 1 where MG voltage is maintained strictly at 170 V, case 2 examines the system stability given that the voltage reference declines while the CPL maintains at a comparatively high level (550 W). As shown in Figure 9.5, when v_{C1r} steps down from 170 V to 160 V, the MG is stable under both DCC and PI controller. However, transient oscillations of the bus voltage and the inductor current are found in the PI regulated system. In the situation that v_{C1r} reduces from 170 V to 150 V, as recorded in Figure 9.6, the system with the PI unfortunately collapses, whereas the DCC stabilizes the DC bus voltage at the targeted value. Following case 1, case 3 studies the worst scenario that the CPL abruptly surges from 50 W to 650 W, and experimental results have been shown in Figure 9.7. For this atrocious case, with the proposed DCC, although a voltage drop of 10 V occurs at the instant of a load change,

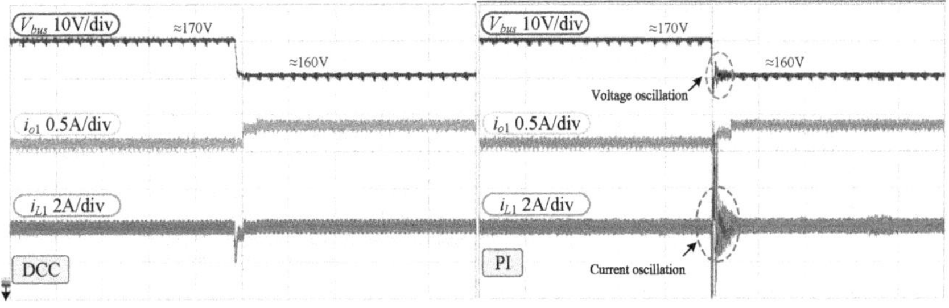

Figure 9.5 Experimental results of case 2: CVM with v_{C1r} from 170 V to 160 V.

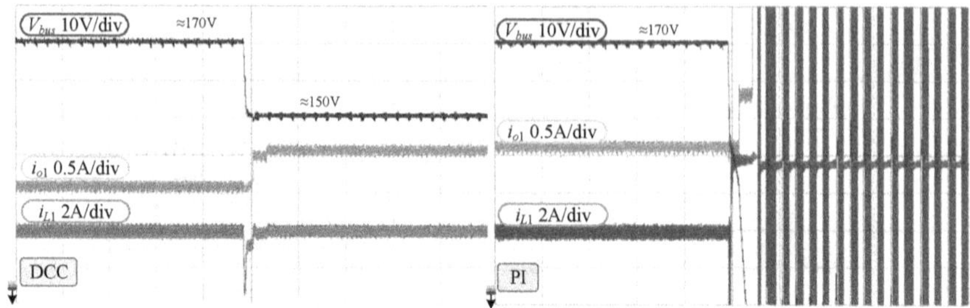

Figure 9.6 Experimental results of case 2: CVM with v_{C1r} from 170 V to 150 V.

the bus voltage quickly recovers to 170 V in a short time, which means voltage regulation has been realized. In contrast, the PI controller fails to survive this huge CPL increase, and the entire MG system is destabilized.

Summarily, the results in cases 2 and 3 experimentally consolidate the theoretical analysis that, the DCC can extend the system operating margin and attain the large-signal stability. By means of the DCC, considerable equilibrium changes are permitted by the MG, and thus, operational flexibility can be markedly raised.

Droop mode test:

Figure 9.8 shows the experimental results with the droop coefficients of two converters set as 0.01 when the CPL steps up from 100 W to 700 W. In this case, DU ratings in the MG are assumed to be identical, and they would evenly share CPL variations. For DCC and PI controller, the same voltage drops of around 7 V are observed when the load increase is triggered. The DC bus reaches a new level which is 166.5 V. Differences lie on that the DCC spends 9 ms on the bus voltage transition, whereas the PI controller takes estimated 50 ms. Similar to case 1, inductor overshoots are found during system transitions by employing the DCC. these phenomena could be well explained by exigent power transfer

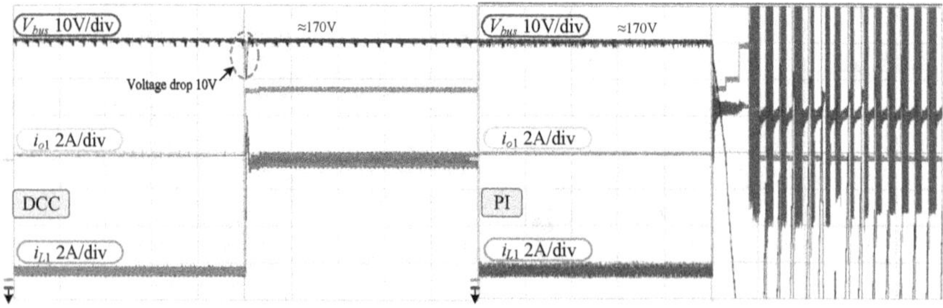

Figure 9.7 Experimental results of case 3: CVM with CPL from 50 W to 650 W.

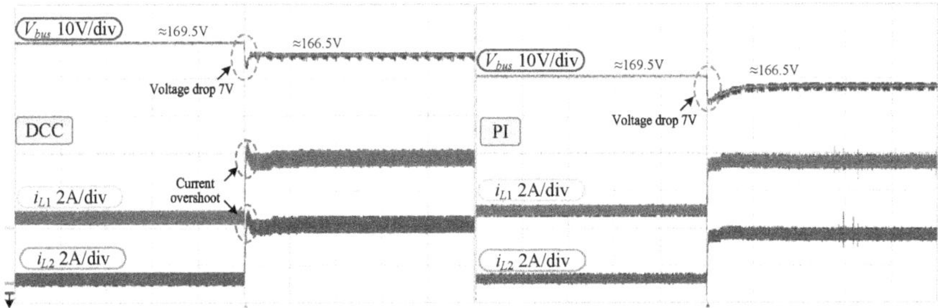

Figure 9.8 Experimental results of case 4: DM with m_1=m_2=0.01 and CPL from 100 W to 700 W.

enforced by the DCC from the source to the load. As a result, more rapid system dynamics can be acquired. It is important to reiterate that droop controllers intentionally enlarge the output impedances of the converters, which is a compromised solution to sidestep the voltage conflictions of multiple sources. The increased droop gains may induce more power interactions between CPL and sources, thus possibly aggravating the MG stability. For this reason, different from ordinary operations shown by case 4, case 5 would explore the impacts of droop gain fluctuations on the MG performances.

At first, droop gains are chosen as 0.02 for the two converters. The bus voltage fixes at 169 V when the CPL is scheduled as 100 W (see Figure 9.9). When the CPL rises to 700 W, both DCC and PI controller can stably regulate the bus voltage at around 163 V. It should be noted that the PI controller causes transient voltage and current oscillations, while the proposed DCC allows a smoother system transition. Continuing to increase the droop gains to 0.04, relevant results have been shown in Figure 9.10. The PI controller-based MG is unstable under the CPL step-up, whereas the DCC regulates the bus voltage stabilized at a new

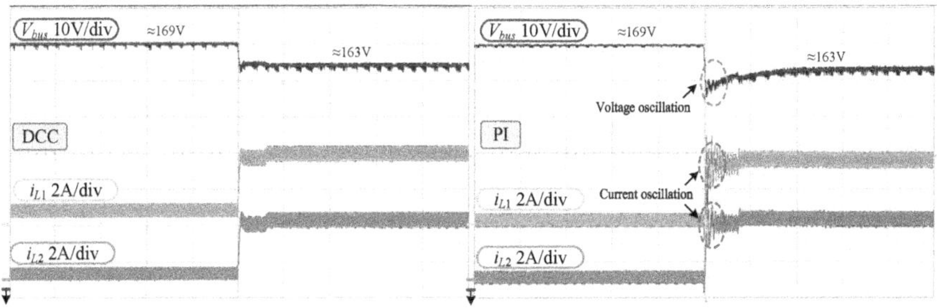

Figure 9.9 Experimental results of case 5: DM with m_1=m_2=0.02 and CPL from 100 W to 700 W.

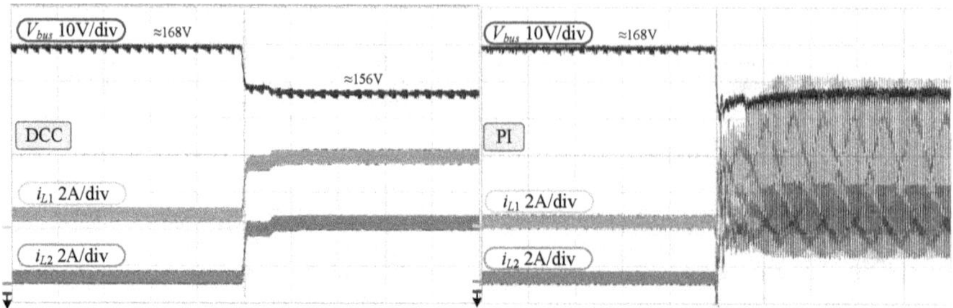

Figure 9.10 Experimental results of case 5: DM with m_1=m_2=0.04 and CPL from 100 W to 700 W.

value that is 156 V. For the three scenarios in cases 4 and 5, the two converters are assigned with identical droop coefficients for easy operations. The consolidated droop gains are 0.005, 0.01, and 0.02 respectively. When the total droop coefficient grows, the destabilizing effects on the DUs exerted by CPLs will increase accordingly. Without loss of generality, it should be noted that the superiorities of the proposed DCC uncovered by the above comparisons can also be found in the case that the converter droop gains are differently set. For example, given that m_1 and m_2 are set as 0.06 and 0.03, the total droop gain is $(m_1m_2)/(m_1+m_2)=0.02$, which is equivalent to the second situation of case 5. In this sense, it can be inferred that the corresponding results would be similar to that in Figure 9.10.

Subsequent to case 4, in this case, droop gains are maintained invariant as 0.01. The overall system stabilities regulated by both DCC and PI controller are compared when the CPL brutally grows to 1000 W. As plotted in Figure 9.11, although voltage drop of 8 V is inspected in the transient state, the DCC stabilizes the DC bus at 165 V after the CPL step-up is enabled. Unluckily, this cruel load increase cannot be tolerated under the PI control case, and thus, the whole MG is

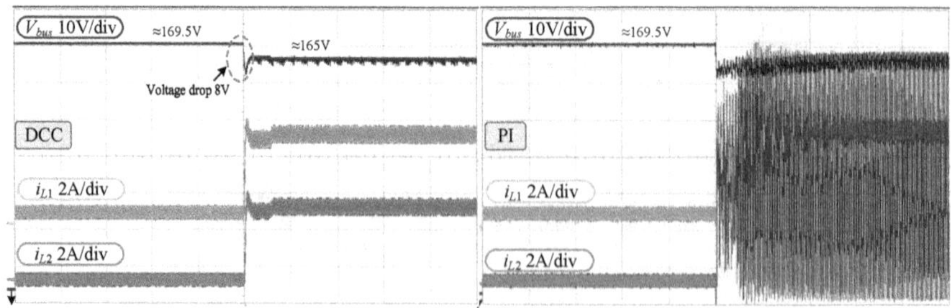

Figure 9.11 Experimental results of case 6: DM with m_1=m_2=0.01 and CPL from 100 W to 1000 W.

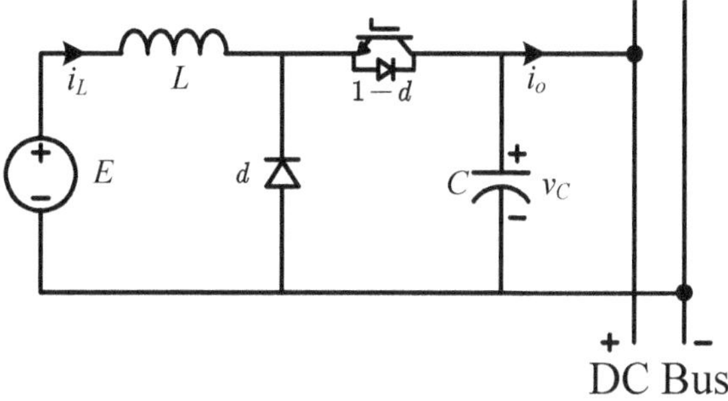

Figure 9.12 Topology of a typical buck converter.

unstable, as depicted in Figure 9.11. Conclusively, the observations reported by cases 5 and 6 again validate that the DCC proposed in this section is a large-signal regulator. It can weather large system disturbances and ensure stable operations in a wider range.

9.2.4 Extension to other types of converters

In this subsection, DCC design procedures are applied to buck converters and buck-boost converters working in CVM. If these DU converters intend to work in DM, one may simply change the voltage reference generating pattern, while DCC remain unaffected at all.

As shown in Figure 9.12, the averaged model of a buck converter can be given as follows:

$$\begin{cases} L\dot{i}_L = Ed - v_C, \\ C\dot{v}_C = i_L - i_o, \end{cases} \tag{9.22}$$

where L, C, i_L, i_o, v_C, d, and E are respectively the inductance value, capacitance value, instantaneous inductor current, instantaneous output current, instantaneous capacitor voltage, duty cycle, and DC source voltage of the buck converter.

Then, system (9.22) can be transformed by the following coordinate transformations:

$$\begin{aligned} z_1 &= 0.5Cv_C^2, \\ z_2 &= v_C i_L. \end{aligned} \tag{9.23}$$

The first-order derivative of the last two equations can be obtained:

$$\dot{z}_1 = Cv_C\dot{v}_C = Cv_C\left(\frac{i_L - i_o}{C}\right) = v_C i_L - v_C i_o,$$
$$\dot{z}_2 = \dot{v}_C i_L + \dot{i}_L v_C = \left(\frac{i_L - i_o}{C}\right) i_L + v_C\left(\frac{E\mu - u_C}{L}\right). \tag{9.24}$$

Combing (9.23) with (9.24), the following relationship can be obtained:

$$\begin{cases} \dot{z}_1 = z_2 + \varsigma_1, \\ \dot{z}_2 = u + \varsigma_2, \end{cases} \tag{9.25}$$

where $u = \frac{v_C E\mu}{L} - \frac{v_C^2}{L} + \frac{i_L^2}{C}$, $\varsigma_1 = -v_C i_o$, $\varsigma_2 = -\frac{i_L i_o}{C}$.

In (9.25), ς_1 and ς_2 are the disturbance terms that impact the dynamic states z_1 and z_2. Both these disturbances, as elaborated previously, can be regarded as the electrical couplings of buck converter with other devices. Resembling (9.7), high gain observers can be designed to estimate the necessary information of ς_1 and ς_2 respectively.

$$\begin{cases} \dot{\hat{z}}_{11} = z_2 + \hat{z}_{12} + l_{11}\sigma(z_1 - \hat{z}_{11}), \\ \dot{\hat{z}}_{12} = \hat{z}_{13} + l_{12}\sigma^2(z_1 - \hat{z}_{11}), \\ \dot{\hat{z}}_{13} = l_{13}\sigma^3(z_1 - \hat{z}_{11}), \end{cases} \tag{9.26}$$

$$\begin{cases} \dot{\hat{z}}_{21} = u + \hat{z}_{22} + l_{21}\sigma(z_2 - \hat{z}_{21}), \\ \dot{\hat{z}}_{22} = l_{22}\sigma^2(z_1 - \hat{z}_{21}), \end{cases} \tag{9.27}$$

where $\hat{z}_{11}, \hat{z}_{12}, \hat{z}_{13}$ are the estimates of $z_1, \varsigma_1, \dot{\varsigma}_1$. $\hat{z}_{21}, \hat{z}_{22}$ are the estimates of z_2, ς_2. $\sigma > 1$ is a positive scaling gain. col (l_{11}, l_{12}, l_{13}) and col (l_{21}, l_{22}) are observer gain vectors with their components being the coefficients of Hurwitz polynomials.

Then, the estimated quantities can now be employed to accomplish the composite controller design. To this end, referring to (9.1), intermediate states could be declared first,

$$\xi_1 = z_1 - z_{1r},$$
$$\xi_2 = (z_2 - z_{2r})/\beta,$$
$$v = (u - u_r)/\beta^2, \tag{9.28}$$

where $\beta > 1$ is a positive scaling gain. v is an auxiliary control input. z_{1r}, z_{2r}, and u_r, denote the reference signals for z_1, z_2, and u respectively; they can be recursively computed from (9.25)–(9.27),

$$z_{1r} = 0.5Cv_{Cr}^2,$$
$$z_{2r} = \dot{z}_{1r} - \hat{z}_{12},$$
$$u_r = \ddot{z}_{1r} - \hat{z}_{13} - \hat{z}_{22}, \tag{9.29}$$

where v_{Cr} is the output voltage reference. Note that in (9.28), buck converter internal states (z_1, z_2) and control input u will approximate to the respective references if the intermediate states are regulated to zeros. Then the output voltage control for the buck converter is achieved.

Taking the differentiation of ξ_1 and ξ_2 leads to the following dynamic equations:

$$\begin{aligned} \dot{\xi}_1 &= \beta\xi_2 + \sigma e_{12}, \\ \dot{\xi}_2 &= \beta v + l_{12}\sigma^2 e_{11}/\beta + \sigma e_{22}/\beta. \end{aligned} \tag{9.30}$$

To stabilize the above states, a linear state feedback control law that contains both ξ_1 and ξ_2 can be designed

$$v = -k_1\xi_1 - k_2\xi_2, \tag{9.31}$$

where k_1 and k_2 correspond to the Hurwitz polynomial coefficients. Then the equivalent controller u could be identified in the following form:

$$u = -\beta^2(k_1\xi_1 + k_2\xi_2) + u_r. \tag{9.32}$$

As shown in Figure 9.13, the averaged model of a buck-boost converter can be given as follows:

$$\begin{cases} L\dot{i}_L = Ed - (1-d)v_C, \\ C\dot{v}_C = (1-d)i_L - i_o, \end{cases} \tag{9.33}$$

where L, C, i_L, i_o, v_C, d, and E are the inductance value, capacitance value, instantaneous inductor current, instantaneous output current, instantaneous capacitor voltage, duty cycle, and DC source voltage of the buck-boost converter,

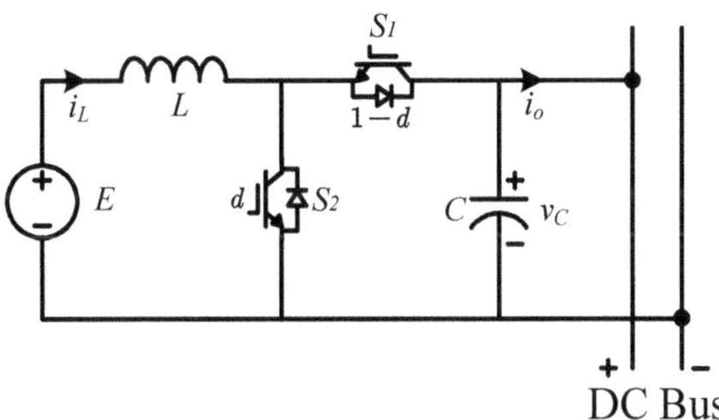

Figure 9.13 Topology of a typical buck-boost converter.

respectively. Then, the system (9.33) can be transformed by the following coordinate transformations:

$$z_1 = 0.5Li_L^2 + 0.5Cv_C^2 + CEv_C,$$
$$z_2 = Ei_L. \tag{9.34}$$

The first-order derivative of the last two equations can be obtained:

$$\dot{z}_1 = Li_L\dot{i}_L + Cv_C\dot{v}_C + CE\dot{v}_C = Ei_L - (E + v_C)i_o,$$
$$\dot{z}_2 = E\dot{i}_L = \frac{E^2d - (1-d)Ev_C}{L}. \tag{9.35}$$

Combing (9.34) with (9.35), the following relationship can be obtained:

$$\begin{cases} \dot{z}_1 = z_2 + \varsigma, \\ \dot{z}_2 = u, \end{cases} \tag{9.36}$$

where $u = \frac{E^2d-(1-d)Ev_C}{L}$, $\varsigma = -(E + v_C)i_o$.

Inspecting (9.36), it now has the identical form to the boost converter canonical model as in (9.4). In this sense, the disturbance observer (9.7) and the composite controller (9.21) can be directly extended to (9.36) without any additional modifications.

9.3 Finite-Time Performance Recovery and Decentralized Control for DC Microgrids

In this section, we consider a typical multi-source autonomous DC microgrid as shown in Figure 9.1, where the power electronic interfaces are boost type converters as they are most widely used in DC microgrid systems. To proceed with theoretical analysis, we treat the system as a simplified structure including n-th boost converters linked in a parallel with the DC bus, while the loads are categorized into a lumped resistive load R and a lumped CPL P_{CPL}.

9.3.1 *Large-signal modeling process*

The governing equations for the i-th converter can be given as follows:

$$\begin{cases} L_i\dot{i}_{Li} = -(1-\mu_i)v_{Ci} + E_i, \\ C_i\dot{v}_{Ci} = (1-\mu_i)i_{Li} - \frac{v_{Ci}}{R} - \frac{P_{CPL}}{v_{Ci}}, \end{cases} \tag{9.37}$$

where L_i, C_i, and R are the inductance, capacitance, and resistance respectively, i_{Li} and v_{Ci} are the instantaneous inductor current and capacitor voltage, respectively, μ_i is the duty cycle generated by the controller, E_i is the input voltage of each DC source. According to [2], we can transfer system (9.37) into a controllable canonical form via the following change of coordinates:

$$x_{i,1} = 0.5L_i i_{Li}^2 + 0.5C_i v_{Ci}^2, \tag{9.38}$$

$$x_{i,2} = E_i i_{Li} - \frac{v_{Ci}^2}{R_0}, \tag{9.39}$$

where R_0 is the nominal value of the resistance.

Calculating the derivative of $x_{i,1}$ and $x_{i,2}$ yields

$$\dot{x}_{i,1} = L_i i_{Li} \dot{i}_{Li} + C_i v_{Ci} \dot{v}_{Ci} = E_i i_{Li} - \frac{v_{Ci}^2}{R} - P_{CPL},$$

$$\dot{x}_{i,2} = \frac{E_i^2}{L_i} + \frac{2v_{Ci}^2}{R_0^2 C_i} - \left(\frac{E_i v_{Ci}}{L_i} + \frac{2i_{Li} v_{Ci}}{R_0 C_i}\right)(1-\mu_i) + \frac{2}{R_0 C_i}\left(P_{CPL} - \frac{v_{Ci}^2}{R_0} + \frac{v_{Ci}^2}{R}\right).$$

Define the following variables:

$$\begin{aligned}
\Delta_{i,1} &:= -P_{CPL} + \frac{v_{Ci}^2}{R_0} - \frac{v_{Ci}^2}{R}, \\
u_i &:= \frac{E_i^2}{L_i} + \frac{2v_{Ci}^2}{R_0^2 C_i} - \left(\frac{E_i v_{Ci}}{L_i} + \frac{2i_{Li} v_{Ci}}{R_0 C_i}\right)(1-\mu_i), \\
\Delta_{i,2} &:= \frac{2}{R_0 C_i}\left(P_{CPL} - \frac{v_{Ci}^2}{R_0} + \frac{v_{Ci}^2}{R}\right),
\end{aligned} \tag{9.40}$$

then system (9.37) can be rewritten as

$$\begin{cases} \dot{x}_{i,1} = x_{i,2} + \Delta_{i,1}, \\ \dot{x}_{i,2} = u_i + \Delta_{i,2}. \end{cases} \tag{9.41}$$

With (9.40), the duty cycle of each converter can be accordingly derived as

$$\mu_i = 1 - \left(\frac{E_i^2}{L_i} + \frac{2v_{Ci}^2}{R_0^2 C_i} - u_i\right) \Big/ \left(\frac{E_i v_{Ci}}{L_i} + \frac{2i_{Li} v_{Ci}}{R_0 Ci}\right).$$

From the above analysis, it is concluded that the control objective can be transformed into designing a control signal u_i such that $x_{i,1}$ can track its reference x_{i1r} asymptotically, which is depicted by

$$\begin{aligned}
x_{i1r} &= 0.5L_i i_{Lir}^2 + 0.5C_i v_{Cir}^2 \\
&= 0.5L_i \left(\frac{P_{CPL} + v_{Cir}^2/R}{E_i}\right)^2 + 0.5C_i v_{Cir}^2,
\end{aligned} \tag{9.42}$$

where $v_{Cir} = V^* + m_i \Delta_{i,1}$ with V^*, m_i representing the nominal DC bus voltage and the droop coefficient for i-th converter, respectively.

Up to now, one obvious hurdle appears: how to identify the two lumped terms, i.e., $\Delta_{i,1}$ and $\Delta_{i,2}$ in order to appoint a precise tracking objective v_{Cir} for each converter. By recalling that a classical double closed-loop PI control would result in a serious reaction delay and thereafter an adverse effect imposed on transient-time performance. In this section, we will investigate a novel composite control strategy by integrating a finite-time feedforward decoupling procedure with a feedback control loop.

9.3.2 Composite controller construction

Following the proposed control design procedure in the above section, we are able to obtain a composite decentralized control scheme for each converter of the DC microgrid system.

First, we denote the following auxiliary functions

$$\begin{aligned}
\chi_{i,1} &= 0.5L_i\left(\Delta_{i,1} - \frac{v_{Ci}^2}{R_0}\right)^2 / E_i^2 + 0.5C_i v_{Cir}^2, \\
\chi_{i,2} &= \frac{d\chi_{i,1}}{dt} - \Delta_{i,1}, \\
\chi_{i,3} &= \frac{d\chi_{i,2}}{dt} - \Delta_{i,2}.
\end{aligned} \tag{9.43}$$

By recalling that the load information R and P_{CPL} are inaccessible, the following HOSM observers are hence essentially required:

$$\begin{aligned}
&1)\begin{cases}
\dot{z}_{i,1,0} = x_{i,2} + \hbar_{i,1,0},\ \dot{z}_{i,1,1} = \hbar_{i,1,1},\ \dot{z}_{i,1,2} = \hbar_{i,1,2} \\
\hbar_{i,1,0} = -l_{i,1,0}\lambda_{i,1}^{1/3}\lfloor z_{i,1,0} - x_{i,1}\rceil^{2/3} + z_{i,1,1} \\
\hbar_{i,1,1} = -l_{i,1,1}\lambda_{i,1}^{1/2}\lfloor z_{i,1,1} - \hbar_{i,1,0}\rceil^{1/2} + z_{i,1,2} \\
\hbar_{i,1,2} = -l_{i,1,2}\lambda_{i,1}\lfloor z_{i,1,2} - \hbar_{i,1,1}\rceil^{0};
\end{cases} \\
&2)\begin{cases}
\dot{z}_{i,2,0} = u_i + \hbar_{i,2,0},\ \dot{z}_{i,2,1} = \hbar_{i,2,1},\ \dot{z}_{i,2,2} = \hbar_{i,2,2} \\
\hbar_{i,2,0} = -l_{i,2,0}\lambda_{i,2}^{1/3}\lfloor z_{i,2,0} - x_{i,2}\rceil^{2/3} + z_{i,2,1} \\
\hbar_{i,2,1} = -l_{i,2,1}\lambda_{i,2}^{1/2}\lfloor z_{i,2,1} - \hbar_{i,2,0}\rceil^{1/2} + z_{i,2,2} \\
\hbar_{i,2,2} = -l_{i,2,2}\lambda_{i,2}\lfloor z_{i,2,2} - \hbar_{i,2,1}\rceil^{0}.
\end{cases}
\end{aligned} \tag{9.44}$$

Second, by replacing the variables $\Delta_{i,1}, \Delta_{i,1}^{(1)}, \Delta_{i,2}$ in $\chi_{i,j}$, $j = 1,2,3$ with their corresponding estimates $z_{i,1,1}, z_{i,1,2}, z_{i,2,1}$, one can obtain the following steady-state function of each states as

$$x_{i,j}^* = \chi_{i,j}(z_{i,1,1}, z_{i,1,2}, z_{i,2,1}),\ j = 1,2,3. \tag{9.45}$$

With (9.45) in mind, using a change of coordinates:

$$\xi_{i,1} = x_{i,1} - x_{i,1}^*,\ \xi_{i,2} = \frac{x_{i,2} - x_{i,2}^*}{L},$$

where $L \geq 1$ is a design parameter, the following decentralized composite controller for $i-$th converter could be constructed

$$u_i = -L^2\left(k_{i,1}\xi_{i,1} + k_{i,2}\xi_{i,2}\right) + x_{i,3}^*, \tag{9.46}$$

where $k_{i,1}$, $k_{i,2}$ are control gains.

9.3.3 Simulation studies

To validate the proposed control strategy, simulation tests are first conducted in Matlab/Simulink. The classical double closed-loop PI controller is chosen to compare with the proposed controller so as to manifest the superiorities of the proposed method. Detailed parameters are provided in Table 9.2. For the sake of fair comparison, it is noted that the benchmark PI control gain parameters are well selected according to reference [42] in order to meet an optimal control performance.

Case 1. Input Voltage Variation Test: In this case, the converter input voltage is changed to examine the stabilization performance. Two DU subsystems are involved and their droop coefficients are both set as 0.01. At the beginning, the bus voltage is regulated at 167.7 V and a 450 W CPL is connected to the DC bus. As 0.3 s, the converter input voltage of DU1 steps down from 100 V to 80 V and DU2 remains unchanged. In order to keep the stability of the DC bus, DU1 needs to release more current. As is shown in Figure 9.15, the DU1 controlled by the proposed approach responses immediately and reaches to its desired value quickly. Owing to its preeminent transient characteristics, DU2 can be immune to this sudden change and keeps constant current output. The transient process of DU1 lasts almost 30ms and thus, the DU2 has to release additional current to make up the lacking power. The performance on bus voltage in Figure 9.14

Table 9.2 System parameters configuration

Parameters	Description	Value
V^*	nominal bus voltage	170 V
E_1, E_2	converter input voltage	100 V
L_1, L_2	nominal inductance value	2 mH
C_1, C_2	nominal capacitance value	470 μF
$l_{i,j,0}, l_{i,j,1}, l_{i,j,2}$	observer gains $(i, j = 1, 2)$	4, 2, 1
$\lambda_{i,j}$	observer scaling gains $(i, j = 1, 2)$	1e9
$k_{i,1}, k_{i,2}$	controller gains $(i = 1, 2)$	30, 20
L	controller scaling gain	100
f_{sw}	switching frequency	20 kHz
k_{cp}, k_{ci}	PI gains for current control loop	0.13,35
k_{vp}, k_{vi}	PI gains for voltage control loop	0.3,35

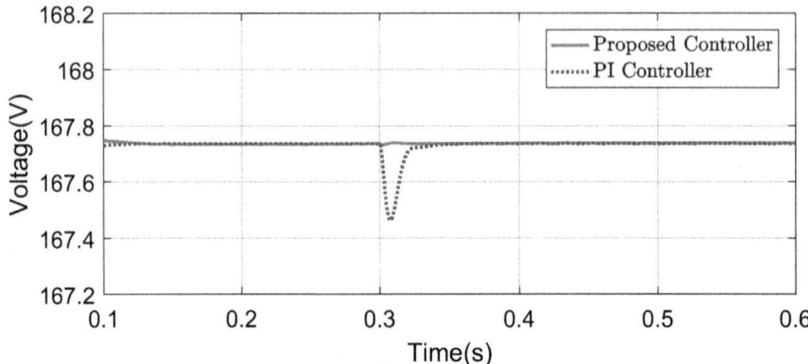

Figure 9.14 Bus voltage responses with a converter input voltage variation from 100 V to 80 V.

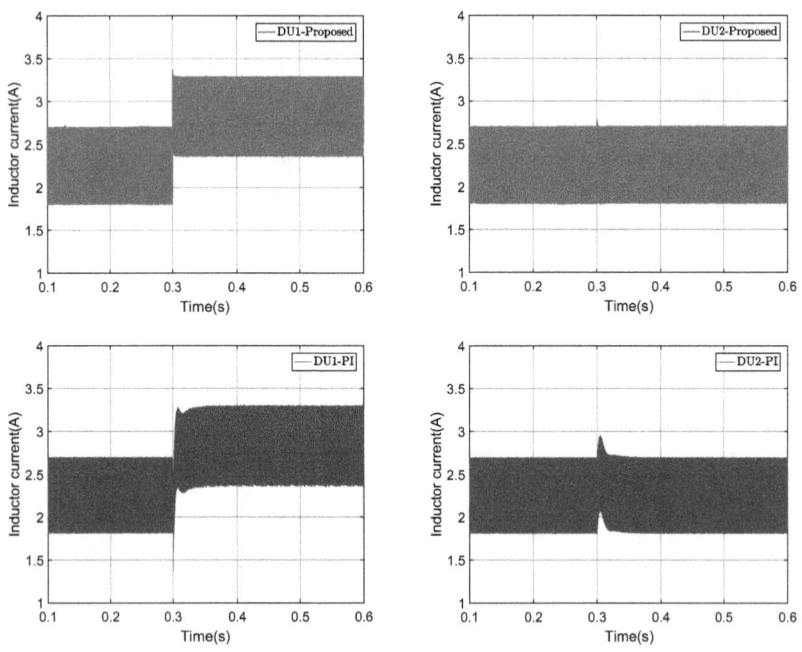

Figure 9.15 Current responses with a converter input voltage variation from 100 V to 80 V.

also reflects the differences between two candidate controllers. It is not difficult to find that the recovery capability of bus voltage gets markedly promoted under the proposed control strategy.

Case 2. CPL Variation Test: The CPL is changed in this case to further explore the advantages which have been brought by the proposed control method. The basic setting is identical with the former case. Initially, the bus voltage is stable at 169 V and a 200 W CPL is connected. Then, the CPL increases to 400 W in 0.25s and decreases to 200 W in 0.45 s. The transient response of bus voltage is shown in Figure 9.16. With the same load change, obvious voltage overshoot can be observed under the PI controller. The amplitude of the voltage overshoot goes down by 2.2 V and the duration of the recovery process is 30 ms. Meanwhile, the proposed controller demonstrates excellent dynamic performance consistently. Upon the disturbances happening, the proposed controller can adjust its operating voltage and reach a steady state in a short period of time. Besides, by further looking into the current output shown in Figure 9.17, the proposed strategy not only shows fast dynamic performance as always, but also realizes the accurate power sharing as we set before.

Case 3. Plug-and-Play Property Test: In what follows, the PnP property is tested under the proposed controller. DU3 is introduced in this case and its basic setting is identical with previous DUs. As is shown in Figure 9.18, three DUs are considered in this test, whose droop coefficients are set as 0.01, 0.02, and 0.015, respectively. At first, DU1 and DU3 work together to supply 750 W CPL. Meanwhile, DU2 operates independently with 200 W CPL. At 0.2 s, another 250 W CPL is added into the double DU system and the power of DU1 and DU3 increases to 600 W and 400 W (3:2). In order to take the load off of the DU1 and DU3, DU2 is connected to the DC bus at 0.3 s. The power is reallocated according to their respective droop coefficients, that is 553 W, 369 W, and 278 W (6:3:4). Subsequently, the 250 W CPL is removed at 0.4s and the output power of each DU decreases to 438 W, 220 W, 292 W (6:3:4). At 0.5 s, DU3 is disconnected. The bus voltage response curve is shown in Figure 9.18(c). The above tests prove the plug-and-play property of the proposed strategy is well built.

9.3.4 Experimental verification

In what follows, an experimental platform is built to verify the proposed control strategy in the application of DC microgrids, as shown in Figure 9.3.

To examine the stability of the system, the programmable electronic load operates in constant power mode to emulate the CPL. In the meantime, the DC microgrid is operated in the droop mode and the droop coefficients of DU1 and DU2 are set as 0.01. The parameters of converter components are identical with Table 9.2 while the gain parameters $l_{i,j,0}$, $l_{i,j,1}$, $l_{i,j,2}$, $\lambda_{i,j}$, L, $k_{i,1}$, $k_{i,2}$ are redesigned as 200, 400, 200, 10, 650, 1, 2 for optimal practical control performance. In what follows, experimental results are provided to verify the proposed control strategy.

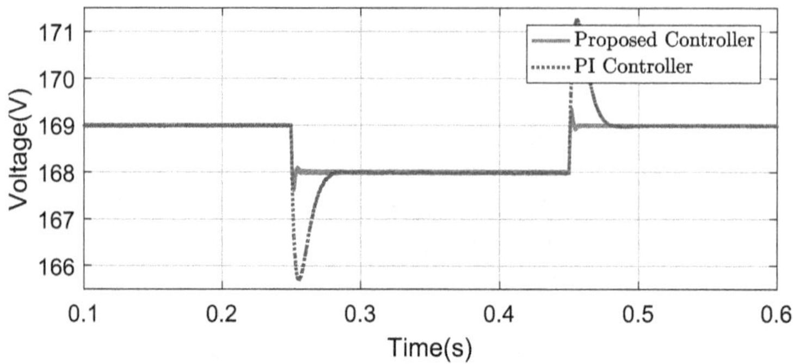

Figure 9.16 Bus voltage responses with load variations from 200 W to 400 W, and from 400 W to 200 W.

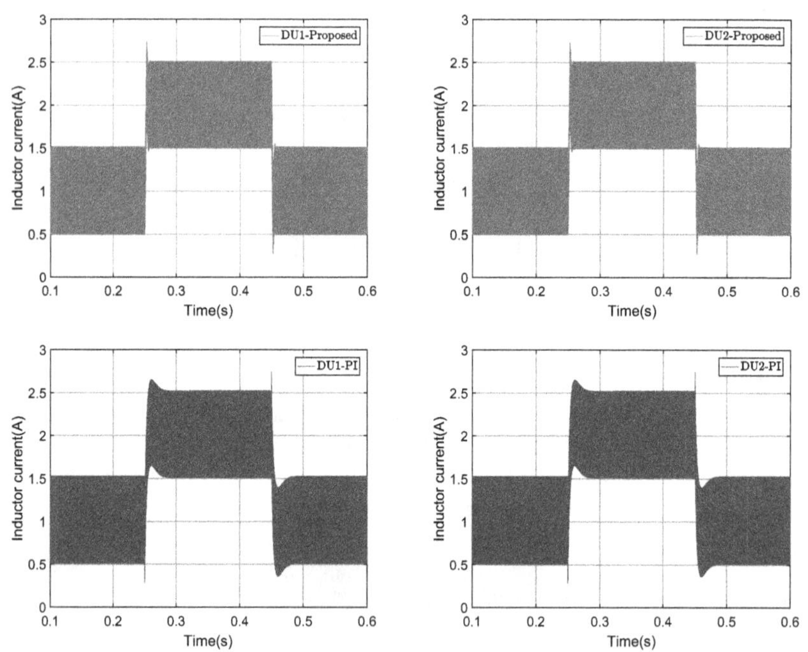

Figure 9.17 Current responses with load variations from 200 W to 400 W, and from 400 W to 200 W.

The input voltage changes and the variations of the loads are both considered in the experimental studies.

First, the converter input voltage is changed to examine the stabilization issue of the system. DC bus is rigorously regulated at 167.7 V and a 650 Ω resistive

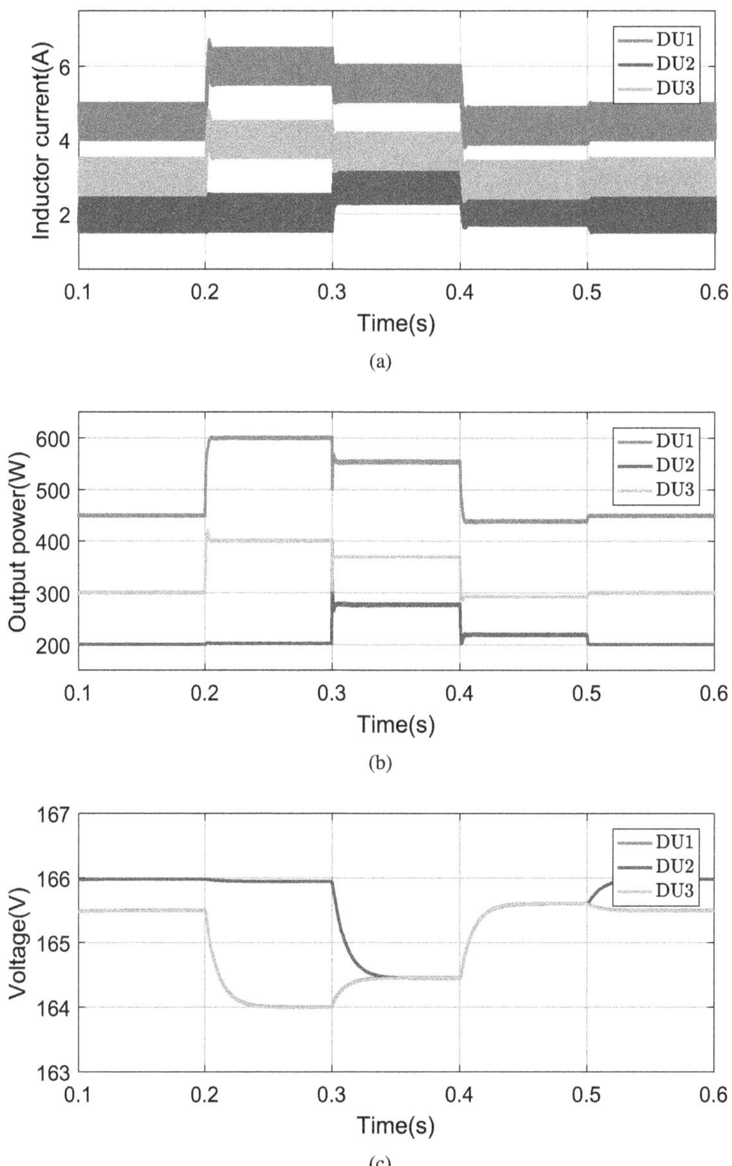

Figure 9.18 Performances of PnP test. (a) Current responses; (b) Power responses; (c) Voltage responses.

load is connected to the DC bus. As shown in Figure 9.19, the converter input voltage of DU1 steps down from 100 V to 80 V and DU2 remains unchanged. In order to keep the constant voltage of the DC bus, the current amplitude of

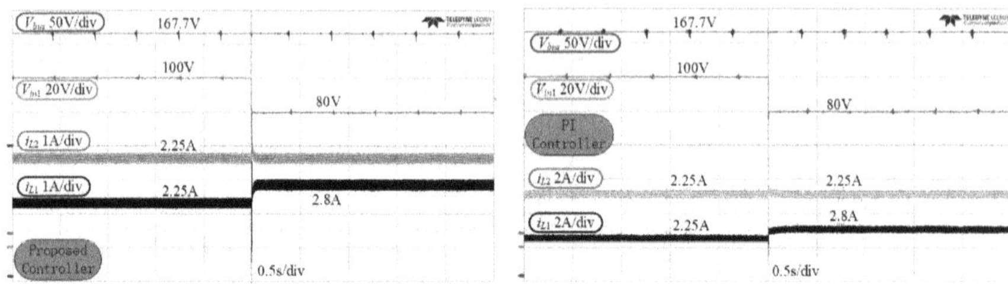

Figure 9.19 Experimental comparison results with a converter input voltage variation from 100 V to 80 V.

DU1 rises from 2.25 A to 2.8 A in a short period of time. It can be observed that both candidate controllers could achieve similar performances, which is beneficial from the effort we spent in tuning the parameters for both controllers, aiming for a fair control performance comparison in the subsequent cases.

Second, only CPL is connected to the DC bus in order to test the microgrid system stability under different variation conditions of the CPL. Figure 9.20 shows the experimental results when the CPL steps up from 50 W to 650 W. It demonstrates that in this case, the PI controller can also achieve the control objective and maintain system stability. Thereafter, to compare the stability margin and outstand the advantages of the proposed controller, as shown in Figure 9.21, a larger CPL variation case is conducted when the CPL suddenly changes from 50 W to 1100 W. Although the voltage has a slight drop, the proposed controller is able to maintain the stability of the DC bus at 154.5 V eventually. With regard to the PI controller under the same control parameters setup as in the former cases, it is conspicuous that large oscillation occurs and increases until the system collapses. Consequently, it is confident to reach that the proposed controller is able to stabilize the system toward a large-signal stability, and therefore provides a wider operating range for DC microgrid system.

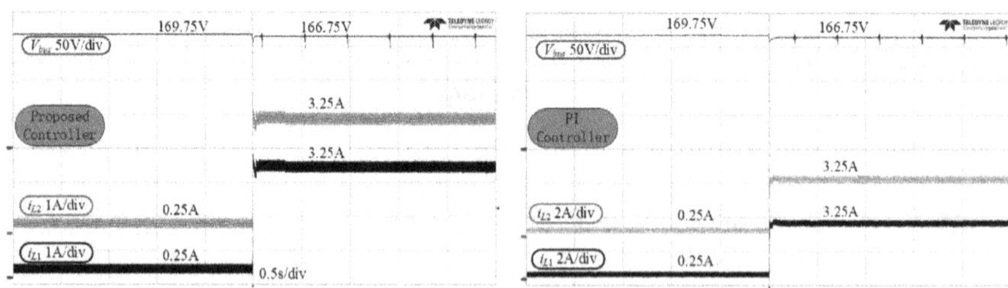

Figure 9.20 Experimental comparison results with a CPL variation from 50 W to 650 W.

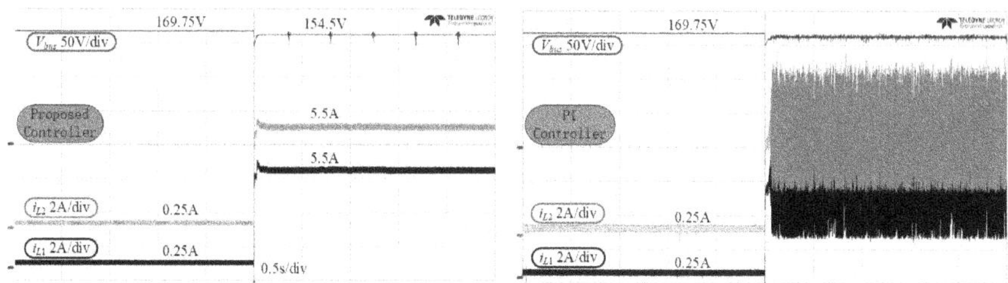

Figure 9.21 Experimental comparison results with a large CPL variation from 50 W to 1100 W.

9.4 Decentralized Adaptive Controller Design

As is well acknowledged that most of the existing robust control approaches are worst case based design and they aim to design controllers with fixed gain parameters to cope with all possible situations. However, the robustness is obtained at the price of sacrificing the nominal control performance. During real-life operations, DC microgrids may operate around the nominal working conditions in most of the operating time and operate in some extreme conditions occasionally. Designing control parameters with too much worst-case thinking may lead to a degraded dynamic performance in most of the operation conditions. To be specific, although strong robustness can ensure system stability when large-signal disturbances happen, the consequent voltage/current overshoots and even oscillations adverse to smooth operation of DC microgrids are normally observed in small-signal disturbance cases.

In this section, a novel decentralized adaptive control law is proposed to achieve a balance between system robustness and adaptivity. Specifically, by employing an online scaling gain update mechanism which is subject to the real-time working conditions, both the large-signal stability and desirable nominal performance can be promised.

9.4.1 Pre-treatment and system modeling

According to the philosophy of disturbance observer-based control, system disturbances and uncertainties are packaged together, and then, the lumped disturbances are estimated by an observation mechanism [9]. This thought can also be applied to the modeling process of DC microgrids. Besides, from the perspective of power flow, RESs can be emulated as constant power sources (CPSs). In our methodology, resistive loads, CPLs and CPSs power are classified as a lumped unknown term

$$P_{lumped} = \sum_{i=1}^{n} P_{oi} = P_{RLs} + P_{CPLs} - P_{CPSs}, \tag{9.47}$$

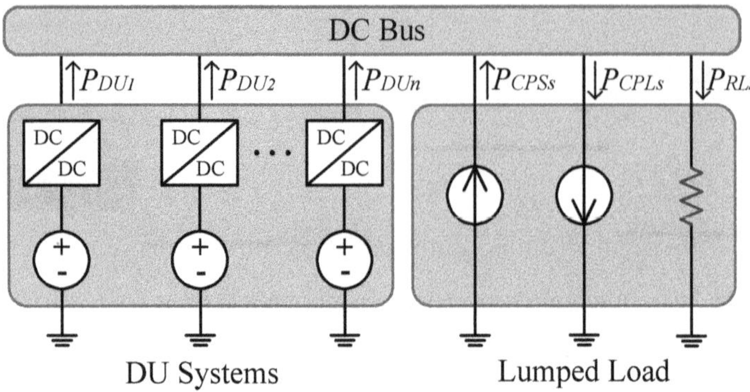

Figure 9.22 Equivalent diagram of a DC microgrid.

where P_{lumped} is the lumped power of the system, P_{oi} is the output power assigned by each subsystem, P_{RLs} and P_{CPLs} are the power consumed by the resistive loads (RLs) and CPLs, P_{CPSs} is the power of CPSs. Equivalent diagram is detailed in Figure 9.22.

This section is meant to propose a universal control scheme for both single DU subsystem and multiple DU subsystems. By following a similar modeling process in (9.41), the following relationship can be obtained:

$$\begin{cases} \dot{\chi}_{i,1} = \chi_{i,2} + \psi_{i,1}, \\ \dot{\chi}_{i,2} = u_i + \psi_{i,2}, \end{cases} \tag{9.48}$$

where $u_i = \frac{E_i^2}{L_i} - \frac{E_i v_{Ci}}{L_i}(1-\mu_i)$, $\psi_{i,1} = -P_{oi}$, $\psi_{i,2}$ represents parameter uncertainties, unmodeled dynamics, and external disturbances, respectively.

Through the above analysis, the reference value is obtained and the control objective of μ_i is enable $\chi_{i,1}$ track its reference value $\chi_{i,1r}$, which is expressed as

$$\begin{aligned} \chi_{i,1r} &= 0.5L_i i_{Lir}^2 + 0.5C_i v_{Cir}^2 \\ &= 0.5L_i(\psi_{i,1}/E_i)^2 + 0.5C_i(V^* + m_i\psi_{i,1})^2, \end{aligned} \tag{9.49}$$

where V^* and m_i represent the nominal DC bus voltage and the droop coefficient of the ith DU subsystem.

Remark 9.1 It is worthy of pointing out that the proposed controller can switch between constant voltage mode and droop mode by modifying the droop coefficient m_i. The descriptions of the above two modes will be discussed in details in the later section. The proposed controller can also be applied to buck converter, buck-boost converter, and AC/DC rectifier with some modifications on particular coordinate transformations as depicted in the previous section. Besides, the nonlinear models of CPLs and CPSs could be linearized at a given operating point ($I = P/V$). Then, the obtained equivalent circuit models of RLs, CPLs, and CPSs can incorporate into the averaged model of the DC/DC boost converter.

9.4.2 Decentralized adaptive controller design

In what follows, the detailed design procedure of the proposed decentralized adaptive controller is presented along with rigorous system stability analysis.

In order to facilitate the practical applications, a schematic diagram of the proposed adaptive approach that includes both power stage and control stage is provided in Figure 9.23.

As previously discussed, both $\psi_{i,1}$ and $\psi_{i,2}$ have their actual implications in practical systems, but these values cannot be measured directly by measurement devices. To achieve this aim, the higher-order sliding mode observation technology provides a soft measurement technique to estimate the lumped uncertainties.

Considering the practical physical characteristics of DC microgrids, a reasonable assumption condition can be given: the unknown disturbances $\psi_{i,1}$, $\psi_{i,2}$, and their (4-j)th-order derivatives are bounded, where j=1, 2, 3.

By referring to the HOSM design in Chapter 5, two finite-time disturbance observers are constructed as the following form:

$$1)\begin{cases}\dot{z}_{i,1,0} = \chi_{i,2} + \hbar_{i,1,0},\\ \dot{z}_{i,1,j} = \hbar_{i,1,j}, j = 1,2,3,\\ \hbar_{i,1,0} = -l_{i,1,0}\lambda_{i,1}^{1/4}\lfloor z_{i,1,0} - \chi_{i,1}\rceil^{3/4} + z_{i,1,1},\\ \hbar_{i,1,1} = -l_{i,1,1}\lambda_{i,1}^{1/3}\lfloor z_{i,1,1} - \hbar_{i,1,0}\rceil^{2/3} + z_{i,1,2},\\ \hbar_{i,1,2} = -l_{i,1,2}\lambda_{i,1}^{1/2}\lfloor z_{i,1,2} - \hbar_{i,1,1}\rceil^{1/2} + z_{i,1,3},\\ \hbar_{i,1,3} = -l_{i,1,3}\lambda_{i,1}\lfloor z_{i,1,3} - \hbar_{i,1,2}\rceil^{0};\end{cases}$$

$$2)\begin{cases}\dot{z}_{i,2,0} = u_i + \hbar_{i,2,0},\\ \dot{z}_{i,2,j} = \hbar_{i,2,j}, j = 1,2,\\ \hbar_{i,2,0} = -l_{i,2,0}\lambda_{i,2}^{1/3}\lfloor z_{i,2,0} - \chi_{i,2}\rceil^{2/3} + z_{i,2,1},\\ \hbar_{i,2,1} = -l_{i,2,1}\lambda_{i,2}^{1/2}\lfloor z_{i,2,1} - \hbar_{i,2,0}\rceil^{1/2} + z_{i,2,2},\\ \hbar_{i,2,2} = -l_{i,2,2}\lambda_{i,2}\lfloor z_{i,2,2} - \hbar_{i,2,1}\rceil^{0}\end{cases}$$

where $l_{i,1,0}$, $l_{i,1,1}$, $l_{i,1,2}$, $l_{i,1,3}$, $l_{i,2,0}$, $l_{i,2,1}$ and $l_{i,2,2}$ are design parameters, $\lambda_{i,1}$ and $\lambda_{i,2}$ are properly selected to satisfy $\lambda_{i,1} > 0$ and $\lambda_{i,2} > 0$, respectively. $\hbar_{i,1,0}$, $\hbar_{i,1,1}$, $\hbar_{i,1,2}$, $\hbar_{i,1,3}$, $\hbar_{i,2,0}$, $\hbar_{i,2,1}$ and $\hbar_{i,2,2}$ are selected as auxiliary intermediate variables, $z_{i,1,0}$, $z_{i,1,1}$, $z_{i,1,2}$, $z_{i,1,3}$, $z_{i,2,0}$, $z_{i,2,1}$, $z_{i,2,2}$ represent the estimates of $\chi_{i,1}$, $\psi_{i,1}$, $\psi_{i,1}^{(1)}$, $\psi_{i,1}^{(2)}$, $\chi_{i,2}$, $\psi_{i,2}$, $\psi_{i,2}^{(1)}$, respectively.

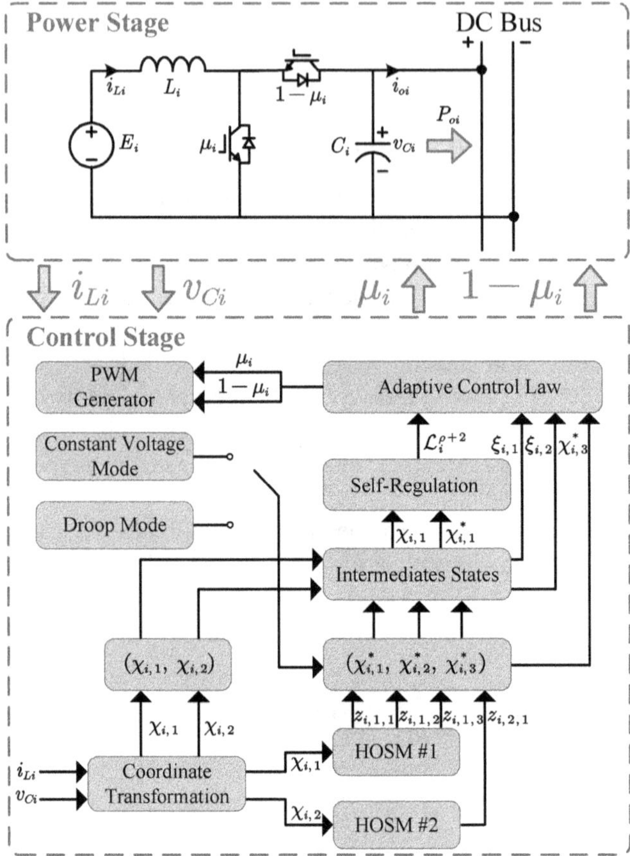

Figure 9.23 Schematic diagram of the ith DU subsystem under the proposed adaptive control approach.

On the basis of the previous preparation step, the proposed decentralized adaptive controller can be designed in this subsection.

To begin with, the following auxiliary functions should be denoted:

$$
\begin{aligned}
x_{i,1} &= 0.5L_i\psi_{i,1}^2/E_i^2 + 0.5C_i v_{Cir}^2,\\
x_{i,2} &= \frac{dx_{i,1}}{dt} - \psi_{i,1},\\
x_{i,3} &= \frac{dx_{i,2}}{dt} - \psi_{i,2}.
\end{aligned}
\tag{9.50}
$$

Then, owing to the HOSM observers, the original value can be replaced by their corresponding estimated value, and hence the reference states can be derived as

$$\begin{aligned}\chi_{i,1}^{*} &= 0.5L_i z_{i,1,1}^2/E_i^2 + 0.5C_i(V^* + m_i z_{i,1,1})^2,\\ \chi_{i,2}^{*} &= L_i z_{i,1,1} z_{i,1,2}/E_i^2 + C_i(V^* + m_i z_{i,1,1}) m_i z_{i,1,2} - z_{i,1,1},\\ \chi_{i,3}^{*} &= L_i(z_{i,1,2}^2 + z_{i,1,1} z_{i,1,3})/E_i^2 - z_{i,1,2} - z_{i,2,1}\\ &\quad + C_i(V^* + m_i z_{i,1,1}) m_i z_{i,1,3} + C_i m_i^2 z_{i,1,2},\end{aligned} \tag{9.51}$$

where $\chi_{i,2}^{*}$ and $\chi_{i,3}^{*}$ are the reference value of $\chi_{i,2}$ and u_i, respectively.

Inspired by the design concept of the universal adaptive control, a dynamic scaling gain ρ is introduced by the following inequality: $\rho > \max\left\{0, -\frac{\lambda_{\min}(\Theta Q_i + Q_i\Theta)}{2\lambda_{\min}(Q_i)}\right\}$, where $\Theta \triangleq [0,0;0,1]$, $Q_i \in \mathbb{R}^{2\times 2}$ is a positive definite and symmetrical matrix which satisfies $(A - BK_i)^\top Q_i + Q_i(A - BK_i) \leq -I_2$, $A = [0,1;0,0]$, $B = [0,1]^\top$, $K_i = \left[k_{i,1}, k_{i,2}\right]$is a coefficient vector of a Hurwitz polynomial $H(s) = s^2 + k_{i,2}s + k_{i,1}$, I_2 is the second order identity matrix.

In what follows, several intermediate states $\xi_i = \left[\xi_{i,1}, \xi_{i,2}\right]^\top$ are established by utilizing the dynamic scaling gain ρ and the adaptive gain $\mathcal{L}_i$, which can be written as

$$\begin{cases}\xi_{i,1} = (\chi_{i,1} - \chi_{i,1}^{*})/\mathcal{L}_i^{\rho},\\ \xi_{i,2} = (\chi_{i,2} - \chi_{i,2}^{*})/\mathcal{L}_i^{\rho+1},\\ v_i = (u_i - \chi_{i,3}^{*})/\mathcal{L}_i^{\rho+2},\end{cases} \tag{9.52}$$

where v_i is an auxiliary control input, which can be designed as the combination of $\xi_{i,1}$ and $\xi_{i,2}$, that is, $v_i = -k_{i,1}\xi_{i,1} - k_{i,2}\xi_{i,2}$, $\mathcal{L}_i$ is a dynamic self-regulation gain function which needs further elaboration in the next subsection.

In what follows, the compact expression of the control law is designed as

$$u_i = -\mathcal{L}_i^{\rho+2}(k_{i,1}\xi_{i,1} + k_{i,2}\xi_{i,2}) + \chi_{i,3}^{*}. \tag{9.53}$$

Up to now, we are able to construct an adaptive control law as follows:

$$\dot{\mathcal{L}}_i = \mathcal{C}_i(\chi_{i,1} - \chi_{i,1}^{*})^2/\mathcal{L}_i^{2\rho},\ \mathcal{L}_i(t_0) = 1, \tag{9.54}$$

where $\mathcal{C}_i \in \mathbb{R}_+$ is a design parameter.

In addition, the control design procedure of the proposed decentralized adaptive algorithm is presented in Figure 9.24 to provide a detailed and visualized expression.

9.4.3 Simulation studies

In this section, the feasibility and superiority of the proposed controller are verified by simulation studies. The simulation studies are conducted in Matlab/Simulink environment.

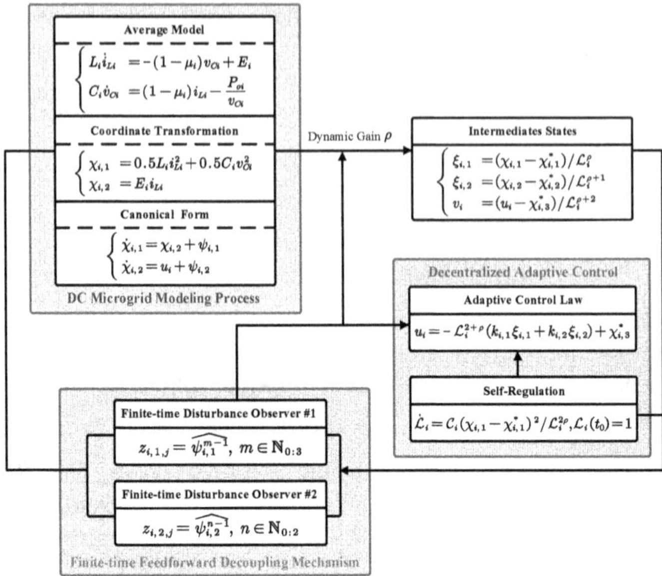

Figure 9.24 Control design procedure of the proposed adaptive control algorithm.

In what follows, we mainly consider two typical bus voltage control modes: CVM and DM. The CVM is preferred when a single DU has enough capacity to cope with all possible load changes. However, under the circumstance of coordinated operation of multiple DUs, DM can be employed to realize the power allocation by regulating corresponding droop coefficient m_i, which is depicted in (9.49). Just as its name implies, the reference voltage in droop mode fluctuates as the load changes. To be specific, heavier loads comes with larger drop and vice versa.

In order to present the actual control effects, the parameters of simulations and experiments remain consistent. The detailed system parameters for DC microgrid system and proposed adaptive controller are summarized in Table 9.3.

In simulation studies, a DC microgrid operated in CVM, which is shown in Figure 9.25, is considered to verify the proposed control strategy.

Simulation Case 1. Constant power load tests: The variation of CPLs is considered in this test. A robust backstepping controller in [69] is chosen to compare with the proposed controller. As shown in Figure 9.26, the systems separately controlled by the two controllers are running steadily and the voltage is precisely stabilized at 160 V. In 0.4 s, CPLs vary from 100 W to 300 W. The proposed controller reacts immediately once the loads are increased. More currents are released to compensate the required power and the bus voltage can return to

Table 9.3 System parameters configuration

Description and Parameters	Value
DC/DC Converters	
Converter input voltages E_1, E_2	80 V
Nominal bus voltage V^*	160 V
Nominal inductance values L_1, L_2	6.8m H
Nominal capacitance values C_1, C_2	1800 μF
Switching frequency f_{sw}	10 kHz
Observer and Controller	
Observer gains for observer #1 $l_{i,1,s}$, s=0,1,2,3	5, 4, 2, 1
Observer gains for observer #2 $l_{i,2,s}$, s=0,1,2	4, 2, 1
Observer scaling gains $\lambda_{i,j}$ (i, j=1,2)	1e9
Dynamic scaling gain ρ	0.25
Design parameter C_i (i=1,2)	5, 5
Controller gains $k_{i,1}, k_{i,2}$ (i=1,2)	50, 50

its nominal value quickly. By contrast, obvious voltage/current overshoots can be founded in the system controlled by the robust backstepping controller due to the worst-case-based design in the robust control, which leads to the degraded nominal performance of controlled system. The well observation effect of the HOSM observer and the response curve of adaptive gain $\mathcal{L}$ are also plotted in Figure 9.26. It is worth noting that the initial value of the adaptive gain $\mathcal{L}$ is set to 1 and it could make real-time adjustments according to the condition of operation.

Remark 9.2 It should be noted that due to the ideal environment of simulation system, the controlled system has a relatively fast convergence speed, and hence the

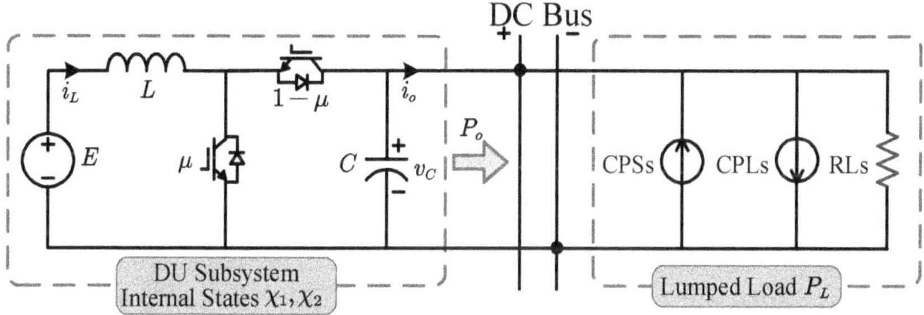

Figure 9.25 A DC microgrid operated in constant voltage mode.

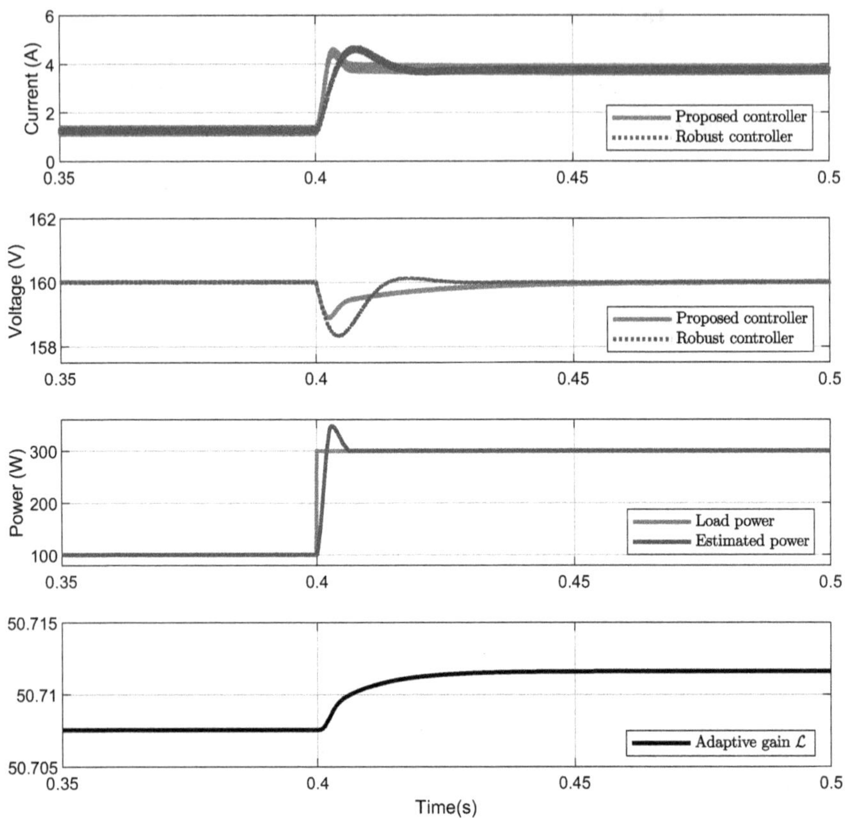

Figure 9.26 Simulation results for the proposed adaptive control algorithm and robust controller with CPL changes from 100 W to 300 W.

increase of $\mathcal{L}$ is not obvious. In the experimental tests, the proposed controller will be further examined by the actual cases while the system parameters remain the same.

Simulation Case 2. Renewable energy source tests: As shown in Figure 9.27, the intermittency of RESs is considered to further validate the effectiveness of the proposed controller. In this test, both the CPL and RES are connected to the DC bus. The power of CPL is set as 200 W and the power of RES changes every 0.1 s. With the variation of RES power, the DU subsystem can make timely adjustments to its inductor current and the bus voltage can be regulated at the nominal value within a short time. The HOSM observer can accurately estimate the lumped power of CPL and RES. The simulation results in Figure 9.27 demonstrate the proposed control strategy can cope well with the frequent power change

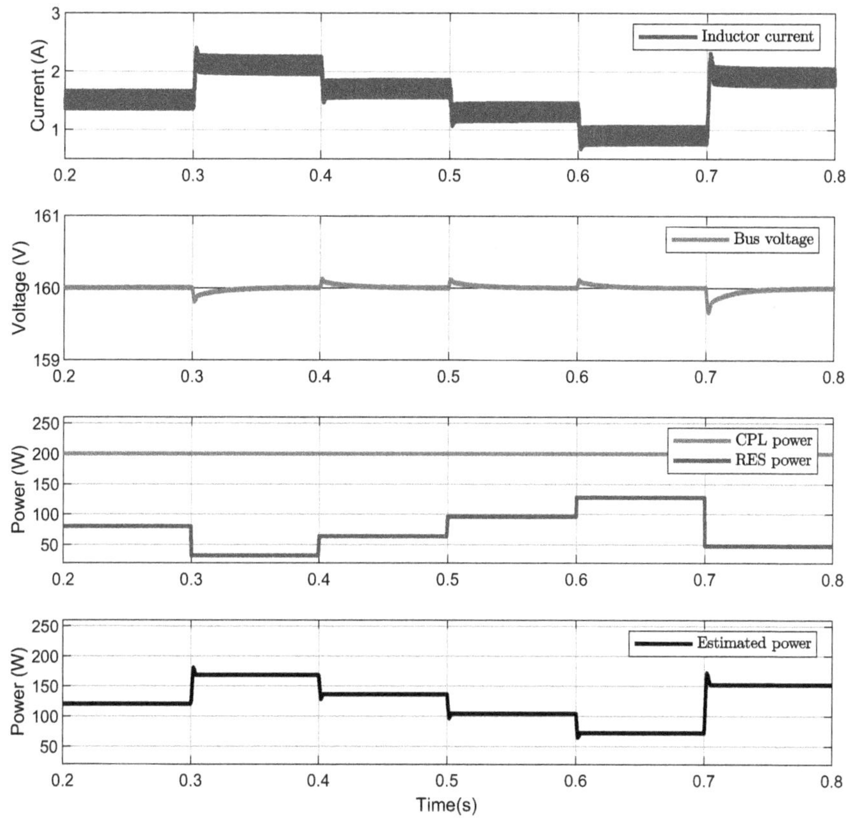

Figure 9.27 Simulation results with the intermittency of RESs.

of RESs. The universality of the proposed method in dealing with loads and RESs is validated.

9.4.4 Experimental tests

In order to verify the proposed control strategy, a scale-down DC microgrid is built in the laboratory, which is shown in Figure 9.3. The designed control algorithm is embedded in dSPACE 1202 to generate PWM signals to the DC/DC converters with a switching frequency of 10 kHz.

In addition, the classical PI controller is chosen to compare with the proposed method. For a relatively fair comparison, the parameters of the PI controller are not obtained by those "trial and error" methods. They are carefully tuned by the standard design guideline recommended by [42] and finally determined as (k_{pv}, k_{iv}, k_{pc}, k_{ic})=(0.3, 0.15, 0.275, 0.275).

In this test, both CVM and DM are considered and the performance of the two candidate controllers will be examined by different degrees of disturbances.

Experimental case 1. Constant voltage mode tests: In the first case, both of the two controllers work in the CVM and only one DU system operates. As shown in Figure 9.28, the bus voltage of the DU system stabilizes at 160 V and the power of CPL is 50 W. When the CPL increases to 200 W, different dynamic performances can be observed. According to the DU subsystem controlled by the PI controller, a slight voltage drop happens and it takes about 3 s to reach its reference value. And the similar performance occurs with the inductor current. Different from the former one, although a certain amount of voltage drop can be noticed, both the bus voltage and inductor current only take 0.5 s to reach their corresponding steady values.

Similar experimental results can also be found in Figures 9.29 and 9.30. Through the above results, we can find that although obvious voltage drop and current overshoot can be observed, the proposed controller has a rapid dynamic response speed and that sort of validates our algorithm.

Remark 9.3 From the experimental results in the first case, we can find that the two candidate controller exhibit different dynamic performances and the PI controller seems to perform better from the general point of view. This distinct contrast indicates that the parameters for PI controller are well selected to deal with the small-signal disturbances. The variation rate of $\mathcal{L}$ is limited, which is summarized in Table 9.4. Hence the proposed controller does not have enough energy to enable the actual voltage to track its reference value in a short time. In the next case, the decentralized cooperate control is adopted and the large-signal disturbances are tested to conduct a further comparison between the two controllers.

Experimental Case 2. Droop mode tests: In the second case, the DM is employed in order to realize the power allocation among the two DU subsystems.

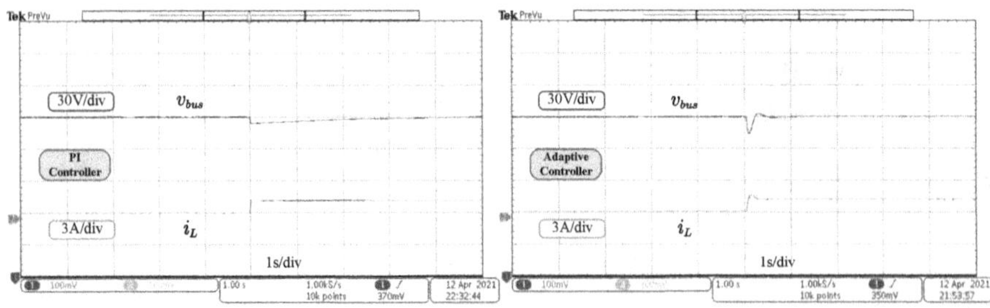

Figure 9.28 Experimental results in constant voltage mode with CPL changes from 50 W to 200 W.

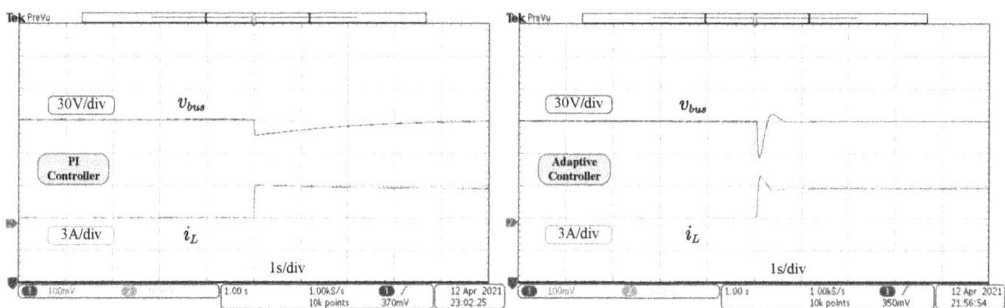

Figure 9.29 Experimental results in constant voltage mode with CPL changes from 50 W to 400 W.

The droop coefficients of DU1 and DU2 are both set at 0.1 and hence the load power can be equally allocated. Figure 9.31 shows the dynamic performance of the two controllers in DM when CPL changes from 50 W to 200 W. Different from the constant voltage case, the proposed controller exhibits better performances than the PI controller whether in voltage or current.

Proceed to Figures 9.32 and 9.33 in which the CPL is increased to 400 W and 800 W respectively. Slow voltage response and obvious voltage drop can be noticed in the DC system controlled by the PI controller. Turning to another one, the system controlled by the proposed controller can quickly adjust its voltage to the desired value and meanwhile, the inductor current also responses immediately. It attributes to timely dynamic changes of $\mathcal{L}_i$ and large-signal disturbances can accelerate and intensify its rate of changes.

In order to further test the dynamic performance and the system stability, the CPL is changed from 50 W to 1000 W in the last case. As can be observed from Figure 9.34, the system controlled by the PI controller appears to manifest voltage and current oscillations and which eventually causes the system to crash. In contrast, under the circumstance of large-signal disturbance, the proposed controller can online adjust its adaptive gains and eventually determine at 53.32 and 53.09. Although a slight current overshoot can be observed,

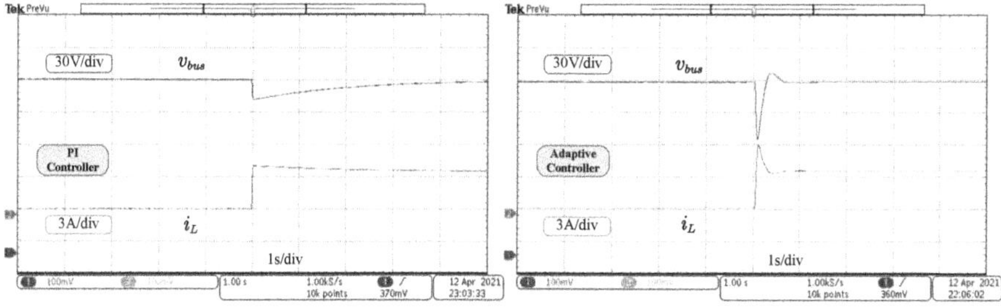

Figure 9.30 Experimental results in constant voltage mode with CPL changes from 50 W to 500 W.

Table 9.4 Adaptive gain values

Case	Load Change	Adaptive Gains	
Constant Voltage Mode Tests			
Case 1	CPL: 50 W–200 W	$\mathcal{L}$: 26.13	
Case 2	CPL: 50 W–400 W	$\mathcal{L}$: 34.46	
Case 3	CPL: 50 W-500 W	$\mathcal{L}$: 39.22	
Droop Mode Tests			
Case 4	CPL: 50 W–200 W	$\mathcal{L}_1$: 29.80	$\mathcal{L}_2$: 28.86
Case 5	CPL: 50 W–400 W	$\mathcal{L}_1$: 32.91	$\mathcal{L}_2$: 32.86
Case 6	CPL: 50 W–800 W	$\mathcal{L}_1$: 45.64	$\mathcal{L}_2$: 45.92
Case 7	CPL: 50 W–1000 W	$\mathcal{L}_1$: 53.32	$\mathcal{L}_2$: 53.09

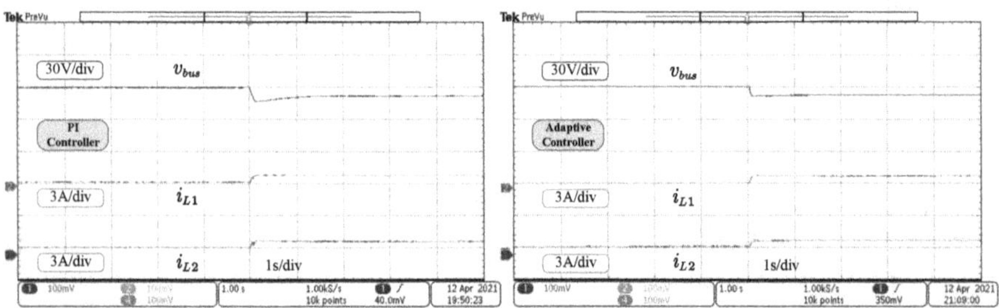

Figure 9.31 Experimental results in droop mode with CPL changes from 50 W to 200 W.

the large-signal disturbance can be well coped with by enlarging the controller energy and the stable operation of the system can be ensured. This effectively demonstrates the improvements in dynamic performance and stability margin brought by the proposed method.

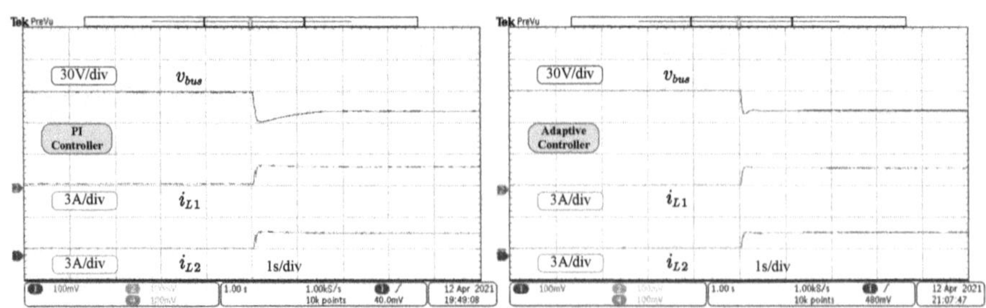

Figure 9.32 Experimental results in droop mode with CPL changes from 50 W to 400 W.

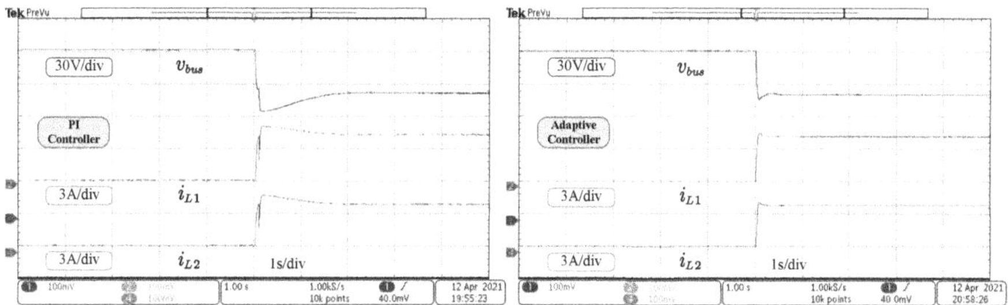

Figure 9.33 Experimental results in droop mode with CPL changes from 50 W to 800 W.

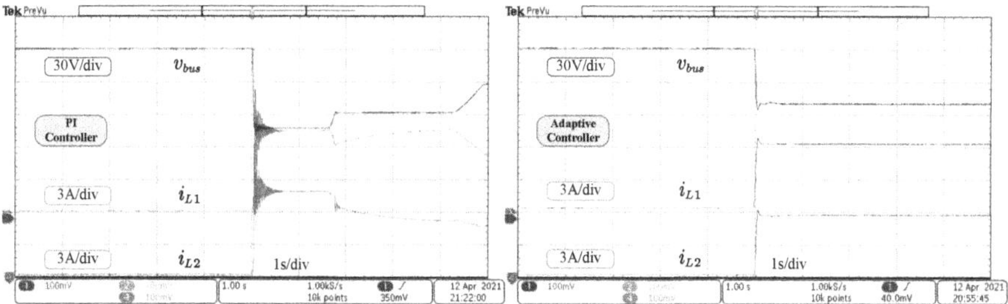

Figure 9.34 Experimental results in droop mode with CPL changes from 50 W to 1000 W.

9.5 Nonsmooth Control Design for Multi-Bus DC Microgrids

A representative multi-bus DC MG has been plotted in Figure 9.35. For a certain bus, DGs will contribute to regulating the bus voltage and limit the voltage within an allowable range. RESs like PVs and wind turbines are normally working maximum power point tracking modes. Resistive loads (RLs) and CPLs would absorb powers from their feeding buses. Noting that all RESs, resistive loads and CPLs are not participating in the bus regulation, the stability of DC voltage will ultimately count on those distributed generators (DGs). In Figure 9.35, the lumped load P_{il} of the ith bus is defined as

$$P_{il} = P_{i\text{Loads}} - P_{i\text{RESs}} = P_{i\text{RLs}} + P_{i\text{CPLs}} - P_{i\text{RESs}},\ i = 1,2,3,4. \tag{9.55}$$

Concerning DGs, three commonly used DC/DC converters (boost, buck, buck-boost) have been shown in Figure 9.36. For all that they are initially devised for different voltage levels, without loss of generality, all three converter topologies are considered here as they will more than likely appear in a generic DC architecture. To standardize the design process of the stabilizer, regardless of

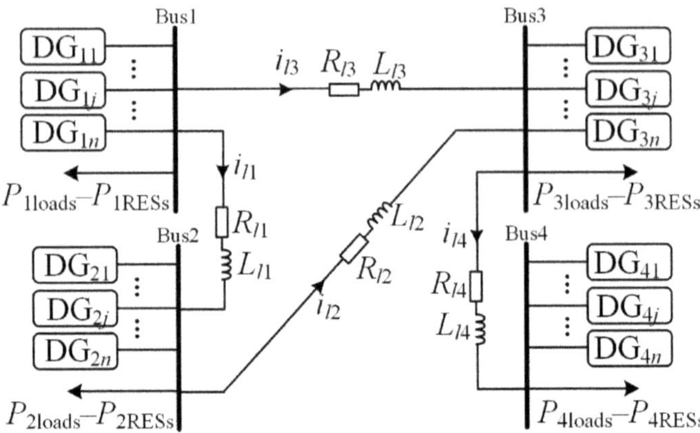

Figure 9.35 A typical multi-bus (4 buses) DC MG with ring and radical topologies. No communication between any two of DGs is allowed.

model differences, proper coordinate transformations will be performed on the converters in Figure 9.36. Hence, they can eventually be expressed in a uniform dynamic form. For the boost type converter, its average model in a switching period can be written as

$$\begin{cases} \dot{i}_L = E/L-(1-d)v_C/L,\ L=L_0+\Delta_L, \\ \dot{v}_C=(1-d)i_L/C-P_o/(Cv_C),\ C=C_0+\Delta_C \end{cases} \tag{9.56}$$

where L_0 and C_0 stand for the nominal inductance and nominal capacitance, respectively. ΔL and ΔC denote parameter uncertainties. E, i_L, v_C, and P_o are input voltage, inductor current, capacitor voltage, and output power, respectively. d is the duty cycle. Then writing (9.56) into the one with respect to L_0, C_0, ΔL, and ΔC yields,

$$\begin{cases} \dot{i}_L = \dfrac{E}{L_0}-\dfrac{(1-d)v_C}{L_0}+\left[-\dfrac{E\Delta_L}{L_0(L_0+\Delta_L)}+\dfrac{(1-d)v_C\Delta_L}{L_0(L_0+\Delta_L)}\right], \\ \dot{v}_C=\dfrac{(1-d)i_L}{C_0}-\dfrac{P_o}{C_0v_C}+\left[-\dfrac{(1-d)i_L\Delta_C}{C_0(C_0+\Delta_C)}+\dfrac{P_o\Delta_C}{C_0v_C(C_0+\Delta_C)}\right] \end{cases} \tag{9.57}$$

Observing (9.57) indicates that the model is nonlinear, entangled with uncertain terms. Two new states can be selected for (9.57) as

$$\begin{aligned} x_1 &= 0.5L_0i_L^2+0.5C_0v_C^2, \\ x_2 &= Ei_L. \end{aligned} \tag{9.58}$$

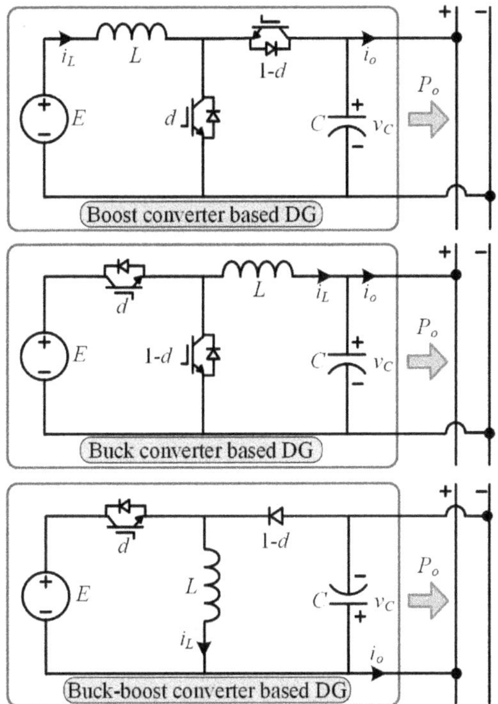

Figure 9.36 DGs based on three commonly used DC/DC converters: boost, buck, and buck-boost, respectively.

Computing the derivatives of x_1 and x_2 respectively results in,

$$\begin{cases} \dot{x}_1 = L_0 i_L \dot{i}_L + C_0 v_C \dot{v}_C = E i_L + \vartheta_1 \\ \dot{x}_2 = E \dot{i}_L = E^2/L_0 - (1-d)E v_C/L_0 + \vartheta_2 \end{cases} \Rightarrow \begin{cases} \dot{x}_1 = x_2 + \vartheta_1 \\ \dot{x}_2 = u + \vartheta_2, \end{cases} \tag{9.59}$$

where disturbances ϑ_1, ϑ_2, and the equivalent control input u can be identified as follows,

$$\begin{aligned} \vartheta_1 &= -P_o - \frac{E i_L \Delta_L}{(L_0+\Delta_L)} + \frac{(1-d) i_L v_C \Delta_L}{(L_0+\Delta_L)} - \frac{(1-d) i_L v_C \Delta_C}{(C_0+\Delta_C)} + \frac{P_0 \Delta_C}{(C_0+\Delta_C)}, \\ \vartheta_2 &= -\frac{E^2 \Delta_L}{(L_0+\Delta_L) L_0} + \frac{(1-d) E v_C \Delta_L}{(L_0+\Delta_L) L_0}, \\ u &= \frac{E^2}{L_0} - \frac{1-d}{L_0} E v_C \Rightarrow d = 1 - \frac{E}{v_C} + \frac{L_0 u}{E v_C}. \end{aligned} \tag{9.60}$$

It is noticed that, in (9.59), non-minimum phase situations can be favorably eluded. Comparing (9.57) and (9.59), it suggests that the nonlinear dynamics of the boost converter have been delicately transformed into the linear ones in the new coordinates (x_1, x_2), and the differential equations of x_1 and x_2 governed by (9.59) are referred as the uniform model. Proceeding to the buck converter in Figure 9.36, its average model involving parameter uncertainties can be delineated as

$$\begin{cases} \dot{i}_L = \dfrac{Ed}{L_0} - \dfrac{v_C}{L_0} + \left[-\dfrac{Ed\Delta_L}{L_0(L_0+\Delta_L)} + \dfrac{v_C\Delta_L}{L_0(L_0+\Delta_L)} \right], \\ \dot{v}_C = \dfrac{i_L}{C_0} - \dfrac{P_o}{C_0 v_C} + \left[-\dfrac{i_L\Delta_C}{C_0(C_0+\Delta_C)} + \dfrac{P_0\Delta_C}{C_0 v_C(C_0+\Delta_C)} \right]. \end{cases} \tag{9.61}$$

Two new states could be formulated for (9.61) as

$$x_1 = 0.5C_0 v_C^2,\ x_2 = v_C i_L. \tag{9.62}$$

The time derivatives of these two states are

$$\begin{cases} \dot{x}_1 = C_0 v_C \dot{v}_C = v_C i_L + \vartheta_1 \\ \dot{x}_2 = \dot{v}_C i_L + v_C \dot{i}_L = Edv_C/L_0 - v_C^2/L_0 + i_L^2/C_0 + \vartheta_2 \end{cases} \Rightarrow \begin{cases} \dot{x}_1 = x_2 + \vartheta_1 \\ \dot{x}_2 = u + \vartheta_2, \end{cases} \tag{9.63}$$

where ϑ_1, ϑ_2, and u could be figured out in the following form:

$$\begin{aligned} \vartheta_1 &= -P_o - \frac{v_C i_L \Delta_C}{(C_0+\Delta_C)} + \frac{P_0\Delta_C}{(C_0+\Delta_C)}, \\ \vartheta_2 &= -\frac{i_L P_o}{v_C C_0} - \frac{i_L^2\Delta_C}{C_0(C_0+\Delta_C)} + \frac{i_L P_0 \Delta_C}{C_0 v_C(C_0+\Delta_C)} - \frac{Edv_C\Delta_L}{(L_0+\Delta_L)L_0} + \frac{v_C^2\Delta_L}{(L_0+\Delta_L)L_0}, \\ u &= \frac{Edv_C}{L_0} - \frac{v_C^2}{L_0} + \frac{i_L^2}{C_0} \Rightarrow d = \left(u + \frac{v_C^2}{L_0} - \frac{i_L^2}{C_0}\right)\frac{L_0}{Ev_C}. \end{aligned} \tag{9.64}$$

The buck-boost converter in Figure 9.36 has the following expressions,

$$\begin{cases} \dot{i}_L = \dfrac{Ed}{L_0} - \dfrac{(1-d)v_C}{L_0} + \left[-\dfrac{Ed\Delta_L}{L_0(L_0+\Delta_L)} + \dfrac{(1-d)v_C\Delta_L}{L_0(L_0+\Delta_L)} \right] \\ \dot{v}_C = \dfrac{(1-d)i_L}{C_0} - \dfrac{P_o}{C_0 v_C} + \left[-\dfrac{(1-d)i_L\Delta_C}{C_0(C_0+\Delta_C)} + \dfrac{P_0\Delta_C}{C_0 v_C(C_0+\Delta_C)} \right]. \end{cases} \tag{9.65}$$

Similar to (9.58) and (9.62), two new states should also be carefully chosen for (9.65) to attain the uniform model, given as

$$\begin{aligned} x_1 &= 0.5L_0 i_l^2 + 0.5C_0 v_c^2 + C_0 E v_c, \\ x_2 &= E i_L. \end{aligned} \tag{9.66}$$

The dynamics of x_1 and x_2 could be described as

$$\begin{cases} \dot{x}_1 = L_0 i_L \dot{i}_L + C_0 v_C \dot{v}_C + C_0 E \dot{v}_C = E i_L + \vartheta_1 \\ \dot{x}_2 = E\dot{i}_L = E^2 d/L_0 - (1-d)Ev_C/L_0 + \vartheta_2. \end{cases} \Rightarrow \begin{cases} \dot{x}_1 = x_2 + \vartheta_1, \\ \dot{x}_2 = u + \vartheta_2. \end{cases} \tag{9.67}$$

Corresponding variables ϑ_1, ϑ_2, and u in (9.67) are derived as

$$\begin{aligned}
\vartheta_1 &= -\frac{(v_C+E)P_0}{v_C} - \frac{Edi_L\Delta_L}{L_0+\Delta_L} + \frac{(1-d)v_Ci_L\Delta_L}{L_0+\Delta_L} - \frac{(v_C+E)(1-d)i_L\Delta_C}{C_0+\Delta_C} \\
&\quad + \frac{(v_C+E)P_o\Delta_C}{(C_0+\Delta_C)v_C} \\
g_2 &= -\frac{(1-d)Ei_L\Delta_C}{(C_0+\Delta_C)C_0} + \frac{EP_0\Delta_C}{(C_0+\Delta_C)C_0v_C} \\
u &= \frac{E^2d}{L_0} - \frac{(1-d)Ev_C}{L_0} \Rightarrow d = \frac{L_0u+Ev_C}{E^2+Ev_C}.
\end{aligned}$$

It is obvious that the transformed equations of boost, buck, and buck-boost converters have the same mathematical expressions, as in (9.59), (9.63), and (9.67). These uniform converter modeling approaches make it easier to develop a decentralized composite stabilizer that uniformly applies disregarding converter model discrepancies. The stabilizer will be minutely elaborated in the next section.

9.5.1 Controller design

The previous section has explicated that the three converter topologies could be unexceptionally represented by a uniform dynamic form. In a generic multi-bus DC MG, a boost converter could appear at any DC bus, and so do its peer converters (buck and buck-boost). For easy explanations, uniform dynamics in (9.59), (9.63), and (9.67) are assigned with subscript "ij" and retrofitted as,

$$\begin{aligned}
\dot{x}_{ij1} &= x_{ij2} + \vartheta_{ij1}, \\
\dot{x}_{ij2} &= u_{ij} + \vartheta_{ij2},
\end{aligned} \tag{9.68}$$

where "ij" indicates that the associated state variables are in the jth converter at the ith DC bus. Although the control input u_{ij} and x_{ij1} are in the linear relation, the dual-integrator system is unfortunately coupled with ϑ_{ij1} and ϑ_{ij2} which are imposed by other DGs, the impedance network and parameter uncertainties, etc. The presence of ϑ_{ij1} and ϑ_{ij2} may adversely affect the reference tracking performances and even cause instabilities to DGs. It is hence essential to know the exact disturbances and eliminate them in the controller. To this end, a decentralized nonsmooth disturbance observer is constructed in (9.69) for fast and precise ϑ_{ij1} estimation. The observer for ϑ_{ij2} can be derived from (9.69) with slight modifications as the two equations in (9.68) share the same format.

$$\begin{cases}
\dot{\hat{x}}_{ij11} = x_{ij12} + \hat{x}_{ij12} + l_{ij11}\sigma_{ij1}\,\mathrm{sig}^{1+\tau}(x_{ij1} - \hat{x}_{ij11}), \\
\dot{\hat{x}}_{ij12} = \hat{x}_{ij13} + l_{ij12}\sigma_{ij1}^2\,\mathrm{sig}^{1+2\tau}(x_{ij1} - \hat{x}_{ij11}), \\
\dot{\hat{x}}_{ij13} = l_{ij13}\sigma_{ij1}^3\,\mathrm{sig}^{1+3\tau}(x_{ij1} - \hat{x}_{ij11}).
\end{cases} \tag{9.69}$$

where $\hat{x}_{ij11}$, $\hat{x}_{ij12}$, and $\hat{x}_{ij13}$ are the estimations of x_{ij1}, ϑ_{ij1}, and $\dot{\vartheta}_{ij1}$ respectively. The denotation of $\text{sig}^a(\cdot)$ is defined as $\text{sign}(\cdot)|\cdot|a$. $\sigma_{ij1} > 1$ is a scaling gain. τ should be within $(-\frac{1}{3},0)$, and it acts as a fine-tuning factor which will be shown contributing to dynamic enhancements in the subsequent context. l_{ij11}, l_{ij12} and l_{ij13} correspond to the coefficients of a Hurwitz polynomial $h(s) = s^3 + l_{ij11}s^2 + l_{ij12}s + l_{ij13}$. In practice, all components in ϑ_{ij1} and ϑ_{ij2} have their physical meanings, and it does allow the following assumption that

$$\max\left\{\sup\left|\dot{\vartheta}_{ij1}\right|, \sup\left|\ddot{\vartheta}_{ij1}\right|, \sup\left|\dot{\vartheta}_{ij2}\right|\right\} \leq D_{ij}, \tag{9.70}$$

where D_{ij} is a positive real number.

As for σ_{ij2}, a nonsmooth disturbance observer can be designed similar to (9.69) while the observer could reduce to a second order system.

$$\begin{cases} \dot{\hat{x}}_{ij21} = u_{ij} + \hat{x}_{ij22} + l_{ij21}\sigma_{ij2}\lfloor x_{ij2} - \hat{x}_{ij21}\rceil^{1+\tau}, \\ \dot{\hat{x}}_{ij22} = l_{ij22}\sigma_{ij2}^2\lfloor x_{ij2} - \hat{x}_{ij21}\rceil^{1+2\tau}, \end{cases} \tag{9.71}$$

where $\hat{x}_{ij21}$ and $\hat{x}_{ij22}$ are the estimations of x_{ij2} and ϑ_{ij2}, respectively. $\sigma_{ij2} > 1$ is a scaling gain. l_{ij21} and l_{ij22} are the coefficients of a second order Hurwitz polynomial, $h(s) = s^2 + l_{ij21}s + l_{ij22}$.

Observers (9.69) and (9.71) have estimated the disturbances in the DG uniform representation given by (9.68). These disturbances should now be neutralized by a nonsmooth feedback controller which simultaneously ensures the local DG large signal stability and achieves the desired reference tracking. For this purpose, intermediate states in relation to x_{ij1}, x_{ij2} could be defined of the following form:

$$\chi_{ij1} = x_{ij1} - x_{ij1r}, \chi_{ij2} = (x_{ij2} - x_{ij2r})/\beta_{ij}, v_{ij} = (u_{ij} - u_{ijj})/\beta_{ij}^2, \tag{9.72}$$

where $\beta_{ij} > 1$ is a scaling gain. v_{ij} is the provisional control input for the stabilization of χ_{ij1} and χ_{ij2}. v_{ij} is hence designed as

$$v_{ij} = -k_{ij1}\lfloor\chi_{ij1}\rceil^{1+2\tau} - k_{ij2}\lfloor\chi_{ij2}\rceil^{(1+2\tau)/(1+\tau)}, \tag{9.73}$$

where k_{ij1} and k_{ij2} are the coefficients of a Hurwitz polynomial $p(s) = s^2 + k_{ij2}s + k_{ij1}$. x_{ij1r}, x_{ij2r} and u_{ijr} are reference signals of x_{ij1}, x_{ij2} and u_{ij} respectively.

For the DGs based on boost, buck, and buck-boost converters, their voltage references could be from droop controllers for realizing power sharing, or the references are directly consigned with constants to minimize output voltage deviations. In the multi-bus MG in Figure 9.35, constant voltages at all buses would cause no power flow on the impedance network. Then linking cables in DC buses can be equivalently removed while not affecting individual bus regulations, which violates the initial intention of establishing a multi-bus MG. As such, it is preferable to configure all DGs in droop patterns and the power interactions among buses would thus be possible. Once the voltage reference v_{Cr}

for a DG is correctly determined, the reference signal x_{ij1r} for the transformed uniform system (9.68) can be accordingly computed referring to (9.58), (9.62), and (9.67),

$$\begin{aligned}&\text{Boost}: x_{ij1r} = 0.5L_0\left(P_o/E\right)^2 + 0.5C_0 v_{Cr}^2,\\&\text{Buck}: x_{ij1r} = 0.5C_0 v_{Cr}^2,\\&\text{Buck-boost}: x_{ij1r} = 0.5L_0\left(P_o/E\right)^2 + 0.5C_0 v_{Cr}^2 + C_0 E v_{Cr}.\end{aligned} \tag{9.74}$$

After knowing x_{ij1r}, x_{ij2r} and u_{ijr} are identified as,

$$\begin{aligned}x_{ij2r} &= \dot{x}_{ij1r} - \hat{x}_{ij12},\\u_{ijr} &= \ddot{x}_{ij1r} - \hat{x}_{ij13} - \hat{x}_{ij22}.\end{aligned} \tag{9.75}$$

By virtue of (9.72), the original control input u_{ij} is calculated as $u_{ij} = \beta_{ij2}v_{ij} + u_{ijr}$. The intermediate state dynamics of χ_{ij1}, χ_{ij2} would have the following close-loop form:

$$\dot{\chi}_{ij} = \beta_{ij} g_{ij} + \Phi_{ij}, g_{ij} = \left[g_{ij1}, g_{ij2}\right]^{\top}, \Phi_{ij} = \left[\Phi_{ij1}, \Phi_{ij2}\right]^{\top}, \tag{9.76}$$

where $\chi_{ij} = \left[\chi_{ij1}, \chi_{ij2}\right]^{\top}, g_{ij1} = \chi_{ij2}$,

$$\begin{aligned}g_{ij2} &= -k_{ij1}\lfloor\chi_{ij1}\rceil^{1+2\tau} - k_{ij2}\lfloor\chi_{ij2}\rceil^{(1+2\tau)/(1+\tau)},\\\Phi_{ij1} &= \sigma_{ij1}e_{ij12},\ \Phi_{i2} = \left(l_{ij12}\sigma_{ij1}^2\lfloor e_{ij11}\rceil^{1+2\tau} + \sigma_{ij2}e_{ij22}\right)/\beta_{ij}.\end{aligned} \tag{9.77}$$

By the composition of the nonsmooth disturbance observer and the nonsmooth composite control, disturbances in the uniform expression (9.68) can be fully neutralized. Besides, the proposed controller does allow decentralized stabilizer deployment into individual DGs among which no communication is incurred, but large signal stability of the entire multi-bus MG is achieved. This novel stabilizing concept categorically strengthens the MG PnP functionality, reliability, and scalability.

9.5.2 Discussions

1) Justification of using multi-bus DC system

On the one hand, in rural areas and wild battlefields, it has remote change to introduce AC power from urban generation stations, but a possible choice is to build up islanded DC MGs to incorporate sources and loads together, for the reduction of energy conversion stages, to achieve better system management [47]. When it comes to multiple DC MGs, as suggested by [66], MG buses could be interconnected by tie lines such that the DC systems can provide support to each other and enhance overall DC system operational reliability.

On the other hand, for a given MG, it is not always feasible to maintain only a single DC bus. Electrical devices are coupled together to form a complete DC system through cable lines which indisputably have their own line impedances. Although many existing results simply assume that all DGs are collectively linked to a certain bus. The scenario that makes more general sense is to allow the DG output terminals to be interlinked via the line impedance network of arbitrary topology. These DGs together with their loads establish local DC buses, and therefore, there would exist multiple DC buses in the MG. On these bases, the topology in Figure 9.35 is capable of representing the two multi-bus concepts discussed above. For either a DC system with multiple MGs or a multi-bus DC MG, they unexceptionally arouse wide academic attention and do make practical sense in real applications.

2) Grid connected operations

DC MG can be coupled to the utility grid through a three-phase AC/DC inverter which can be controlled under power-regulated mode. In this sense, the inverter, an interface to the utility, actually serves as a constant power source (CPS) or a constant power load (CPL) provided that the power command for the inverter is invariant. Specifically, if the inverter detects that the average value of multiple bus voltages exceeds the preset threshold, which indicates the multi-bus DC system is lightly loaded, then the grid-connected inverter will transfer power from the MG to the grid, such that the bus voltages would be regulated within allowable ranges. Inversely, if it is perceived that the DC MG is heavily loaded, then the inverter will extract power from the grid and inject it into the MG system.

9.5.3 Simulation studies

The results presented above have shown how to formulate decentralized composite stabilizers for individual DGs. To make the stabilizer easier to implement by engineering practitioners, key parameter selection guidance for observers in (9.69), (9.71) and controller in (9.72), (9.73) will be provided in this subsection. The impacts of fine-tuning factor τ on converter dynamics will also be illustrated.

Note that DG converter models can be expressed as a uniform model in (9.68). Without loss of generality and for clear explanations, a boost type converter with its output voltage always regulated at 170 V is used to demonstrate the parameter selection procedures. The selecting results can be directly used in buck and buck-boost converters without further changes because all three types of converters in Figure 9.36 share the same dynamic model under the proposed UNCS. Overviewing the structure of composite stabilizer, the observer would convey necessary estimations to the controller.

Thus, it is preferred that the observer dynamics are much faster than the controller. Taking the observer expressed in (9.69) as an example, a Hurwitz polynomial, which indicates all its roots are in the left-half plane, can be selected as $h(s) = (s+1)^3 = s^3 + 3s^2 + 3s + 1$. Then l_{ij11}, l_{ij12}, and l_{ij13} in (9.69) can be accordingly identified as 3, 3, and 1. Similarly, in observer (9.71), a second order polynomial with repeated roots of 1 could be formulated, $h(s) = (s+1)^2 = s^2 + 2s + 1$. By coefficients matching, l_{ij21} and l_{ij22} are respectively assigned as 2 and 1, and k_{ij1} and k_{ij2} in the controller (9.73) can also be designed as 1 and 2, respectively.

Figure 9.37 shows the dynamics of $\hat{x}_{ij12}$ in observer (9.69) with various σ_{ij1} when CPL changes from 200 W to 400 W, β_{ij} is differently set and τ is temporarily given as -0.02. From Figure 9.37(a), the transient duration of $\hat{x}_{ij12}$ are around 0.017s, 0.009s, 0.005s, 0.003s, and 0.002s with σ_{ij1} varying from 500 to 2500 in step of 500. The smaller σ_{ij1} helps to increase the observer dynamic response. Comparing Figure 9.37(a) and Figure 9.37(b) further suggests that the β_{ij} variation in the controller hardly affects the observer transient performances. Since the observer described by (9.71) has the similar form to (9.69), the parameter σ_{ij2} can be set the same as σ_{ij1}. Afterward, it is possible to fix σ_{ij1} and σ_{ij2} and scrutinize controller dynamics along with the changes of β_{ij}, as illustrated in Figure 9.38. All the converging speeds of converter output voltage, internal states x_{ij1} and x_{ij2} are improved by the increase of β_{ij}. For example, in Figure 9.38(a), in the case of CPL step-up, the settling times of the voltage can be read

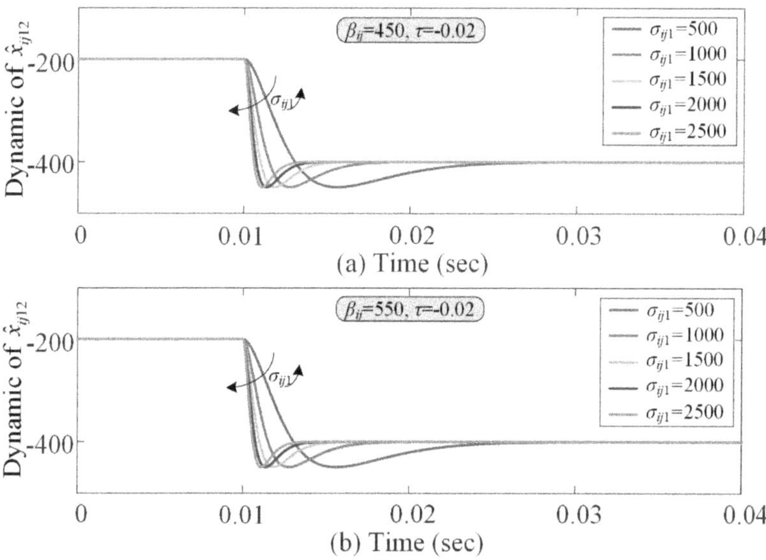

Figure 9.37 The dynamics of nonsmooth high gain observer with various σ_{ij1} when CPL changes from 200 W to 400 W, β_{ij} is differently set and the fine-tuning factor τ is given as -0.02.

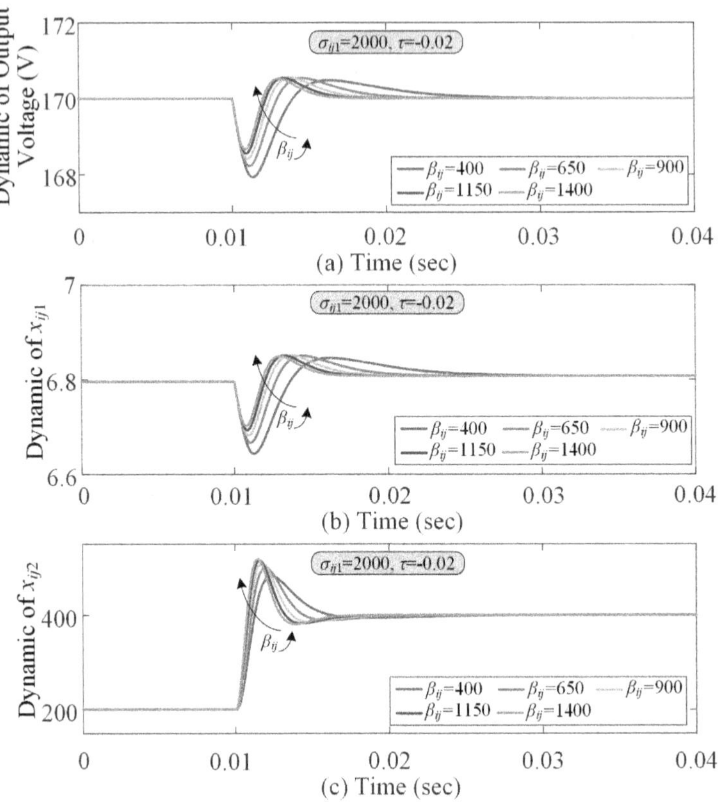

Figure 9.38 The dynamics of boost converter output voltage, x_{ij1} and x_{ij2} with various β_{ij} when CPL changes from 200 W to 400 W, σ_{ij1} and σ_{ij2} are both set as 2000, and τ is given as -0.02.

as 0.02 s, 0.015 s, 0.011 s, 0.01 s, and 0.009 s respectively, when β_{ij} is designed as 400, 650, 900, 1150, and 1400. In this simulation, $\sigma_{ij1}, \sigma_{ij2}$, and β_{ij} are respectively selected as 2000, 2000, and 650 such that the observer could respond over 5 times faster than the controller. This set of parameters is subsequently used to demonstrate how the fine-tuning factor τ refines the composite stabilizer dynamics, and the parameters will be straightly utilized by experimental tests.

It is clear that the highest dynamic order appears in either the observer or the controller is 3, and according to the aforesaid homogeneity theory, τ can only vary in the range $(-\frac{1}{3}, 0)$. On this point, transient performances of $\hat{x}_{ij12}$, output voltage, states x_{ij1} and x_{ij2} subjected to τ variations are all displayed in Figure 9.39. The figure explicitly illustrates that τ provides an extra control degree of freedom additional to $\sigma_{ij1}, \sigma_{ij2}$, and β_{ij}, but not impairing the large signal

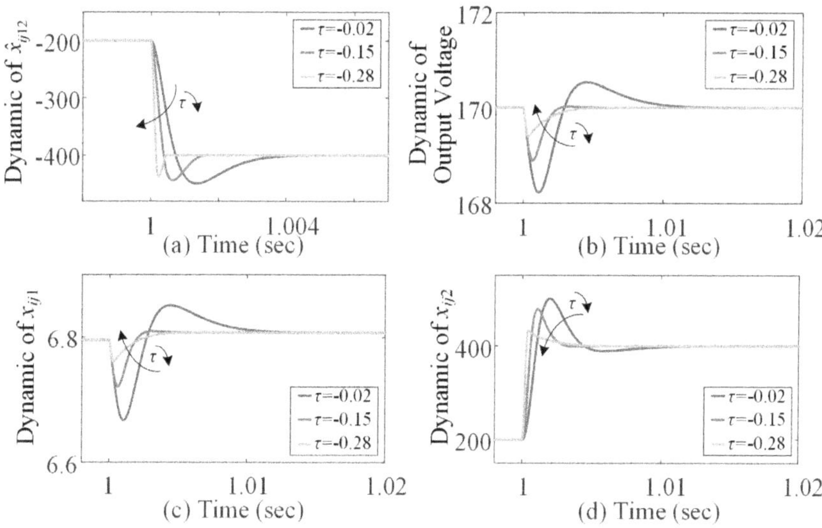

Figure 9.39 Refining dynamic responses in the observer and the controller by changing fine-tuning factor τ within permissible range when both σ_{ij1} and σ_{ij2} are 2000, and β_{ij} is 650.

stability. A lowered τ brings about better dynamic responses, i.e., less undershoots and shorter transient durations. In this section, τ is chosen as the smallest one (-0.28), so as to make full use of refining functionality endowed by the fine-tuning factor.

9.5.4 Experimental tests

A representative multi-bus DC MG for experiments is plotted in Figure 9.40. A boost converter, a buck converter, and a buck-boost converter are DGs separately distributed at three buses for voltage regulations. The PWM signals are generated from dSPACE 1006. Line impedances of dissimilar values interlink the three buses to constitute a multi-bus system as intended. Each bus is loaded by a programmable electronic load. All DGs are under droop modes, with the droop controllers written as

$$V_{Cri} = V_n - m_i P_{oi},\ i = 1,2,3, \tag{9.78}$$

where V_{Cri} denotes the voltage reference for the i-th DG. P_{oi} is the output power and m_i is droop coefficient. More details of the hardware system have been summarized in Table 9.5.

A. Plug and Play Test

For the configuration in Figure 9.40, the voltages at three DC buses equal v_{C1}, v_{C2}, and v_{C3} respectively. All electronic loads in this PnP test are configured

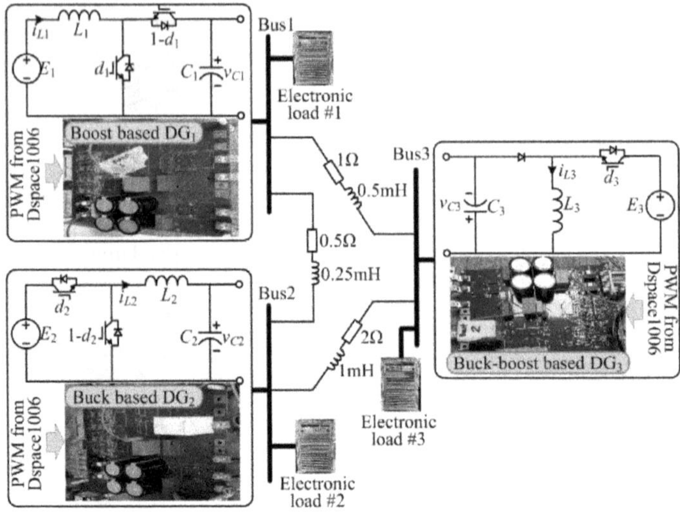

Figure 9.40 Hardware experimental platform of a representative multi-bus DC MG.

as CPLs, and the experimental results are recorded in Figure 9.41. At the beginning, the CPLs of DG1, DG2, and DG3 are separately scheduled as 300 W, 400 W, and 500 W. As seen in Figure 9.41(a), before DG2 plugs in, the output powers of DGs are 300, 400 W, and 500 W. The DC bus voltages can be estimated as 167 V, 166 V, and 165 V. At this stage, v_{C1} is the highest since DG1 has the lightest load. Its output voltage deviation from the nominal value (V_n) would be less than other DGs due to the existence of droop controller. When DG2 is connected to DG1 via line impedance (0.5 Ω, 0.25 mH), DG1 takes up a part of the load of DG2; hence the power of DG2 decreases by around 44.3 W. In Figure 9.41(b), DG3 is further plugged into the system to form a complete three-bus MG. The three DGs will simultaneously share all the loads across the MG. The DG powers can thus be read as 386.4 W, 393.5 W, and 420.3 W. This means DG1

Table 9.5 System parameters

Parameter	Description	Value
V_n	Nominal DC voltage level	170 V
E_1, E_2, E_3	Converter input voltages	100 V, 220 V, 120 V
L_1, L_2, L_3	Converter inductors	2 mH
C_1, C_2, C_3	Converter capacitors	470 uF
m_1, m_2, m_3	Droop coefficients	0.01

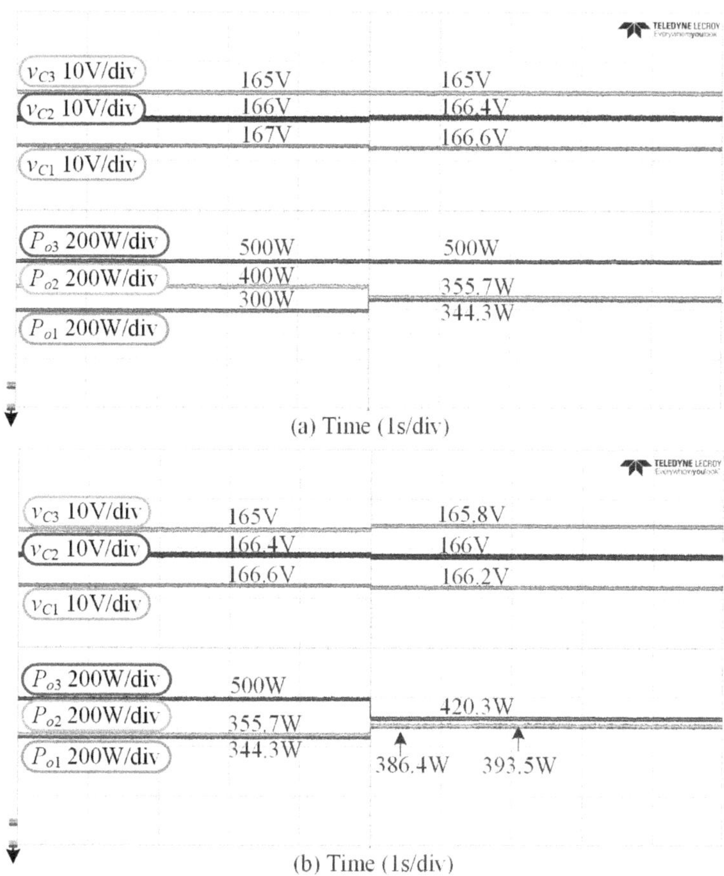

Figure 9.41 Plug and play tests for the experimental multi-bus DC MG wherein each DG is integrated with only a CPL.

and DG2 help to alleviate the loading stresses of DG3 as its output power significantly decreases while the power supplied by DG1 and DG2 increases. From the testing results, it is safe to conclude that the proposed composite stabilizer in UNCS does allow PnP functions, which certainly enhance the operational flexibility of multi-bus MGs.

B. Stability Comparison By using the proposed composite stabilizer, it is true that large-signal stabilization has been achieved. It is now ready to experimentally justify that the proposed UNCS is superior to the conventional PI control schemes which are also accomplished in a decentralized way. According to standard design of double PI loops given in [2], the parameters for the boost, buck, and buck-boost based DGs are tabularized in Table 9.6. For either voltage loops

Table 9.6 PI parameters for three converters

Parameter	Description	Value
k_{pv1}, k_{iv1}	PI parameters of voltage and current loops	0.66, 55.79
k_{pc1}, k_{ic1}	in boost DG1	0.098, 82.16
k_{pv2}, k_{iv2}	PI parameters of voltage and current loops	0.39, 32.82
k_{pc2}, k_{ic2}	in boost DG2	0.076, 63.48
k_{pv3}, k_{iv3}	PI parameters of voltage and current loops	0.94, 79.32
k_{pc3}, k_{ic3}	in boost DG3	0.058, 48.16

or current loops, the stability margins are uniformly designed as 84.29° which is supposed to be sufficient for handling possible worse scenarios.

To compare stabilization capability between the UNCS and PI control, differing from the preceding PnP tests, electronic loads are all programmed as pure CPLs to emulate the worst operating condition. Figure 9.42 shows the experimental outcomes with three CPLs concurrently stepping up from 100 W to 700 W by 100 W. At each step, the multiple DC buses and the DG output powers are stable. However, for the PI regulated MG, although PI parameters are obtained following the standard design criteria in [2], as in Figure 9.43(a)–(b), the entire MG is unfortunately destabilized when CPLs simultaneously grow to 600 W. These observations can be supported by eigenvalue analysis in Figure 9.43(c) where dominant eigenvalues traverse into the right half plan when CPLs are 600 W. From the above comparisons, it could be discerned that the decentralized composite stabilizer withstands a wider operating range than PI control does. Moreover, the proposed stabilizers are in DGs without communication, which substantially strengthens the MG scalability. Theoretically, as long as MG operation is within its nominal rating, the multi-bus system can be functioning at any desired equilibrium, which markedly characterizes the large signal stabilization of UNCS.

Figure 9.44 shows the experimental results in the situation that all resistances in line impedances are reduced to 0.5 Ω and MG is with UNCS. This maneuver will mitigate the stabilizing effects of passive components. In the final stage, it is clear that all CPLs climb to 800 W and the entire CPL power is $800 \times 3 = 2400$ W. The DC MG is consistently stable, which attests to the effectiveness of the proposed UNCS method.

9.6 Nonsmooth Adaptive Controller Design for DC Microgrids

In the above sections, composite controllers applied into DC microgrids integrating disturbance observation with output regulation are explicitly shown. The

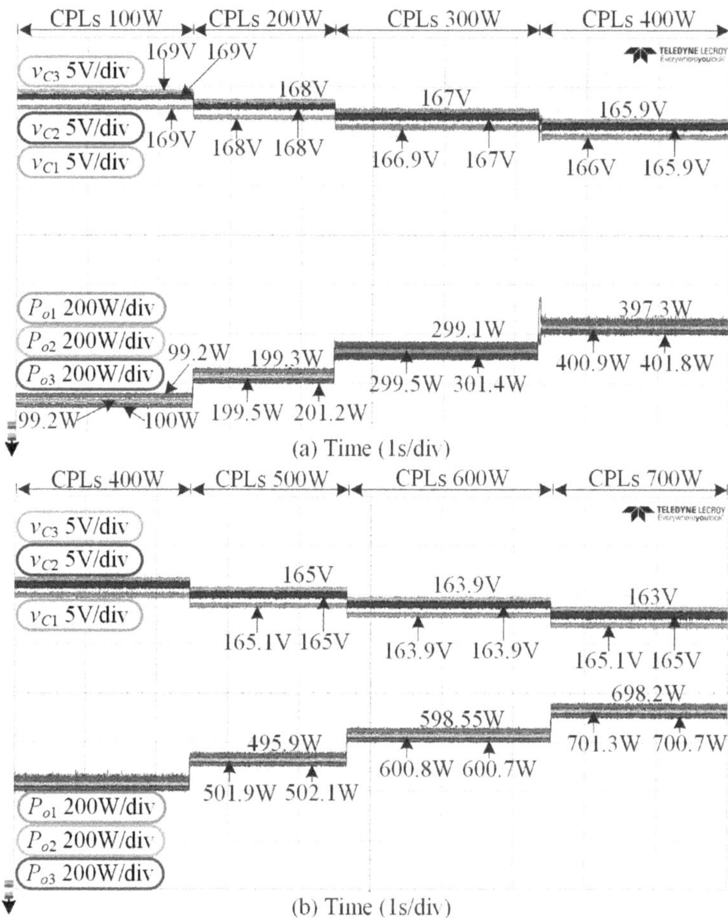

Figure 9.42 System-level operation of the experimental multi-bus DC MG wherein each DG loaded by a pure CPL is with the proposed composite robust controller.

distinguishing feature can be summarized as the fixed high gain design in a non-recursive manner. Therefore, a phenomenon can be observed in engineering experiments to DC-DC boost converter that the fixed high gain design can indeed stabilize the bus voltage, however, with the wide range of fluctuations of constant power loads (CPLs), the transient-time performance, i.e., overshoot value, tends to be influenced by the fixed value of scaling gain. Moreover, the settling time varies significantly between large-signal and small-signal disturbances, with longer settling time for large-signal disturbances which may probably not be a satisfactory result. In this section, we set the control mode for DC microgrids as

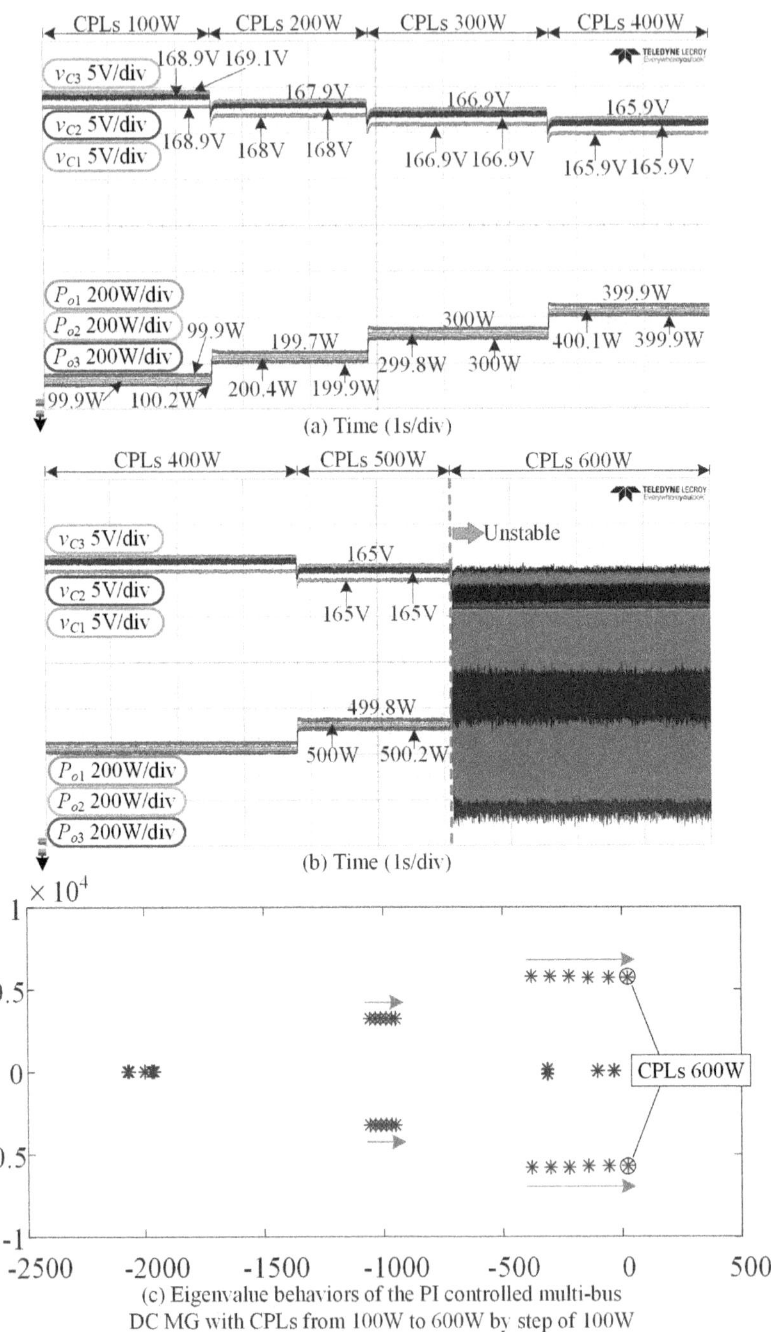

Figure 9.43 System-level operation of the experimental multi-bus DC MG wherein each DG loaded by a pure CPL is with the standardly designed PI controllers.

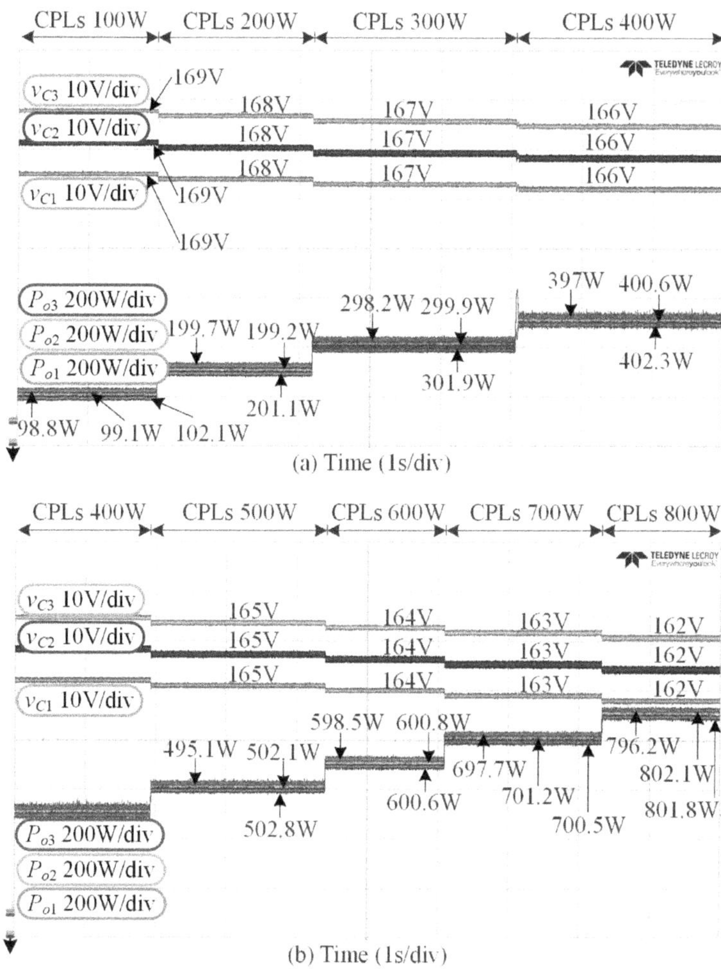

Figure 9.44 Experimental results of multi-bus DC MG under UNCS with each CPL from 100 W to 800 W, and all resistances in line impedances being set as 0.5 Ω.

the constant voltage mode, a novel nonsmooth adaptive controller design strategy is proposed, aiming to further improve the control performance while the control system could include a homogeneous degree to exquisitely tune the control performances.

9.6.1 Controller design

Without loss of generality, we are able to simplify the bus stabilization issue for the DC microgrid to a constant voltage regulation problem for a DC-DC boost converter. By following the standardized modeling process in Subsection 9.2.1, one could straightforwardly arrive at the transformed equivalent system of the following form:

$$\begin{aligned}\dot{x}_1 &= x_2 + d_1,\\ \dot{x}_2 &= u + d_2,\\ y &= x_1,\end{aligned} \tag{9.79}$$

by viewing $-P_{CPL}, \frac{E^2}{L} - \frac{Ev_C}{L}(1-\mu)$ as a mismatched disturbance d_1 and the control input u, respectively. d_2 is regarded as the lumped matched disturbance including external disturbances, internal uncertainties, etc.

On the basis of the proposed nonsmooth adaptive controller in Chapter 6, the nonsmooth dynamic bus voltage regulation law for this case can be constructed as:

$$\begin{cases} u = -\ell^{2+\rho}\left(k_1\lfloor\eta_1\rceil^{1+2\kappa} + k_2\lfloor\eta_2\rceil^{\frac{1+2\kappa}{1+\kappa}}\right) + u^* \\ \dot{\ell} = \lambda\left(|\eta_1|^2 + |\eta_2|^{\frac{2}{1+\kappa}}\right)\left(1 + \mathrm{sign}(|e_s| - \delta)\right), \end{cases} \tag{9.80}$$

where $\eta_1 = (x_1 - x_1^*)/\ell^\rho$, $\eta_2 = (x_2 - x_2^*)/\ell^{\rho+1}$, $x_1^* = 0.5L\hat{z}_{1,2}^2/E^2 + 0.5Cv_{Cr}^2$, $x_2^* = L\hat{z}_{1,2}\hat{z}_{1,3}/E^2 - \hat{z}_{1,2}$, $u^* = L(\hat{z}_{1,3} + \hat{z}_{1,2}\hat{z}_{1,4})/E^2 - \hat{z}_{1,3} - \hat{z}_{2,2}$, and $\hat{z}_{1,2}, \hat{z}_{1,3}, \hat{z}_{1,4}, \hat{z}_{2,2}$ are derived from (6.14).

9.6.2 Experimental tests

The experimental setup is depicted in Figure 9.3. The proposed algorithm is embedded in dSPACE to generate PWM signals for DC-DC boost converter with a switching frequency of 20 kHz. The analog working condition switching is achieved by switching the power of the Chroma DC electrical load in the CPL mode. The involved parameters are listed in Table 9.7.

The experimental design is described as follows: first, the classical PI controller, the nonsmooth composite controllers with fixed and self-tuning scaling gain are chosen respectively for control performance comparison. Second, the working conditions are mainly divided into two scenarios: CPL variation changes from 40 W to 200 W and from 40 W to 400 W that correspond to small-signal and large-signal stability comparisons. Third, in order to better demonstrate the updating discipline of self-tuning scaling gain, the CPL variation is also carried out from 40 W to 100 W to 200 W to 300 W and to 400 W, respectively.

Table 9.7 Parameters of the DC-DC boost converter control system

Description	Parameter	Value
Converter Input Voltage	E	75 V
Nominal Bus Voltage	V_{ref}	150 V
Nominal Inductance Value	L	2 mH
Nominal Capacitance Value	C	1.1 mF
Switching Frequency	f_{sw}	20 kHz
Observer Gains for Observer # 1	$\beta_1, l_{1,1} \cdots l_{1,4}$	80, 4, 6, 4, 1
Observer Gains for Observer # 2	$\beta_2, l_{2,1}, l_{2,2}$	40, 2, 1
Controller Gains	k_1, k_2	0.5, 0.3
Design Parameters	ρ, λ, δ	0.1, 300, 0.7
Homogeneous Degrees	τ, κ	−0.08, −0.12
PI Gains of Voltage Control Loop	k_{vp}, k_{vi}	1, 23
PI Gains of Current Control Loop	k_{cp}, k_{ci}	0.01, 0.5

The nonsmooth robust controller is achieved by setting the scaling gain ℓ as a sufficiently large constant value, which is expressed as

$$u = -\ell^{2+\rho}(k_1 \lfloor \eta_1 \rceil^{1+2\kappa} + k_2 \lfloor \eta_2 \rceil^{\frac{1+2\kappa}{1+\kappa}}) + u^*, \ \ell \in \mathbb{R}_+.$$

In this experimental verification, we set a fixed ℓ as 950.

Scenario 1: CPL varies from 40 W to 200 W.

In this scenario, the results of conducted experiments can be seen in Figure 9.45 (a), in which one can observe that for similar voltage deviation, the longest settling time occurs in the PI controlled system (0.063 s). The control system behaves as smoother steady-state performance under the controller proposed in this section than the nonsmooth robust controller, and the dynamic scaling gain $\ell(t)$ will converge from 1 to 604.84 to 608.12 with working condition switching, which implies that the value of ℓ as 608.12 tends to be more suitable for the current working condition. In other words, the fixed scaling gain has to be set as a sufficiently large value aiming to maintain the stability of the bus voltage with the presence of operating condition switching. Conversely, if the gain selection is over-conservative, the possible robustness redundancy may deteriorate the steady-state performance to a certain extent.

Scenario 2: CPL varies from 40 W to 400 W.

Figure 9.45 (b) presents the detailed results. Obviously, compared with *Scenario 1*, the transient-time performance of all types of methods has deteriorated under large-signal switching. However, the overall trends are still consistent with the transient-time performance under small-signal case, i.e., the settling time of the PI controller is obviously longer than the nonsmooth controllers, and the steady-state performance of the nonsmooth robust controller has a slight voltage oscillation.

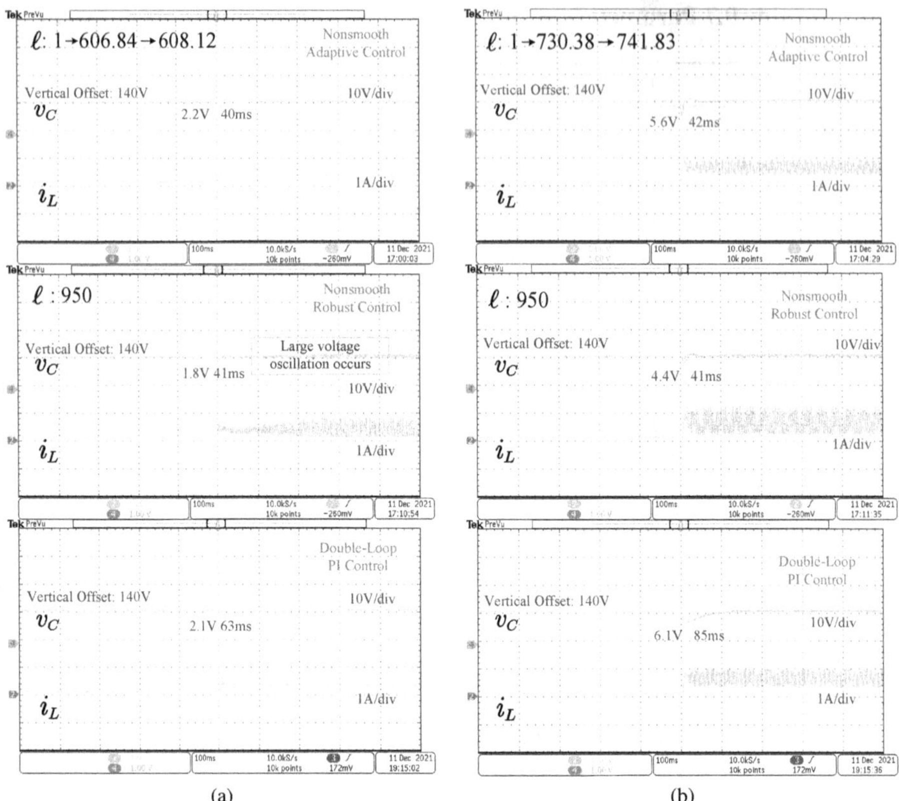

Figure 9.45 Experimental comparison results. (a) CPL changes from 40 W to 200 W; (b) CPL changes from 40 W to 400 W.

Scenario 3: CPL varies from 40 W to 100 W, 40 W to 200 W, 40 W to 300 W, 40 W to 400 W, respectively.

We present four working conditions to demonstrate the dynamic gain updating mechanism. As depicted in Figure 9.46, the switching of working conditions becomes larger and larger, the increasing rate of dynamic gain becomes larger and larger (0.65, 1.28, 7.04, 11.45). To see why, the larger working condition switching causes more intense bus voltage fluctuations, which provides greater energy to trigger the updating mechanism of ℓ.

The set of data with the desirable experimental results can be viewed in Table 9.8 and the experimental results are basically consistent with the design expectation. Furthermore, the performance improvement of the proposed nonsmooth dynamic controller is verified by comparing the performance index of the three controllers, mainly settling time and steady-state performance.

Table 9.8 Experimental performance indexes of Scenarios 1 and 2

Controllers	CPL Variation Level	Maximum Deviation	Settling Time	ISE	MSE
The Proposed Controller	CPL: 40 W→200 W	2.2 V	40 ms	0.23	0.93
	CPL: 40 W→400 W	5.6 V	42 ms	1.42	1.13
Nonsmooth Robust Controller	CPL: 40 W→200 W	1.8 V	41 ms	0.37	8.46
	CPL: 40 W→400 W	4.4 V	41 ms	2.44	7.88
PI Controller	CPL: 40 W→200 W	2.1 V	63 ms	0.18	0.91
	CPL: 40 W→400 W	6.1 V	85 ms	2.5	2.64

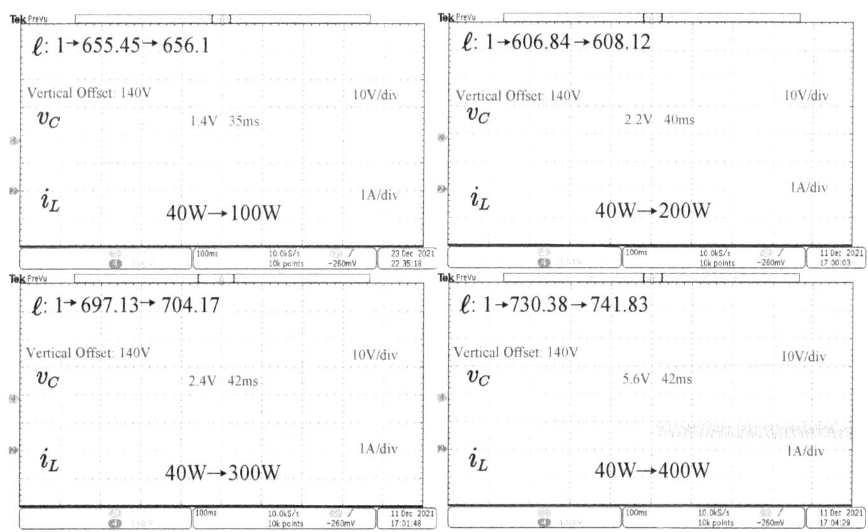

Figure 9.46 Experimental comparison results of CPL changes from 40 W to 100 W, 40 W to 200 W, 40 W to 300 W, and 40 W to 400 W under the controller (9.80).

References

[1] A. Astolfi and R. Ortega. Immersion and invariance: A new tool for stabilization and adaptive control of nonlinear systems. *IEEE Transactions on Automatic Control*, 48(4):590–606, 2003.

[2] S. Bacha, I. Munteanu, and A. I. Bratcu. Power electronic converters modeling and control. In *Power Electronic Converters Modeling and Control*. Springer, London, U.K., 2014.

[3] S. Bhat and D. Bernstein. Continuous finite-time stabilization of the translational and rotational double integrators. *IEEE Transactions on Automatic Control*, 43(5):678–682, 1998.

[4] S. Bhat and D. Bernstein. Geometric homogeneity with applications to finite-time stability. *Mathematics of Control, Signals and Systems*, 17(2):101–127, 2005.

[5] R. Brockett and C. Byrnes. Multivariable nyquist criteria, root loci, and pole placement: A geometric viewpoint. *IEEE Transactions on Automatic Control*, 26(1):271–284, 1981.

[6] J. Cai, C. Wen, H. Su, Z. Liu, and L. Xing. Adaptive backstepping control for a class of nonlinear systems with non-triangular structural uncertainties. *IEEE Transactions on Automatic Control*, 62(10):5220–5226, 2017.

[7] F. Castaños and L. Fridman. Analysis and design of integral sliding manifolds for systems with unmatched perturbations. *IEEE Transactions on Automatic Control*, 51(5):853–858, 2006.

[8] W-H Chen, J. Ballance, and J. Gawthrop. Optimal control of nonlinear systems: a predictive control approach. *Automatica*, 39(4):633–641, 2003.

[9] W-H Chen, J. Yang, L. Guo, and S. Li. Disturbance-observer-based control and related methods–an overview. *IEEE Transactions on Industrial Electronics*, 63(2):1083–1095, 2016.

[10] C. Dai, T. Guo, J. Yang, and S. Li. A disturbance observer-based current-constrained controller for speed regulation of PMSM systems subject to unmatched disturbances. *IEEE Transactions on Industrial Electronics*, 68(1):767–775, 2020.

[11] J. Davila. Exact tracking using backstepping control design and high-order sliding modes. *IEEE Transactions on Automatic Control*, 58(8):2077–2081, 2013.

[12] S. Ding, A. Levant, and S. Li. Simple homogeneous sliding-mode controller. *Automatica*, 6:22–32, 2016.

[13] D. Kwon and W. Book. A time-domain inverse dynamic tracking control of a single-link flexible manipulator. *Journal of Dynamical Systems, Measurement, and Control*, 116(193), 1994.

[14] H. Du, C. Qian, S. Li, and Z. Chu. Global sampled-data output feedback stabilization for a class of uncertain nonlinear systems. *Automatica*, 99:403–411, 2019.

[15] H. Du, C. Qian, S. Yang, and S. Li. Recursive design of finite-time convergent observers for a class of time-varying nonlinear systems. *Automatica*, 49(2):601–609, 2013.

[16] H. Du, G. Wen, Y. Cheng, and J. Lv. Design and implementation of bounded finite-time control algorithm for speed regulation of permanent magnet synchronous motor. *IEEE Transactions on Industrial Electronics*, 68(3):2417–2426, 2020.

[17] A. Emadi, A. Khaligh, H. Rivetta, and A. Williamson. Constant power loads and negative impedance instability in automotive systems: Definition, modeling, stability, and control of power electronic converters and motor drives. *IEEE Transactions on Vehicular Technology*, 55(4):1112–1125, 2006.

[18] R. Errouissi, M. Ouhrouche, W-H Chen, and M. Trzynadlowski. Robust nonlinear predictive controller for permanent-magnet synchronous motors with an optimized cost function. *IEEE Transactions on industrial Electronics*, 59(7):2849–2858, 2012.

[19] A. Ferreira, C. Jérôme, H. David, Z. Ali, and F. Leonid. Output tracking of systems subjected to perturbations and a class of actuator faults based on hosm observation and identification. *Automatica*, 59:200–205, 2015.

[20] Q. Gong and C. Qian. Global practical output regulation of a class of nonlinear systems by measurement feedback. *Automatica*, 43:184–189, 2007.

[21] H. Hermes. Homogeneous coordinates and continuous asymptotically stabilizing feedback controls. In S. Elaydi, editor, *Differential Equations, Stability and Control*, pages 249–260. Dekker: New York, 1991.

[22] Y. Hong, J. Wang, and D. Cheng. Adaptive finite-time control of nonlinear systems with parametric uncertainty. *IEEE Transactions on Automatic Control*, 51(5):858–862, 2006.

[23] J. Huang and Z. Chen. A general framework for tackling the output regulation problem. *IEEE Transactions on Automatic Control*, 49(12):2203–2218, 2004.

[24] J. Huang, C. Wen, W. Wang, and Y. Song. Design of adaptive finite-time controllers for nonlinear uncertain systems based on given transient specifications. *Automatica*, 69:395–404, 2016.

[25] X. Huang, Y. Song, and C. Wen. Output feedback control for constrained pure-feedback systems: A non-recursive and transformational observer based approach. *Automatica*, 113:108789, 2020.

[26] A. Isidori. *Nonlinear control systems. II.* Communications and Control Engineering Series. Springer-Verlag London Ltd., London, 1999.

[27] A. Isidori. *Nonlinear control systems*. Springer Science & Business Media, 2013.

[28] A. Isidori and C. Byrnes. Output regulation of nonlinear systems. *IEEE Transactions on Automatic Control*, 35(2):131–140, 1990.

[29] M. Kawski. Stabilization of nonlinear systems in the plane. *Systems and Control Letters*, 12(2):169–175, 1989.

[30] H. Khalil and F. Esfandiari. Semiglobal stabilization of a class of nonlinear systems using output feedback. *IEEE Transactions on Automatic Control*, 38(9):1412–1415, 1993.

[31] K. Khalil. *Nonlinear systems*. Macmillan Publishing Company, New York, 1992.

[32] K. Khalil. High-gain observers in nonlinear feedback control. In *New directions in Nonlinear Observer Design (Geiranger Fjord,1999)*, volume 244 of *Lecture Notes in Control and Inform. Sci.*, pages 249–268. Springer, London, 1999.

[33] M-S Koo, H-L Choi, and J-T Lim. Universal control of nonlinear systems with unknown nonlinearity and growth rate by adaptive output feedback. *Automatica*, 47(10):2211–2217, 2011.

[34] P. Krishnamurthy and F. Khorrami. Dynamic high-gain scaling: State and output feedback with application to systems with ISS appended dynamics driven by all states. *IEEE Transactions on Automatic Control*, 49(12):2219–2239, 2004.

[35] M. Krstic, I. Kanellakopoulos, and P. V. Kokotovic. *Nonlinear and Adaptive Control Design*. John Wiley, 1995.

[36] H. Lei and W. Lin. Universal adaptive control of nonlinear systems with unknown growth rate by output feedback. *Automatica*, 42(10):1783–1789, 2006.

[37] F. Leonid and K. Khalil. Performance recovery of feedback-linearization-based designs. *IEEE Transactions on Automatic Control*, 53(10):2324–2334, 2008.

[38] A. Levant. Higher-order sliding modes, differentiation and output-feedback control. *International Journal of Control*, 76(9-10):924–941, 2003.

[39] S. Li, H. Sun, J. Yang, and X. Yu. Continuous finite-time output regulation for disturbed systems under mismatching condition. *IEEE Transactions on Automatic Control*, 60(1):277–282, 2015.

[40] S. Li, J. Yang, W-H Chen, and X. Chen. *Disturbance Observer-Based Control: Methods and Applications*. CRC Press, 2014.

[41] X. Li, G. Chen, Y. Pan, and H. Yu. Region control for robots driven by series elastic actuators. In *2016 IEEE International Conference on Robotics and Automation*, pages 1102–1107, 2016.

[42] P. Lin, P. Wang, J. Xiao, J. Wang, C. Jin, and Y. Tang. An integral droop for transient power allocation and output impedance shaping of hybrid energy storage system in DC microgrid. *IEEE Transactions on Power Electronics*, 33(7):6262–6277, 2018.

[43] W. Lin and C. Qian. Adding one power integrator: A tool for global stabilization of high-order lower-triangular systems. *Systems and Control Letters*, 39(3):339–351, 2000.

[44] W. Lin and C. Qian. Adaptive control of nonlinearly parameterized systems: A nonsmooth feedback framework. *IEEE Transactions on Automatic Control*, 47(5):757–774, 2002.

[45] W. Lin and C. Qian. Adaptive control of nonlinearly parameterized systems: the smooth feedback case. *IEEE Transactions on Automatic Control*, 47(8):1249–1266, 2002.

[46] Z. Liu, Y. Wu, and J. Li. Universal strategies to explicit adaptive control of nonlinear time-delay systems with different structures. *Automatica*, 89:151–159, 2018.

[47] L. Meng, Q. Shafiee, G. F. Trecate, H. Karimi, D. Fulwani, X. Lu, and J. M. Guerrero. Review on control of DC microgrids and multiple microgrid clusters. *IEEE Journal of Emerging and Selected Topics in Power Electronics*, 5(3):928–948, 2017.

[48] R. Morales, H. Sira-Ramírez, and J. A. Somolinos. Robust control of underactuated wheeled mobile manipulators using GPI disturbance observers. *Multibody System Dynamics*, 32(4):511–533, 2014.

[49] E. Moulay. Stabilization via homogeneous feedback controls. *Automatica*, 44(11):2981–2984, 2008.

[50] J. Na, X. Ren, and D. Zheng. Adaptive control for nonlinear pure-feedback systems with high-order sliding mode observer. *IEEE Transactions on Neural Networks and Learning Systems*, 24(3):370–382, 2013.

[51] S. Oh and K. Kong. High-precision robust force control of a series elastic actuator. *IEEE/ASME Transactions on Mechatronics*, 22(1):71–80, 2017.

[52] C. P. Bechlioulis and G. A. Rovithakis. A low-complexity global approximation-free control scheme with prescribed performance for unknown pure feedback systems. *Automatica*, 50(4):1217–1226, 2004.

[53] N. Paine, S. Oh, and L. Sentis. Design and control considerations for high-performance series elastic actuators. *IEEE/ASME Transactions on Mechatronics*, 19(3):1080–1091, 2014.

[54] W. Perruquetti, T. Floquet, and E. Moulay. Finite-time observers: Application to secure communication. *IEEE Transactions on Automatic Control*, 53(1):356–360, 2008.

[55] J. Polendo and C. Qian. A generalized homogeneous approach for global stabilization of inherently nonlinear systems via output feedback. *International Journal of Robust and Nonlinear Control*, 17(7):605–629, 2007.

[56] L. Praly and Z. Jiang. Linear output feedback with dynamic high gain for nonlinear systems. *Systems and Control Letters*, 53(2):107–116, 2004.

[57] J. Pratt, B. Krupp, and C. Morse. Series elastic actuators for high fidelity force control. *Industrial Robot: An International Journal*, 29(3):234–241, 2002.

[58] C. Qian. A homogeneous domination approach for global output feedback stabilization of a class of nonlinear systems. *Proceedings of the 2005 American Control Conference*, pages 4708–4715, 2005.

[59] C. Qian and W. Lin. Almost disturbance decoupling for a class of high-order nonlinear systems. *IEEE Transactions on Automatic Control*, 45(6):1208–1214, 2000.

[60] C. Qian and W. Lin. A continuous feedback approach to global strong stabilization of nonlinear systems. *IEEE Transactions on Automatic Control*, 46(7):1061–1079, 2001.

[61] C. Qian and W. Lin. Practical output tracking of nonlinear systems with uncontrollable unstable linearization. *IEEE Transactions on Automatic Control*, 47(1):21–36, 2002.

[62] L. Rosier. Homogeneous Lyapunov function for homogeneous continuous vector fields. *Systems and Control Letters*, 19(6):467–473, 1992.

[63] C. Rui, R. Mahmut, I. Kolmanovsky, S. Cho, and H. McClamroch. Nonsmooth stabilization of an underactuated unstable two degrees of freedom mechanical system. In *IEEE Conference on Control and Decision*, volume 4, pages 3998–4003, 1997.

[64] E. Sariyildiz, G. Chen, and H. Yu. An acceleration-based robust motion controller design for a novel series elastic actuator. *IEEE Transactions on Industrial Electronics*, 63(3):1900–1910, 2016.

[65] D-W. Seo, Y. Bak, and K-B. Lee. An improved rotating restart method for a sensorless permanent magnet synchronous motor drive system using repetitive zero voltage vectors. *IEEE Transactions on Industrial Electronics*, 67(5):3496–3504, 2019.

[66] Q. Shafiee, T. Dragicevic, J. C. Vasquez, and J. M. Guerrero. Hierarchical control for multiple DC-microgrids clusters. *IEEE Transactions on Energy Conversion*, 29(4):922–933, 2014.

[67] Y. Shen and X. Xia. Semi-global finite-time observers for nonlinear systems. *Automatica*, 44(12):3152–3156, 2008.

[68] Z. Su, Y. Hao, W. Zha, and C. Qian. Semiglobal stabilization of linearly uncontrollable and unobservable nonlinear systems via sampled-data control. *International Journal of Robust and Nonlinear Control*, 30(13):5290–5304, 2020.

[69] H. Sun, S. Li, J. Yang, and W. Zheng. Global output regulation for strict-feedback nonlinear systems with mismatched nonvanishing disturbances.

International Journal of Robust and Nonlinear Control, 25(15):2631–2645, 2015.

[70] Z. Sun, Y. Zhang, S. Li, and X. Zhang. A simplified composite current-constrained control for permanent magnet synchronous motor speed-regulation system with time-varying disturbances. *Transactions of the Institute of Measurement and Control*, 42(3):374–385, 2020.

[71] Z. Sun, T. Li, and S Yang. A unified time-varying feedback approach and its applications in adaptive stabilization of high-order uncertain nonlinear systems. *Automatica*, 70:249–257, 2016.

[72] Z. Sun, L. Xue, and K. Zhang. A new approach to finite-time adaptive stabilization of high-order uncertain nonlinear system. *Automatica*, 58:60–66, 2015.

[73] Z. Sun, S. Yang, and T. Li. Global adaptive stabilization for high-order uncertain time-varying nonlinear systems with time-delays. *International Journal of Robust and Nonlinear Control*, 27(13):2198–2217, 2017.

[74] D. Swaroop, J. K. Hedrick, P. P. Yip, and J. C. Gerdes. Dynamic surface control for a class of nonlinear systems. *IEEE Transactions on Automatic Control*, 45(10):1893–1899, 2000.

[75] Q. Tan, X. Huang, L. Li, and M. Wang. Magnetic field analysis and flux barrier design for modular permanent magnet linear synchronous motor. *IEEE Transactions on Industrial Electronics*, 67(5):3891–3900, 2019.

[76] A. Teel and L. Praly. Global stabilizability and observability imply semi-global stabilizability by output feedback. *Systems and Control Letters*, 22(5):313–325, 1994.

[77] A. Teel and L. Praly. Tools for semiglobal stabilization by partial state and output feedback. *SIAM Journal on Control and Optimization*, 33(5):1443–1488, 1995.

[78] M. Wang, L. Sun, W. Yin, S. Dong, and J. Liu. Continuous finite-time control approach for series elastic actuator. In *2016 IEEE 55th Conference on Decision and Control*, pages 843–848, 2016.

[79] N. Wang, H. Karimi, H. Li, and S. Su. Accurate trajectory tracking of disturbed surface vehicles: A finite-time control approach. *IEEE/ASME Transactions on Mechatronics*, 24(3):1064–1074, 2019.

[80] C. Wen, J. Zhou, Z. Liu, and H. Su. Robust adaptive control of uncertain nonlinear systems in the presence of input saturation and external disturbance. *IEEE Transactions on Automatic Control*, 56(7):1672–1678, 2011.

[81] J. Wu, W. Chen, and J. Li. Global finite-time adaptive stabilization for nonlinear systems with multiple unknown control directions. *Automatica*, 69:298–307, 2016.

[82] Z. Xi and Z. Ding. Global adaptive output regulation of a class of nonlinear systems with nonlinear exosystems. *Automatica*, 43(1):143–149, 2007.

[83] B. Yang and W. Lin. Semi-global stabilization of nonlinear systems by nonsmooth output feedback. *International Journal of Robust and Nonlinear Control*, 24(16):2522–2545, 2014.

[84] J. Yang, W. Chen, S. Li, L. Guo, and Y. Yan. Disturbance/uncertainty estimation and attenuation techniques in PMSM drives-a survey. *IEEE Transactions on Industrial Electronics*, 64(4):3273–3285, 2016.

[85] J. Yang and Z. Ding. Global output regulation for a class of lower triangular nonlinear systems: A feedback domination approach. *Automatica*, 76:65–69, 2017.

[86] J. Yang, Z. Ding, S. Li, and C. Zhang. Continuous finite-time output regulation of nonlinear systems with unmatched time-varying disturbances. *IEEE Control Systems Letters*, 2(1):97–102, 2017.

[87] J. Yang, W. Zheng, S. Li, B. Wu, and M. Cheng. Design of a prediction-accuracy-enhanced continuous-time mpc for disturbed systems via a disturbance observer. *IEEE Transactions on Industrial Electronics*, 62(9):5807–5816, 2015.

[88] H. Yu, S. Huang, G. Chen, Y. Pan, and Z. Guo. Human-robot interaction control of rehabilitation robots with series elastic actuators. *IEEE Transactions on Robotics*, 31(5):1089–1100, 2015.

[89] C. Zhang and C. Wen. A non-recursive $\mathcal{C}^1$ adaptive stabilization methodology for nonlinearly parameterized uncertain nonlinear systems. *Journal of the Franklin Institute*, 355(12):5099–5114, 2018.

[90] C. Zhang, Y. Yan, A. Narayan, and H. Yu. Practically oriented finite-time control design and implementation: Application to series elastic actuator. *IEEE Transactions on Industrial Electronics*, 65(5):4166–4176, 2017.

[91] C. Zhang, J. Yang, and C. Wen. Global stabilisation for a class of uncertain non-linear systems: A novel non-recursive design framework. *Journal of Control and Decision*, 4(2):57–69, 2017.

[92] C. Zhang, J. Yang, C. Wen, L. Wang, and S. Li. Realization of exact tracking control for nonlinear systems via a non-recursive dynamic design. *IEEE Transactions on Systems, Man and Cybernetics: Systems*, 50(2):577–589, 2020.

[93] C. Zhang, J. Yang, Y. Yan, L. Fridman, and S. Li. Semi-global finite-time trajectory tracking realization for disturbed nonlinear systems via higher-order sliding modes. *IEEE Transactions on Automatic Control*, 60(5):2185–2191, 2020.

[94] X. Zhang, G. Feng, and Y. Sun. Finite-time stabilization by state feedback control for a class of time-varying nonlinear systems. *Automatica*, 48(3):499–504, 2012.

[95] L. Zhao, J. Yu, and H. Yu. Adaptive finite-time attitude tracking control for spacecraft with disturbances. *IEEE Transactions on Aerospace and Electronic Systems*, 54(3):1297–1305, 2017.

[96] Z. Zhao and Z. Jiang. Semi-global finite-time output-feedback stabilization with an application to robotics. *IEEE Transactions on Industrial Electronics*, 66(4):3148–3156, 2018.

[97] J. Zhou and C. Wen. *Adaptive backstepping control of uncertain systems: Nonsmooth nonlinearities, interactions or time-variations*. Springer, 2008.

[98] S. Zhou and Y. Song. Neuroadaptive control design for pure-feedback nonlinear systems: A one-step design approach. *IEEE Transactions on Neural Networks and Learning Systems*, 31(9):3389–3399, 2019.

For Product Safety Concerns and Information please contact our EU representative GPSR@taylorandfrancis.com
Taylor & Francis Verlag GmbH, Kaufingerstraße 24, 80331 München, Germany

www.ingramcontent.com/pod-product-compliance
Lightning Source LLC
Chambersburg PA
CBHW060347220326
41598CB00023B/2834

* 9 7 8 1 0 3 2 5 0 6 0 4 3 *